T0141951

Lecture Notes in Networks and Systems

Volume 43

Series editor

Janusz Kacprzyk, Polish Academy of Sciences, Warsaw, Poland
e-mail: kacprzyk@ibspan.waw.pl

The series "Lecture Notes in Networks and Systems" publishes the latest developments in Networks and Systems—quickly, informally and with high quality. Original research reported in proceedings and post-proceedings represents the core of LNNS.

Volumes published in LNNS embrace all aspects and subfields of, as well as new challenges in, Networks and Systems.

The series contains proceedings and edited volumes in systems and networks, spanning the areas of Cyber-Physical Systems, Autonomous Systems, Sensor Networks, Control Systems, Energy Systems, Automotive Systems, Biological Systems, Vehicular Networking and Connected Vehicles, Aerospace Systems, Automation, Manufacturing, Smart Grids, Nonlinear Systems, Power Systems, Robotics, Social Systems, Economic Systems and other. Of particular value to both the contributors and the readership are the short publication timeframe and the world-wide distribution and exposure which enable both a wide and rapid dissemination of research output.

The series covers the theory, applications, and perspectives on the state of the art and future developments relevant to systems and networks, decision making, control, complex processes and related areas, as embedded in the fields of interdisciplinary and applied sciences, engineering, computer science, physics, economics, social, and life sciences, as well as the paradigms and methodologies behind them.

Advisory Board

More information about this series at http://www.springer.com/series/15179

P. Nagabhushan · D. S. Guru
B. H. Shekar · Y. H. Sharath Kumar
Editors

Data Analytics and Learning

Proceedings of DAL 2018

Editors
P. Nagabhushan
Indian Institute of Information Technology
Allahabad, Uttar Pradesh, India

D. S. Guru
Department of Computer Science and Engineering, CBCS Education
University of Mysore
Mysuru, Karnataka, India

B. H. Shekar
Department of Studies in Computer Science
Mangalore University
Mangalore, Karnataka, India

Y. H. Sharath Kumar
Department of Information Science and Engineering
Maharaja Institute of Technology
Belawadi, Karnataka, India

ISSN 2367-3370 ISSN 2367-3389 (electronic)
Lecture Notes in Networks and Systems
ISBN 978-981-13-2513-7 ISBN 978-981-13-2514-4 (eBook)
https://doi.org/10.1007/978-981-13-2514-4

Library of Congress Control Number: 2018954017

This Springer imprint is published by the registered company Springer Nature Singapore Pte Ltd.
The registered company address is: 152 Beach Road, #21-01/04 Gateway East, Singapore 189721, Singapore

Preface

We write this message with deep satisfaction to the proceedings of the "First International Conference on Data Analytics and Machine Learning 2018 (DAL 2018)" held on March 30 and 31, 2018, at Mysore, Karnataka, India, which has the central theme "Data Analytics and its Application." Our research experiences in related areas for the last decade have inspired us to conduct DAL 2018. This conference was planned to provide a platform for researchers from both academia and industries where they can discuss and exchange their research thoughts to have better future research plans, particularly in the fields of data analytics and machine learning. Soon after we notified a call for original research papers, there has been a tremendous response from the researchers. There were 150 papers submitted, out of which we could accommodate only 50 papers based on the reports of the reviewers. Each paper was blindly reviewed by at least two experts from the related areas. The overall acceptance rate is about 30 %. The conference is aimed at image processing, signal processing, pattern recognition, document processing, biomedical processing, computer vision, biometrics, data mining and knowledge discovery, information retrieval and information coding. For all these areas, we got a number of papers reflecting their right combinations. I hope that the readers will appreciate and enjoy the papers published in the proceedings. We could make this conference a successful one, though it was launched at a relatively short notice. It was because of the good response from the research community and the good effort put in by the reviewers to support us with timely reviews. The authors of all the papers submitted deserve our acknowledgments. The proceedings are published and indexed by Springer-LNEE, which is known for bringing out this type of proceedings. Special thanks to them.

We would also like to thank the help of EasyChair in the submission, review, and proceedings creation processes. We are very pleased to express our sincere thanks to Springer, especially Jayanthi Narayanaswamy, Jayarani Premkumar, Aninda Bose, and the editorial staff, for their support in publishing the proceedings of DAL 2018.

Allahabad, India	Dr. P. Nagabhushan
Mysuru, India	Dr. D. S. Guru
Mangalore, India	Dr. B. H. Shekar
Belawadi, India	Dr. Y. H. Sharath Kumar

Contents

Editors and Contributors

About the Editors

P. Nagabhushan is Director of the Indian Institute of Information Technology, Allahabad, Uttar Pradesh. He has previously served as Professor at the University of Mysore and Amrita University. His areas of specialization are pattern recognition, digital image analysis and processing, document image analysis, and knowledge mining. He has published more than 500 papers in peer-reviewed journals and conferences in these areas. He has received research funding from various organizations, e.g., DRDO, MHRD, AICTE, ICMR, UGC, UGC, UPE, and SAP, and has more than 30 years of teaching experience.

D. S. Guru is Professor in the Department of Studies in Computer Science, University of Mysore. He completed his doctorate in computer science from the University of Mysore in 2000 and has published more than 200 research papers in national and international journals and conference proceedings. He has served as Program Chair for the 2009 International Conference on Signal and Image Processing in Mysore and is currently Reviewer for the International Journal of Pattern Recognition Letters, Journal of Pattern Recognition, Journal of Document Analysis and Recognition, Journal of Image and Vision Computing, and IEEE Transactions on Neural Networks.

B. H. Shekar is Professor in the Department of Studies in Computer Science, Mangalore University, India. His research interests include pattern recognition, image processing, database systems, and algorithms. He is an active researcher and has published more than 100 papers in peer-reviewed journals and conferences. He was granted a DST-DAAD Fellowship in 2002 and was the principal investigator for many research projects funded by DST-RFBR and UGC. He is also a member of many academic societies, e.g., the International Association for Pattern Recognition, IACSIT, and Indian Unit for Pattern Recognition and Artificial Intelligence.

Y. H. Sharath Kumar is Associate Professor in the Department of Information Science and Engineering, Maharaja Institute of Technology, Mysuru. He received his M.Tech. in 2009 and his Ph.D. in 2014, both from the University of Mysore. His research interests include image and video processing, pattern recognition, Sign Language Video Analysis, logo classification, biometrics, and precision agriculture. He has authored a number of papers in reputed journals and conference proceedings.

Contributors

Mostafa Ali Department of Studies in Computer Science, University of Mysore, Manasagangotri, Mysuru, Karnataka, India

Mokhtar A. Alworafi DoS in Computer Science, University of Mysore, Mysuru, India

R. Amarnath Department of Studies in Computer Science, University of Mysore, Mysuru, Karnataka, India

H. Annapurna Department of Studies in Computer Science, University of Mysore, Mysuru, Karnataka, India

C. V. Aravinda NMAM Institute of Technology NITTE, Karakal, Karnataka, India

Adela Arpitha DoS in Computer Science, University of Mysore, Mysuru, India

A. P. Ayshath Thabsheera Department of Computer Science, Central University of Kerala, Kasaragod, Kerala, India

Bharathlal Department of Studies in Computer Science, University of Mysore, Mysuru, India

Shabina Bhaskar Central University of Kerala, Kasaragod, Kerala, India

Nikitha Bhat Department of CSE, NMAM Institute of Technology, Nitte, India

Vikram Bhat Department of CSE, NMAM Institute of Technology, Nitte, India

Umesh P. Chavan School of Computational Sciences, Solapur University, Solapur, India

Venkatratnam Chitturi Department of Instrumentation Technology, GSSSIETW, Mysuru, India

Guesh Dagnew Department of Computer Science, Mangalore University, Karnataka, Mangalore, India

N. C. Dayananda Kumar Department of Electronics and Communication Engineering, Siddaganga Institute of Technology, Tumkur, India

Akshaya Devadiga NMAM Institute of Technology NITTE, Karakal, Karnataka, India

Atyaf Dhari College Education for Pure Science, Thi Qar University, Nasiriyah, Iraq

Rajesh K. Dhumal Department of CS and IT, Dr. Babasaheb Ambedkar Marathwada University, Aurangabad, Maharashtra, India

R. Dinesh Department of Information Science and Engineering, Jain University, Bengaluru, India

Shereen A. El-Booz Department of Computer Science and Engineering, Menoufia University, Shibin El Kom, Egypt

Nagi Farrukh Faculty of Mechanical Engineering, Universiti Tenaga Nasional, Kajang, Selangor, Malaysia

D. S. Guru Department of Studies in Computer Science, University of Mysore, Manasagangotri, Mysuru, India

H. L. Gururaj Vidyavardhaka College of Engineering, Mysuru, India

Mallikarjun Hangarge Department of Computer Science, Karnatak Arts Science and Commerce College, Bidar, India

N. Harivinod Department of Computer Science and Engineering, Vivekananda College of Engineering and Technology, Puttur, Karnataka, India

Maryam Hazman Central Laboratory for Agricultural Experts Systems, Agricultural Research Center (ARC), Giza, Egypt

Ravindra S. Hegadi School of Computational Sciences, Solapur University, Solapur, India

G. Hemantha Kumar Department of Computer Science, University of Mysore, Mysuru, India; High Performance Computing Project, DoS in CS, University of Mysore, Mysuru, India

Swaroop Hople Veermata Jijabai Technological Institute, Mumbai, India

Mohammad Imran Department of Computer Science, University of Mysore, Mysuru, India

Mohammed Javed Department of IT, Indian Institute of Information Technology, Allahabad, Allahabad, India

V. K. Jyothi Department of Studies in Computer Science, University of Mysore, Manasagangotri, Mysuru, India

Karbhari V. Kale Department of CS and IT, Dr. Babasaheb Ambedkar Marathwada University, Aurangabad, Maharashtra, India

V. N. Kamalesh VTU RRC, Belagavi, Karnataka, India; TJIT, Bengaluru, India

Saket Karve Veermata Jijabai Technological Institute, Mumbai, India

Thotreingam Kasar Kaaya Tech Inc., Mysuru, India

Mohammadamir Kavousi Southern Illinois University, Carbondale, USA

P. G. Lavanya DoS in Computer Science, University of Mysore, Mysuru, India

Suresha Mallappa DoS in Computer Science, University of Mysore, Mysuru, India

V. G. Manjunatha Guru GFGC, Honnāli, India; VTU RRC, Belagavi, Karnataka, India

K. S. Manjunatha Maharani's Science College for Women, Mysuru, Karnataka, India

N. Manohar Department of Computer Science, University of Mysore, Mysuru, India

K. Manoj Bengaluru, India

S. C. Mehrotra Department of CS and IT, Dr. Babasaheb Ambedkar Marathwada University, Aurangabad, Maharashtra, India

N. Mohammed Muddasir Department of Information Science and Engineering, VVCE, Mysuru, India; NIE, Mysuru, India

Krishna Chaithanya Movva Indian Institute of Information Technology, Sri City, Chittoor, Andhra Pradesh, India

P. Nagabhushan Department of IT, Indian Institute of Information Technology, Allahabad, Allahabad, India

Ajay D. Nagne Department of CS and IT, Dr. Babasaheb Ambedkar Marathwada University, Aurangabad, Maharashtra, India

Aida A. Nasr Department of Computer Science and Engineering, Menoufia University, Shibin El Kom, Egypt

Dattatray I. Navale School of Computational Sciences, Solapur University, Solapur, India

Tejas Nayak Department of CSE, NMAM Institute of Technology, Nitte, India

S. P. Paramesh Department of Studies in Computer Science and Engineering, U.B.D.T College of Engineering, Davanagere, Karnataka, India

Trupti D. Pawar School of Computational Sciences, Solapur University, Solapur, India

Bharathi Pilar Department of Computer Science, University College, Mangalore, Karnataka, India

H. N. Prakash Rajeev Institute of Technology, Hassan, Karnataka, India

J. Prathima Mabel Department of Information Science and Engineering, Dayananda Sagar College of Engineering, Bengaluru, India

Viswanath Pulabaigari Indian Institute of Information Technology, Sri City, Chittoor, Andhra Pradesh, India

K. Raghuveer Department of Information Science and Engineering, NIE, Mysuru, India

R. Rajesh Department of Computer Science, Central University of Kerala, Kasaragod, Kerala, India

K. N. Rama Mohan Babu Department of Information Science and Engineering, Dayananda Sagar College of Engineering, Bengaluru, India

B. Ramesh Malnad College of Engineering, Hassan, India

Radhika Rani SBRR Mahajana First Grade College, Mysuru, India

C. Rashmi High Performance Computing Project, DoS in CS, University of Mysore, Mysuru, India

Darshan D. Ruikar School of Computational Sciences, Solapur University, Solapur, India

Sepehr Saadatmand Southern Illinois University, Carbondale, USA

M. Shankar Lingam University of Mysore, Manasagangotri, Mysuru, India

B. Sharada Department of Studies in Computer Science, University of Mysore, Mysuru, India

Y. H. Sharath Kumar Department of Computer Science & Engineering and Department of Information Science, Maharaja Institute of Technology, Mandya, Karnataka, India

B. H. Shekar Department of Computer Science, Mangalore University, Mangalore, Karnataka, India

Vasisht Shende Veermata Jijabai Technological Institute, Mumbai, India

Sannidhi Shetty Department of CSE, NMAM Institute of Technology, Nitte, India

E. G. Shivakumar Department of Electrical Engineering, UVCE, Bangalore University, Bengaluru, India

K. S. Shreedhara Department of Studies in Computer Science and Engineering, U.B.D.T College of Engineering, Davanagere, Karnataka, India

G. S. Sindhushree Department of Studies in Computer Science, University of Mysore, Mysuru, Karnataka, India

A. M. Sudhakara University of Mysore, Manasagangotri, Mysuru, India

Mahamad Suhil Department of Studies in Computer Science, University of Mysore, Manasagangotri, Mysuru, Karnataka, India

D. S. Sunil Kumar Department of Computer Science, Mangalore University, Mangalore, Karnataka, India

H. D. Supreetha Gowda Department of Computer Science, University of Mysore, Mysuru, India

K. V. Suresh Department of Electronics and Communication Engineering, Siddaganga Institute of Technology, Tumkur, India

S. N. Sushma Department of Studies in Computer Science, University of Mysore, Mysuru, India

Seyedmahmoud Talebi Department of Computer Science, University of Mysore, Mysuru, India

T. M. Thasleema Department of Computer Science, Central University of Kerala, Kasaragod, Kerala, India

K. A. Vani Department of Information Science and Engineering, Dayananda Sagar College of Engineering, Bengaluru, India

C. Veershetty Department of Computer Science, Gulbarga University, Kalaburagi, India

B. T. Venu Gopal Department of Electrical Engineering, UVCE, Bangalore University, Bengaluru, India

Amol D. Vibhute Department of CS and IT, Dr. Babasaheb Ambedkar Marathwada University, Aurangabad, Maharashtra, India

Recognition of Seven-Segment Displays from Images of Digital Energy Meters

Thotreingam Kasar

Abstract This paper describes a method to localize and recognize seven-segment displays on digital energy meters. Color edge detection is first performed on a camera-captured image of the device which is then followed by a run-length technique to detect horizontal and vertical lines. The region of interest circumscribing the LCD panel is determined based on the attributes of intersecting horizontal and vertical lines. The extracted display region is preprocessed using the morphological black-hat operation to enhance the text strokes. Adaptive thresholding is then performed and the digits are segmented based on stroke features. Finally, the segmented digits are recognized using a support vector machine classifier trained on a set of syntactic rules defined for the seven-segment font. The proposed method can handle images exhibiting uneven illumination, the presence of shadows, poor contrast, and blur, and yields a recognition accuracy of 97% on a dataset of 175 images of digital energy meters captured using a mobile camera.

Keywords Seven-segment displays · Character recognition
Camera-based document image analysis

1 Introduction

The camera provides a great opportunity for input from the physical world. In recent years, it has become hard to define the term document due to the blurring in the distinction between documents and user interfaces. In addition to imaging hard copy documents, cameras are now increasingly being used to capture text present on 3-D real-world objects such as buildings, billboards, road signs, license plates, black/whiteboards, household appliances, or even on a T-shirt which otherwise would be inaccessible to conventional scanner-based optical character recognition (OCR) systems. Pervasive use of handheld digital cameras has immense potential for newer applications that go far beyond what traditional OCR has to offer [1]. Recognizing

T. Kasar (✉)
Kaaya Tech Inc., Mysuru 570017, India
e-mail: kasar.kaayatech@gmail.com

P. Nagabhushan et al. (eds.), *Data Analytics and Learning*,
Lecture Notes in Networks and Systems 43,
https://doi.org/10.1007/978-981-13-2514-4_1

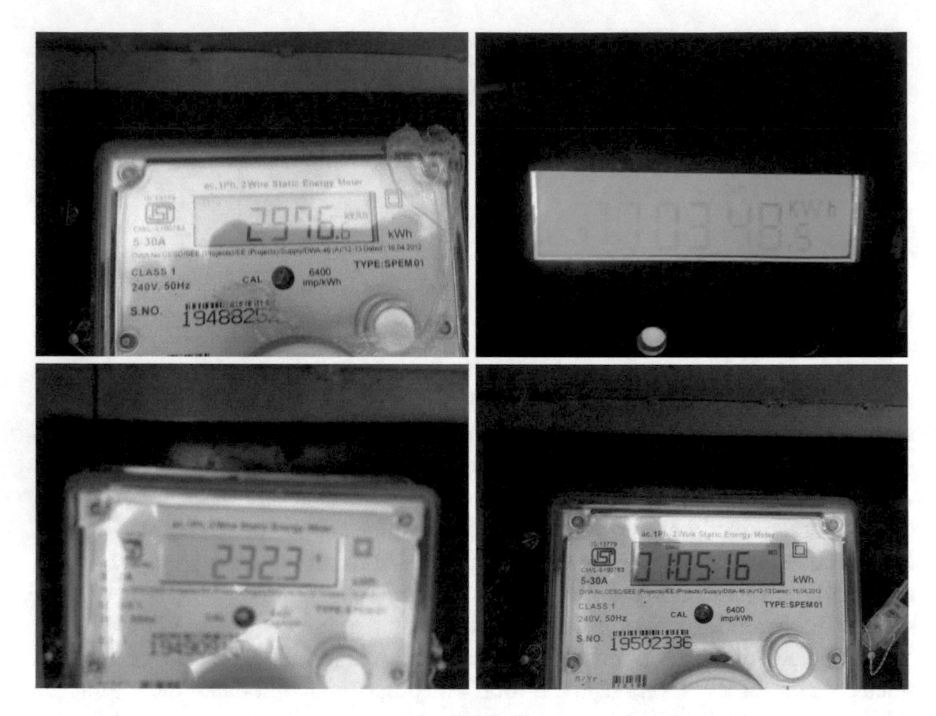

Fig. 1 Typical images of seven-segment displays on digital energy meters exhibiting uneven lighting, shadows, poor contrast, blur, and presence of highlights occluding parts of a digit

text in real-world scenes can be considered as an extension of current OCR technology widely available in the market. The unconstraint mode of document acquisition as well the difference in the target document type calls for a new level of processing for camera-captured images. While it may seem that the low variability of seven-segment displays should make them easy to read, its automatic detection and recognition is in fact a challenging task. A typical image of the display panel on an electronic device contains mostly background clutter and other irrelevant texts. Therefore, a preprocessing step of locating the display area is required before attempting to recognize text in the acquired image. Once the text region is localized, the subsequent recognition task can be performed only on the detected region of interest so as to obviate the effect of background clutter. However, the available technology is still far from being able to reliably separate text from the background clutter. In addition, images of LCD displays often exhibit poor contrast, blur, and may contain highlights and specular reflections from the display surface which make them hard to segment. Figure 1 shows some of these challenges commonly encountered in images of LCD displays.

2 Review of Related Work

While there are a lot of works on recognizing text from natural images [2–4], there has been relatively less work that address the specific problem of recognizing seven-segment displays on electronic devices. The Clearspeech system [5] requires special markers to be affixed to the device to guide the system in localizing the display panel. Shen and Coughlan [6] introduced a phone-based LED/LCD display reader, which do not have such modification of the display. They employ horizontal and vertical edge features in the image and extract the digits using a simple graphical model. Tekim et al. [7] improvized the method in [6] by adopting a connected-component-based approach to detect LED/LCD digits on a Nokia N95 mobile phone that can process up to 5 frames/s allowing the user to overcome issues such as highlights, glare, or saturation by simply varying the camera viewpoint. In [8], the authors address a method to recognize seven-segment displays on multimeters using binary edge and corner features. In this paper, a camera-based system is developed to detect and recognize seven-segment displays in digital energy meters. The method can be applied to images of several other electronic appliances such as calculators, digital blood pressure monitors, digital thermometers, microwave ovens, media players, etc. with minimal or no modification.

3 Proposed Method for Recognition of Seven-Segment Digits

This section describes the proposed method designed for the recognition of seven-segment displays from images of digital energy meters captured using mobile camera phones. The method involves two sub-tasks, namely, (i) localization of the LCD display area and (ii) recognition of the seven-segment digits in the localized area. The LCD display area is localized based on attributes of horizontal and vertical line segments and their intersection, while a support vector machine (SVM) is used for the classification of the digits. A schematic block diagram of the proposed method is shown in Fig. 2.

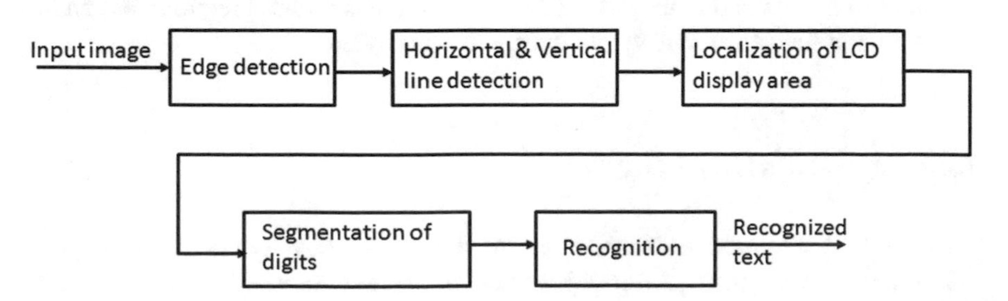

Fig. 2 Block diagram of the proposed seven-segment display recognition system

3.1 Determination of the LCD Display Area

Following image acquisition, the first task is to locate the LCD display area among substantial amounts of background clutter. To this end, Canny edge detection [9] is performed on each color channel of the captured image I. The overall edge map is obtained by taking the union of the individual edge images as follows:

$$E = E_R \cup E_G \cup E_B \tag{1}$$

where $\cup$ denotes the union operation. Following the method in [10], run-length count is performed on the resulting edge image E along the rows and columns to obtain the horizontal and vertical lines, respectively. If the number of runs of the edges starting at a pixel location exceeds a threshold value L, the segment is accepted as a line. Short line segments and other spurious lines with run-lengths less than the specified threshold value are not considered for further processing. The threshold L decides the shortest line that can be detected by the method. This parameter is adaptively set to a fractional proportion of the height of the image. It may be mentioned that the performance of the method is not sensitive to the choice of this parameter since the LCD display panel normally occupies a significant proportion of the image area. The union of the set of validated horizontal and vertical line segments obtained from the two directions yields a composite image I_L. Based on the positions of intersection of horizontal and vertical lines, their heights, and aspect ratios, the rectangular-shaped LCD display area is identified. This step of identifying the region of interest (ROI), i.e., the LCD display region, is an important processing module that removes the background clutter and returns only the relevant display region for further processing. The performance of this module is critical since it serves as the input for the subsequent digit recognition task and affects the overall performance of the system. It may be noted that the run-length method for horizontal and vertical line detection implicitly assumes that orientation of the image is not skewed. However, the method can tolerate a moderate skew angle of up to θ which is given by

$$\theta = \arctan(1/L) \tag{2}$$

For instance, if the minimum detectable line length L is set to eight pixels, the skew tolerance of the method is about $\pm 7.125°$. Thus, there is no strict requirement on the orientation of the camera viewpoint during image capture.

3.2 Segmentation of Digits

Once the ROI is located as described above, the area defined by the ROI is cropped off from the image. Since seven-segment LCD displays are represented in a darker shade with respect to that of the background, the strategy is to look for thin dark structures in the detected ROI. To enhance dark and thin line-like structures, the grayscale morphological bottom-hat operation is performed on the smoothed image obtained by Gaussian filtering the ROI.

$$I_p = (I_\sigma \bullet S_N) - I_\sigma \tag{3}$$

where $I_\sigma = I_g * G_\sigma$ with σ representing the variance of the Gaussian filter, I_g the grayscale image patch defined by the ROI and S_N is a square structuring element of size $N \times N$. The notations "$*$" and "$\bullet$" denote the 2-D convolution and the grayscale morphological closing operation, respectively. The Gaussian filter reduces the effect of noise and helps to maintain the continuity of narrow gaps between line segments. The variance σ of the Gaussian function controls the amount of smoothing. The bottom-hat operation enhances small dark structures while suppressing wide dark structures at the same time. The size N of the structuring element decides the maximum width of the line that can be detected by the system and is empirically set to 15 in this work. The method is not very sensitive to the choice of this parameter and it may be altered without any significant effect on the overall performance.

This intermediate image I_p is then thresholded using a fast implementation of the Niblack's method [11] using integral images [12]. At each pixel location (x, y), the mean $\mu(x, y)$ and the standard deviation $\sigma(x, y)$ within a window $W \times W$ are computed and the gray level at that pixel location is compared with the threshold value $T(x, y)$ given by the following expression:

$$T(x, y) = \mu(x, y) - k\sigma(x, y) \tag{4}$$

The window of size is set to $h/5 \times w/5$ where h and w denote the height and width of the detected ROI, respectively, and the parameter k is set to 0.2. Since the segments of a seven-segment character are not connected, we need to group the individual segments to form the digits before feature extraction and recognition. An eight connected component (CC) labeling is performed on the resulting binary image. Components that touch the image boundary are discarded since it is generally a part of the background. The stroke widths of the remaining CCs are computed using a combination of the distance transform and the skeleton of the CC obtained using a fast thinning algorithm proposed by Zhang and Suen [13]. The maximum stroke width D_{max} is then determined which is used to group the CCs into digits by performing a closing operation with square structuring of size $2D_{max} \times 2D_{max}$. One further step of CC labeling is performed on the resulting image and the bounding box attributes are computed. Since the digits in LCD displays are of similar heights and located horizontally next to each other, the candidate digits are obtained by imposing the height similarity and spatial regularity of the CCs. These filtered CCs are then passed onto the recognition module.

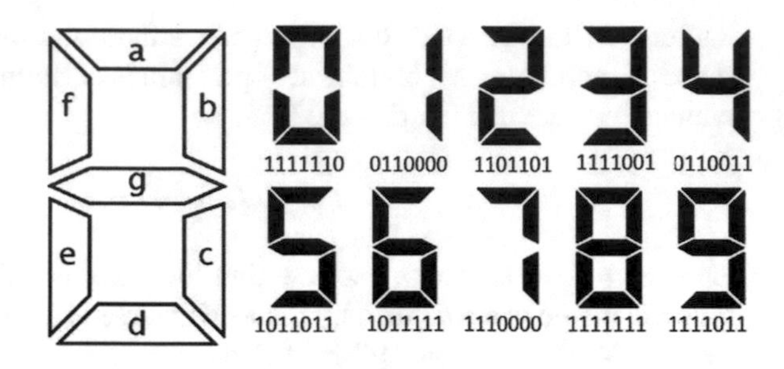

Fig. 3 The seven-segment font and the code (abcdefg) for each digit from 0–9 in terms of its ON/OFF states of the individual segments

3.3 *Digit Recognition*

Figure 3 depicts a typical seven-segment display unit, where any digit can be represented by a 7-bit code depending on the ON or OFF state of the individual segments a, b, ..., g. Since seven-segment displays have a fixed font style, a simple digit recognizer may be formulated based on syntactic rules to recognize the 10 digits [0 1 2 3 4 5 6 7 8 9]. However, such deterministic rules may not work well in practice due to segmentation errors and noise. Here, in this work, an SVM classifier trained on a collection of 500 digits is used. For each segmented CC, the proportion of ON pixels within an area defined by each of individual segments is measured. From seven such measurements over each of the areas defined by the seven segments, a seven-dimensional feature vector is obtained from each candidate CC which is then classified using the trained SVM classifier. While the aspect ratio (width/height) for each digit is around 0.5, the same parameter is much smaller for the digit 1 and hence it can be easily identified. Whenever the aspect ratio of the test CC is less the 0.4, feature extraction is done only for the segments b and c and assign 0 for all the other segments.

3.3.1 Identification of Dots

As a post-processing step, the area within the recognized digits is examined in the presence of dots. Identification of a dot is based on the size and aspect ratio of the CCs. Any small CCs that lie between two segmented digits are separately processed to determine if it is a dot or not. If the size of such a CC is less than t_1 times πD_{max}^2 (t_1 is a scalar, set to 1.5 in this work) and its aspect ratio is close to 1, it is considered as a dot.

3.4 *Experimental Results*

To test the performance of the system, 175 images of energy meters are captured using Apple iPhone 4S and Tinno S4700. These images are captured from a distance

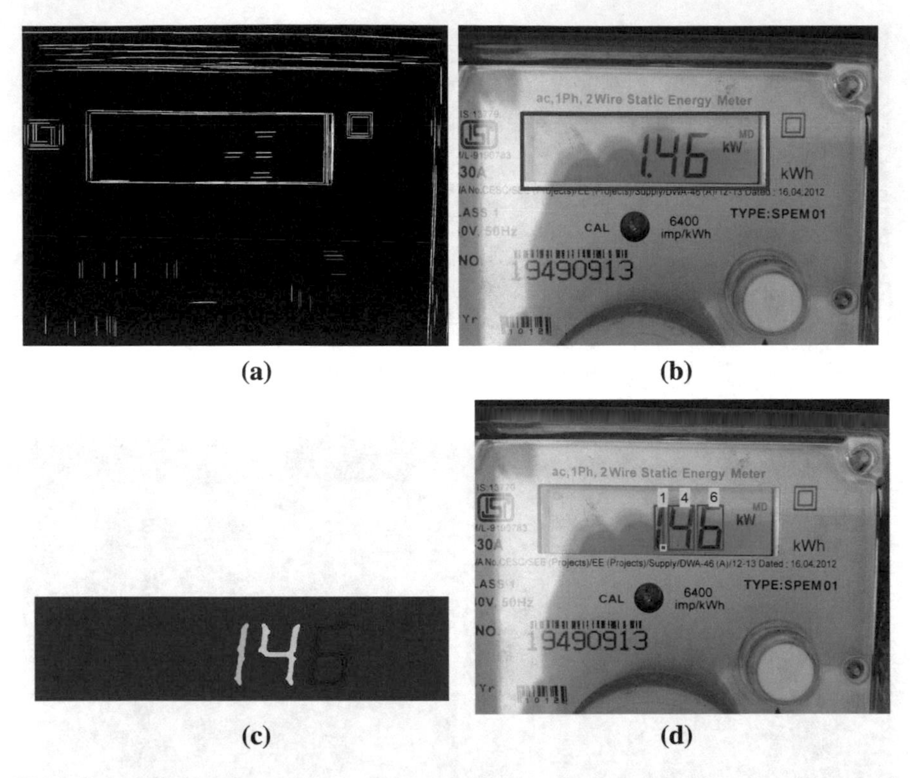

Fig. 4 Intermediate results of the proposed method on a sample test image **a** Detection of horizontal and vertical line segments **b** Localization of the LCD display area **c** Segmentation of the digits **d** Final results of recognition. Note that the decimal point indicated by the cyan rectangle is also identified thereby achieving a correct recognition of the value of the reading on the device

of about 15 to 30 cms from the device and ensuring that the orientation of the captured image is roughly horizontal. However, there is no strict requirement for the position and orientation of the camera viewpoint during image capture. The low and high threshold parameters of the Canny edge detection are set to 0.1 and 0.2, respectively, while the variance of the associated Gaussian function is set to 2. The parameter L for horizontal and vertical line segment detection is adaptively set to 1/30 times the height of the image. Since the LCD display panel of the device is rectangular with a fixed aspect ratio (between 3 to 4), the display area can be identified by subjecting each pair of intersecting horizontal and vertical lines to an aspect ratio test and additionally a lower bound on the length of the vertical line segment to filter out small candidate regions. Whenever there is a detection of nested regions, the overall bounding rectangle of all the detected regions is considered to be the ROI. Following the localization of LCD display area, a preprocessing step of Gaussian smoothing and grayscale morphological black-hat operation is performed to enhance the segments of the seven-segment characters. While conventional thresholding techniques fail to accurately extract the digits, the black-hat operation enables reliable digit

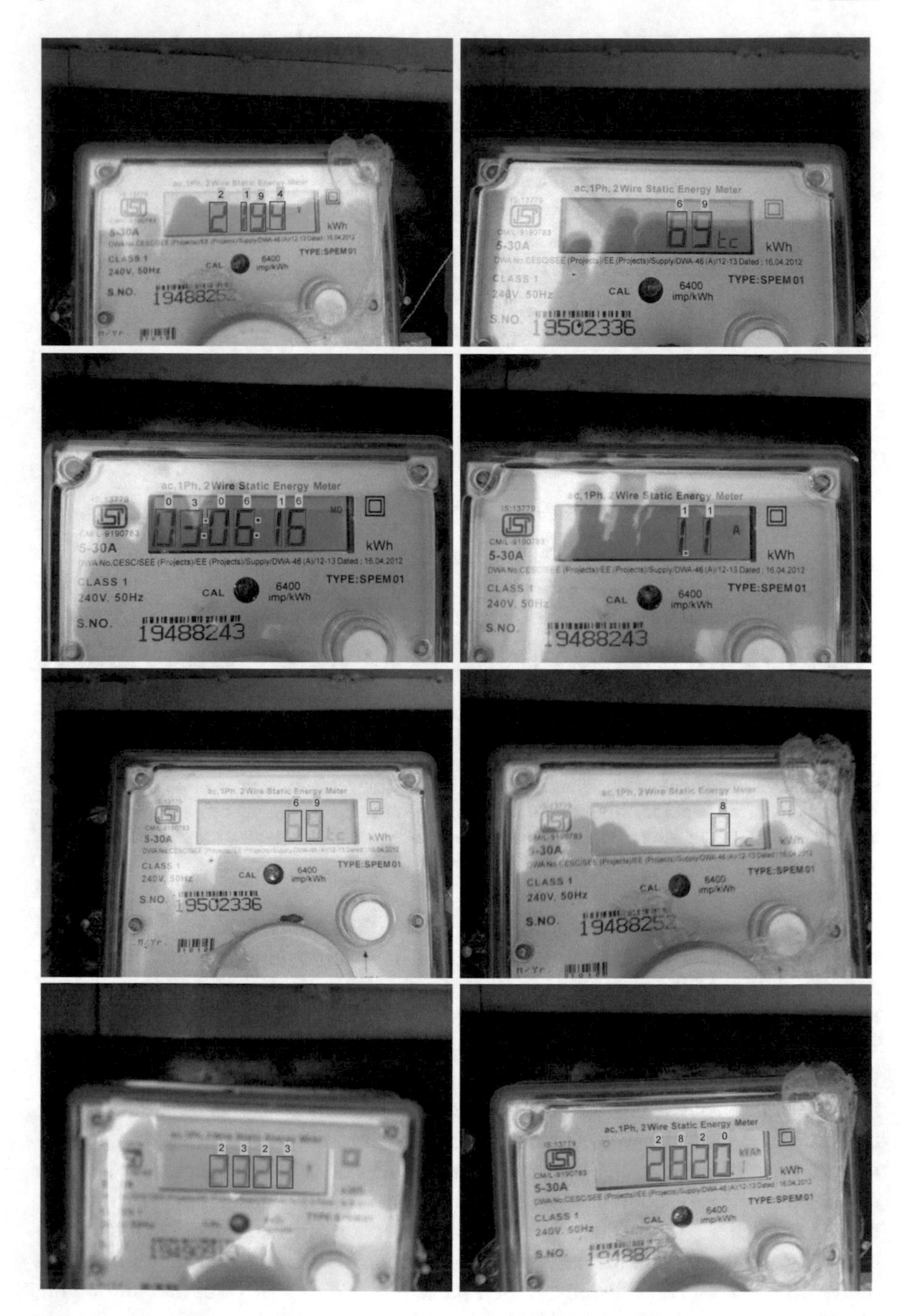

Fig. 5 Representative results of the proposed method. The segmented digits are represented in terms of the bounding boxes and the corresponding recognized results are overlaid above each digit. The method can handle images exhibiting uneven illumination, the presence of shadows, poor contrast, and blur

segmentation even under poor contrast and uneven lighting. Because of the fixed structure of seven-segment font, the SVM classifier can reliably recognize the segmented digits, even when there are instances of one segment segmentation error. Figure 4 shows the intermediate results of various processing stages on a sample test image. It may be noted that the presence of the dot between the first digit 1 and the second digit 4 is also identified that is necessary to infer the correct reading of the display.

Figure 5 illustrates the robustness of the proposed method to uneven lighting, the presence of shadows, low contrast, and blur. The method works well for most images captured under typical imaging conditions. It can also tolerate moderate skew as is evident from the last image of the figure. An overall recognition accuracy of 97% is achieved on a collection of 175 test images. Glare, highlights, and poor contrast are the main sources of recognition error while excessive blur can affect the accuracy of edge detection and consequently the localization of the ROI.

3.5 Conclusions and Future Work

This paper describes a new method to recognize seven-segment displays from images of energy meters captured using a mobile camera. The LCD panel area is determined based on the properties of intersecting horizontal and vertical lines. The morphological black-hat operation enables robust segmentation of the digits even under poor contrast and uneven lighting. The proposed method yields a high performance for images captured under typical imaging conditions and even though the size of dataset used for the evaluation is small, the method is expected to yield a consistent performance on larger test sets too. However, like any other method, it is sensitive to specular highlights, severe blurring, and saturation. The quality of images of LCD displays varies greatly depending on the ambient environment and on the position of the acquisition camera relative to the display. It was observed that glare and reflections that may occlude the digits in the display can be avoided by changing the camera viewpoint. Thus, to address these challenges, the method may be augmented with a continuous feedback system to update the result of recognition to the user in real time. Based on this visual feedback, the user can vary the position the acquisition camera till a perfect recognition is achieved. Such an approach will make it possible to read displays on modern electronic appliances in real time and can lead to a host of new applications.

References

1. Liang, J., Doermann, D., Li, H.: Camera-based analysis of text and documents: a survey. Int. J. Doc. Anal. Recognit. **7**, 83–200 (2005)
2. Wang, K., Babenko, B., Belongie, S.: End-to-End scene text recognition. In: International Conference on Computer Vision, pp. 1457–1464 (2011)

3. Neumann, L., Matas, J.: Real-Time Scene Text Localization and Recognition. In: International Conference on Computer Vision and Pattern Recognition, pp. 3538–3545 (2012)
4. Wang, T., David J. W., Coates, A., Andrew Y. Ng.: End-to-End Text Recognition with Convolutional Neural Networks In: International Conference on Pattern Recognition, pp. 3304–3308 (2012)
5. Morris, T., Blenkhorn, P., Crossey, L., Ngo, Q., Ross, M., Werner, D., Wong, C.: Clearspeech: a display reader for the visually handicapped. IEEE Trans. Neural Syst. Rehabil. Eng. **14**(4), 492–500 (2006)
6. Shen, H., Coughlan, J.: Reading LCD/LED displays with a camera cell phone. In: Conference on Computer Vision and Pattern Recognition Workshop (2006)
7. Tekim, E., Coughlan, J. M., Shen, H.: Real-Time Detection and Reading of LED/LCD Displays for Visually Impaired Persons. In: Workshop Applications of Computer Vision, pp. 181–184 (2011)
8. Rakhi, P. G., Sandip P. N., Mukherji, P., Prathamesh M. K.: Optical character recognition system for seven segment display images of measuring instruments. In: IEEE Region 10 Conference on TENCON, pp. 1–6 (2009)
9. Canny, J.: computational approach to edge detection. IEEE Trans. Pattern Anal. Mach. Intell. **8**(6), 679–698 (1986)
10. Kasar, T., Barlas, P., Adam, S., Chatelain, C., Paquet, T.: Learning to detect tables in scanned document images using line information. In: International Conference on Document Analysis and Recognition, pp. 1185–1189 (2013)
11. Niblack W.: An Introduction to Digital Image Processing. Prentice Hall (1986)
12. Shafait, F., Keysers, D., Breuel, T.M.: Efficient implementation of local adaptive thresholding techniques using integral images. SPIE Electron. Imaging **6815**, 10 (2008)
13. Zhang, T.Y., Suen, C.Y.: A fast parallel algorithm for thinning digital patterns. Commun. ACM **27**(3), 236–239 (1984)

An Enhanced Task Scheduling in Cloud Computing Based on Hybrid Approach

Mokhtar A. Alworafi, Atyaf Dhari, Sheren A. El-Booz, Aida A. Nasr, Adela Arpitha and Suresha Mallappa

Abstract Quality of Services (QoS) has become a more interested research point in cloud computing from the perspectives of cloud users and cloud service providers. QoS mainly concerns minimizing the total completion time of tasks (i.e., makespan), response time, and increasing the efficiency of resource utilization. One of the most investigated techniques to meet QoS requirements in the cloud environment is adopting novel task scheduling strategies. Based on our studies, we found that existing solutions neglect the difference in efficiency of resource performance or the starved processes, which can strongly affect the scheduling solution outcome. In this paper, we consider this difference and propose a Hybrid-SJF-LJF (HSLJF) algorithm, which combines Shortest Job First (SJF) and Longest Job First (LJF) algorithms, while considering the load on resources. To start with, the algorithm sorts the submitted tasks in ascending order. Next, it selects one task according to SJF and another according to LSF. Finally, it selects a VM that has minimum completion time to execute the selected task. The experimental results indicate the superiority of HSLJF in minimizing the makespan, response time, and actual execution time while increasing the resource utilization and throughput when compared to the existing algorithms.

M. A. Alworafi (✉) · A. Arpitha · S. Mallappa
DoS in Computer Science, University of Mysore, Mysuru, India
e-mail: mokhtar119@gmail.com

A. Arpitha
e-mail: adelaarpitha23@gmail.com

S. Mallappa
e-mail: sureshabm20@gmail.com

A. Dhari
College Education for Pure Science, Thi Qar University, Nasiriyah, Iraq
e-mail: atyafcomsinc@gmail.com

S. A. El-Booz · A. A. Nasr
CSE, Menoufia University, Shibin El Kom, Egypt
e-mail: eng.sheren1975@gmail.com

A. A. Nasr
e-mail: aida.nasr2009@gmail.com

P. Nagabhushan et al. (eds.), *Data Analytics and Learning*,
Lecture Notes in Networks and Systems 43,
https://doi.org/10.1007/978-981-13-2514-4_2

Keywords Quality of service · Cloud computing · Makespan · Response time
HSLJF

1 Introduction

Cloud computing is the computing paradigm built above many former computing models, such as distributed computing, cluster computing, utility computing, and autonomic computing, which is considered as an advanced step to grid computing model. Cloud computing allows accessing of a wide range of shared resources which exists in data centers. The benefit of such resources is achieved via virtualization [1], which is the backbone of cloud computing. A Virtual Machine (VM) is created with the resources required for each user's tasks where these tasks will be implemented on this VM [2]. The resources of the cloud computing environment can be shared among the VMs (like processing cores and memory). These resources are offered for each VM based on total available processing power in the cloud [3]. These resources receive a huge amount of demand for tasks coming from different places in the world. So, each cloud needs a scheduling strategy in order to determine and process the execution order of the tasks [4]. Therefore, such a huge conglomeration of tasks and resources, shared among users on subscription basis [5], generates one of the main challenges and a hot scope for research called scheduling issue [6]. Scheduling is the process responsible for assigning tasks submitted to suitable resource provision. There are several objectives related to task scheduling algorithms which can be considered in designing the scheduling algorithm such as minimizing fairness, makespan, reducing energy consumption, minimizing cost, minimizing response time, and so on [2]. In general, there are various task scheduling algorithms utilized in cloud, such as Round Robin, which is a simple algorithm depending on the quantum time concept, meaning the time is portioned into intervals and each VM is given a time interval. The VM will execute its tasks based on this quantum. This algorithm does not consider the load and it selects the resources randomly [7]. Shortest Job First (SJF) is another algorithm in cloud computing, which takes into account the length of task where the tasks are sorted in ascending order, whereas Longest Job First (LJF), the opposite of SJF algorithm orders the tasks in descending order [8]. But both algorithms suffer from the starvation problem, which is one of the main problems that face task scheduling in cloud computing where the task may have to wait for a very long time to get its requested resources served [9].

None of the above algorithms is perfect. Some of these algorithms do not consider the length of the task or the load on resources, while others suffer from starvation. Hence, in this paper, we try to overcome these issues by proposing a Hybrid-SJF-LJF (HSLJF) scheduling algorithm. It fulfills the gaps formed in previous scheduling algorithms. In our proposed HSLJF algorithm, we consider the length of the task and the load on each resource. It will first assign the shortest task followed by the longest. For each task, the load on available resources is calculated and the task will be assigned to resource which has less completion time, and when there is more than one resource returning the same completion time, the resource which has the largest computing processor is selected for maximizing the resource utilization that enhances the system performance. The experimental results demonstrate that the proposed HSLJF algorithm minimizes the total completion time of tasks (makespan), response time, and the actual execution time of each task while maximizing the resource utilization and throughput when compared to the existing algorithms.

The structure of our paper is organized as follows: Sect. 2 presents the related work. Section 3 introduces the system model of scheduling. The proposed algorithm is discussed in Sect. 4. Additionally, we present the simulation environment in Sect. 5. In Sect. 6, the performance evaluation is presented. Finally, Sect. 7 concludes our proposal.

2 Related Work

Elmougy et al. [9] proposed a hybrid task scheduling algorithm called (SRDQ), which combines Shortest Job First (SJF) and Round Robin (RR) algorithms using a dynamic changing task quantum to balance waiting time between short and long tasks. Also, the ready queue is divided into two queues: the first queue includes the short tasks, while the long tasks are put in the second queue. SRDQ assigns two tasks from the first queue followed by another task from the second queue. The simulation results demonstrated that their algorithm outperformed SJF, RR, and Time Slicing Priority Based RR (TSPBRR) algorithms by minimizing the tasks' response time and waiting time, with partly the long tasks being starved.

Yeboah et al. [10] enhanced Round Robin algorithm with Shortest Job First algorithm (RRSJF). This proposal selected the processes depending on the shortest job first in a round robin concept to get optimal selection of job. CloudSim toolkit simulator was used for evaluating the performance of the proposed algorithm and the experimental results proved that their proposed algorithm outperformed Round Robin by minimizing average waiting time, average turnaround, and context switches.

Authors in [11] presented a hybrid algorithm which combined the Shortest Job First and priority scheduling algorithms for reducing the waiting time and response time in addition to turnaround time. The proposed hybrid algorithm solved the problem of deadlock or congestion or increase in latency and enhanced the resources' performance of cloud by reducing the latency and communication overhead.

Suri and Rani [12] introduced three phases in their scheduling algorithm model namely minimization, grouping, ranking, and execution. They took into account some parameters that influenced the algorithm's performance such as average waiting time, makespan, completion time, and turnaround time. The task execution is to have normal distribution and exponential distribution. The tasks were ranked depending on the concept of Shortest Job First algorithm (SJF). The experimental results proved that their proposed algorithm was better than First Come First Serve (FCFS) and the Largest Processing Time First (LPTF) algorithms in improving the defined performance parameters.

Ibrahim et al. [13] proposed an algorithm which calculated the total processing power of the available virtual machine and the total tasks' needed to process the power. Then, the tasks were grouped for every VM according to the requested power ratio of tasks, which corresponded to the total processing power of all VMs. The performance evaluation proved that their proposed algorithm reduced the makespan and increased the resources utilization when compared to GA and PSO algorithms.

Al-maamari and Omara [14] amalgamated the PSO algorithm and the Cuckoo Search (CS) algorithm to present a task scheduling algorithm called PSOCS. Independent tasks were used in CloudSim toolkit for evaluating the performance of their algorithm in comparison to PSO and Random Allocation (RA) algorithms. The simulation results proved that PSOCS reduced the makespan and maximized the utilization ratio better than other algorithms.

Srinivasan et al. [15] improved Shortest Job First Scheduling algorithm by using the Threshold (SJFST), which reduced the response time to process the jobs' requests and this led to reducing the starvation as well as fulfilling the load balancing efficiency.

Ru and Keung [16] focused on reducing the waiting time, makespan, and increasing the resource utilization. The authors amalgamated task grouping, prioritization of bandwidth awareness, and SJF algorithm. CloudSim toolkit simulator was used for implementing the experiments on tasks, which were created by Gaussian distribution and resources creation with random distribution. The simulation results proved that their proposed algorithm outperformed the existing algorithms as per the defined performance parameters under different conditions.

Banerjee et al. [17] introduced a new model of cloudlet allocation that improved the QoS. They took care of distributing the cloudlets to the virtual machines depending upon their capacities so as to make the system more balanced and active. Finally, they proved their model efficiency by reducing the completion time of the cloudlet(s), makespan of the VM(s), and the server(s) of a data center by comparing with various existing allocation models.

3 System Model of Scheduling

Cloud computing depends on two main components: a cloud's users and the service provider. The users submit the tasks to the provider for executing their tasks according to QoS requirements [18]. Typically, the cloud computing environment consists of several Data Centers (DCs), which are interconnected and distributed geographically, and are used effectively in order to get better performance. Every DC includes a number of servers, and each server consists of one or more VMs allotted to that server with share and accessibility to the resources of this server, and they are utilized depending on their capacity. The task(s) which is performed by each VM represents load and each VM has different speed and performance based on processing capacity [17]. This complex environment needs a great deal of management and control of resources. The management of cloud resources and tasks are achieved using a good task scheduling [19]. It is a process initiated by the scheduler which determines the appropriate virtual machine to perform the tasks. In our proposed HSLJF algorithm, there are a total of m virtual machines available in the cloud, which is denoted as $VM = \{VM_1, VM_2, \ldots, VM_m\}$ and the number of n tasks denoted as $T = \{T_1, T_2, \ldots, T_n\}$.

4 Proposed Work

The assigning and ordering of task execution is performed to meet scheduling metrics such as completion time, response time, etc. Therefore, the scheduling process is integrated with some optimization criteria that consider fulfilling QoS parameters [19]. One of the major challenges which must be taken into account in developing task scheduling is starvation problem. In this paper, we propose a hybrid-SJF-LJF scheduling algorithm, which tries to overcome or partially overcome this problem by combining two traditional scheduling algorithms: SJF and LJF. In the proposed algorithm, the task with the shortest length will be selected, followed by the task with the longest length. For maximizing resource utilization and achieving efficient performance, the proposed algorithm will select a VM, which returns less completion time and assigns the task to this VM as shown in Fig. 1.

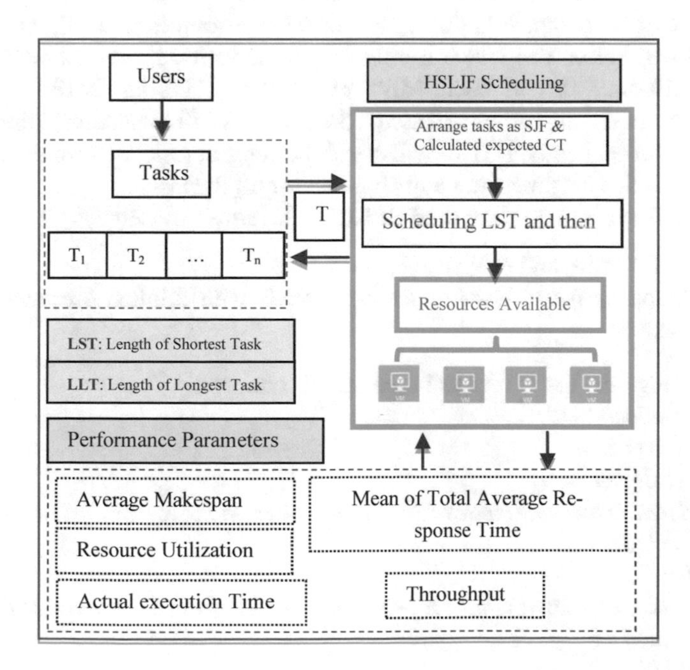

Fig. 1 The proposed HSLJF algorithm

In our proposed HSLJF, the tasks are arranged based on SJF concept. At first, the shortest task will be selected and then the expected completion time in available VMs will be calculated to find the fastest VM. The expected completion time (CT) can be determined by calculating the expected completion time of task in each VM based on Eq. 1 [20].

$$\mathrm{CT} = \mathrm{ET} + load_{VMj} \tag{1}$$

where **ET** refers to the execution time of task that is calculated based on Eq. 2 [20].

$$\mathrm{ET} = \frac{L}{MIPS_j} \tag{2}$$

where **L** indicates the length of the task which is measured with the amount of million instructions. **MIPS** indicates the amount of million instructions per second allocated to $\mathrm{VM_j}$ and $\mathrm{load_{VMj}}$ indicates the previous existing tasks performed by $\mathrm{VM_j}$ as allotted in Eq. 3 [20]:

$$load_{VMj} = \sum_{i=1}^{N} ET_i \tag{3}$$

N refers to the number of tasks into specific VM.

If the expected completion time in two or more resources is equal, our proposed algorithm will select VM which has the largest capacity. If we assume CT of the task in VM0 and VM1 as 2000, MIPS of VM0 as 2000, and MIPS of VM1 as 1000, our proposed algorithm will assign the task to VM0 to maximize the resource utilization. Subsequently, HSLJF will select the longest task and also the lower CT will be selected. This procedure will continue for all the tasks.

The pseudocode of our proposed HSLJF algorithm is as follows:

Input: list of n tasks, and m VMs.
Output: Mapping list of tasks on available VMs with fulfilled user and provider satisfactions.

1. Arrange the list of n tasks based on SJF concept.
2. *Put selected=0*
3. ***For*** *i=1 to n*
4. ***If*** *(selected==0)*
5. *Select the task with index I //select a task from the top of the list*
6. *Selected=1*
7. ***Else***
8. *Select the task with index n − 1 //select a task from the bottom of the list*
9. *Selected=0*
10. ***End if***
11. ***For*** *j=1 to m*

12. *Get the VM with minimum completion time*
13. *If minimum completion time in two VMs is equal*
14. *Select VM with largest MIPS*
15. ***End for**.*
16. *Assign the selected task into the VM with minimum completion time*
17. ***End for**.*

5 Simulation Environment

The proposed HSLJF algorithm has been executed and tested using the CloudSim toolkit 3.0.3, which enables modeling and simulation of cloud computing environment. It is an open source framework, which supports in extending the simulation environment for implementing and evaluating the researchers' algorithms [9].

In our proposed HSLJF, the inner code of CloudSim was extended and modified to test the performance of our proposal in comparison to the traditional algorithms: SJF, LJF, and Round Robin. The basic CloudSim classes that are used for scheduling techniques in cloud environment are relevant to our research in the following manner:

Data center: This class consists of a set of physical machines or servers. It represents an IaaS of cloud provider.

Host: It is a physical machine, which is involved in processing element(s) or core(s), memory, bandwidth, and storage. Various virtual machines numbers can be located on the server that shares the resources of host.

Virtual machine: A virtual machine is a basic unit which performs heterogeneous services to all computer or network resources and shares its resources among several users.

Cloudlet: It refers to a request which is sent from cloud user to the data center. It consists of several characteristics such as length of cloudlet, number of PEs (Processing Elements) required, and output file size [17].

Our algorithm's experiments were performed on a hardware environment involving a laptop, whose configurations are as follows: 2.5 GHz Corei5 CPU, 4 GB memory, and 512 GB hard disk. The simulation environment is described in Table 1.

6 Performance Evaluation

Different metrics are considered to evaluate the performance of task scheduling algorithms, but it is not possible for any algorithm to meet all the metrics in a single solution due to many factors like execution environment or size of the tasks, resource provisioning, etc. [21]. As mentioned earlier, the cloud consists of a cloud service provider and users. They both have their own objectives. While users are focused on the efficiency of the performance in execution, cloud providers are concerned about the efficiency of their resource utilization. These rationales serve as motivations

Table 1 The parameters of CloudSim simulation

Number of datacenter	1
Number of cloud hosts	2
Host MIPS	100,000
Number of CPUs per host	6
Host memory	16 GB
Host storage	1 TB
Host bandwidth	100 GB/s
Virtual machine MIPS	500, 1000, 2000
Virtual machine size	10 GB
Virtual machine memory	0.5 GB
Virtual machine bandwidth	1000
System architecture	x86
Operating system	Linux
Time zone this resource located	10.0
The cost of using processing	0.3
Cloudlets	Generated from a standard formatted workload of a high-performance computing center called HPC2N in Sweden as a benchmark

or objectives of performance criteria while scheduling tasks of users on resources provider. So, the optimization criteria can be distributed into two categories: User satisfaction and provider satisfaction. When scheduling tasks in cloud environment, some of the optimization criteria are:

1. User Satisfaction
   Makespan: It refers to the finishing time of the longest task, as most of the users desire to complete the tasks at the fastest time. Task scheduling should minimize the makespan [22]. It is calculated using Eq. 4.

$$Makespan = max\{CT_i\} \tag{4}$$

   $1 \leq j \leq m$, **m** indicates VM numbers. For all VMs, the average makespan (Avg. Makespan) is calculated based on Eq. 5 [20].

$$Avg.Makespan = \frac{\sum_{j=1}^{m} Makespan}{m} \tag{5}$$

The comparison of our proposed HSLJF algorithm with other algorithms is illustrated in Table 2. It is obvious from all the experiments that the proposed HSLJF returns less average makespan than all other algorithms when the same number of tasks is assigned to different VM numbers. When we consider experiments one and two, HSLJF algorithm has resulted in lesser average makespan. When the number of

Table 2 Average of makespan of HSLJF, SJF, LJF, and Round Robin algorithms

Experiments		Average makespan			
No. of tasks	No. of VMs	**HSLJF**	SJF	LJF	Round robin
500	4	**109,0391**	1,534,173	1,528,533	1,447,505
500	8	**516,290**	683,306	682,665	653,581
1000	8	**677,934**	905,234	923,050	876,210
1000	16	**327,606**	451,778	439,085	468,326
2000	16	**555,931**	749,190	749,182	790,895
2000	32	**273,588**	363,180	366,244	366,497

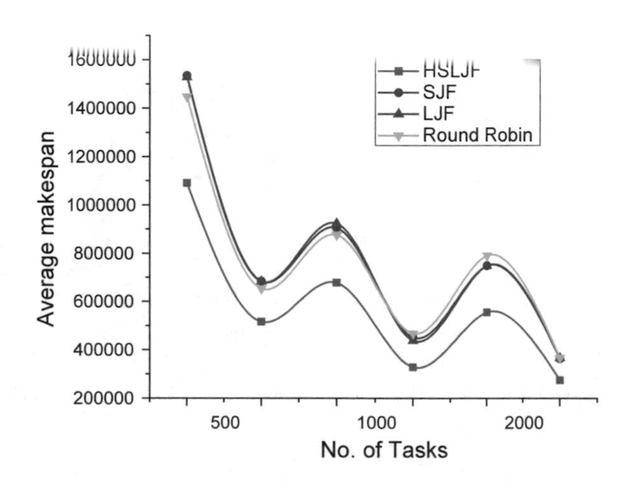

Fig. 2 Comparison of average makespan

tasks is increased on the same VM numbers, such as in experiments two and three, HSLJF algorithm has less incrment of average makespan when compared to the other algorithms.

Figure 2 indicates the average of makespan values in each algorithm. It is clear that HSLJF returns less average makespan, and so achieved better performance than other algorithms when the task number or VM number was maximized.

Response Time (RT): Indicates the total time needed to respond by a cloud computing system and this metric should be minimized. It is calculated based on the Eq. 6.

$$RT = CT_i - SB_i \tag{6}$$

SB: Submission time of task

The average response time (Avg.RT) is obtained as mentioned in Eq. 7.

$$Avg.RT = \frac{\sum_{i=1}^{N} RT}{N} \tag{7}$$

Table 3 Mean of total average response time of HSLJF, SJF, LJF, and Round Robin algorithms

Experiments		Mean of total average response time			
No. of tasks	No. of VMs	**HSLJF**	SJF	LJF	Round Robin
500	4	**325,870**	433,414	432,495	419,602
500	8	**171,651**	194,302	194,290	191,098
1000	8	**173,185**	214,243	218,416	214,022
1000	16	**80,575**	106,607	104,328	114,830
2000	16	**133,997**	168,389	168,024	180,839
2000	32	**66,726**	81,764	82,596	85,519

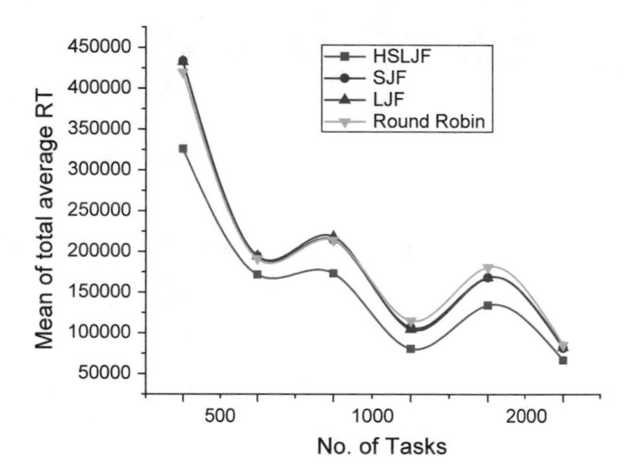

Fig. 3 Comparison of the mean of total average response time

For all VMs, the mean of total average response time (M.Avg.RT) is calculated based on Eq. 8 [20]

$$\text{M.Avg.RT} = \frac{\sum_{j=1}^{m} Avg.RT}{m} \tag{8}$$

The comparison table of another optimization metric which is related to the user is shown in Table 3, where the mean of total average response time of our proposed HSLJF is less than other algorithms in all experimental scenarios.

Figure 3 demonstrates the results of the mean of total average response time of the proposed algorithm when compared to other existing algorithms, where it clearly shows that efficient average response time is represented in HSLJF, while other algorithms got approximate amounts in the mean of total average response time in all experiments.

Table 4 Actual execution time of HSLJF, SJF, LJF, and Round Robin algorithms

Experiments		Actual execution time			
No. of tasks	No. of VMs	**HSLJF**	SJF	LJF	Round Robin
500	4	**15,847,1825**	216,706,843	216,247,473	209,801,456
500	8	**75,372,771**	97,205,197	97,207,053	95,669,524
1000	8	**165,254,666**	214,242,552	218,416,158	214,022,948
1000	16	**78,438,360**	106,634,593	104,372,993	114,809,683
2000	16	**255,094,609**	336,778,584	336,047,149	361,678,942
2000	32	**12,679,7097**	163,565,264	165,217,293	17,131,6101

Actual Execution Time (AT): It refers to the actual execution time that was consumed to implement the task within the available resource. It is defined as in Eq. 9 [23].

$$\text{AT} = \sum_{i=1}^{n} (CT_i - ST_i) \tag{9}$$

where ST_i is the starting time of task.

All users desire the time required to implement their tasks to be as less as possible. The experimental results of actual execution time illustrated in Table 4 indicate that lesser consuming time to execute the tasks is clear in HSLJF algorithm when compared to the existing algorithms, which proves the efficiency of our proposed HSLJF algorithm against SJF, LJF, and Round Robin algorithms.

In addition, Fig. 4 illustrates the actual execution time consumed to execute tasks. Our proposed HSLJF algorithm achieved users' satisfaction by minimizing the total execution time of tasks when compared to the existing algorithms.

2. Cloud Provider Satisfaction

   Resource Utilization (RU): The important optimization metric concerning the provider is maximizing resource utilization. It can be derived from Eqs. 10 and 11 [3].

$$\text{RU} = \frac{\sum_{j=1}^{m} Makespan_j}{m \times Max\_Makespan} \tag{10}$$

   where max_makespan can be expressed as

$$\text{Max\_Makespan} = \max\{\text{Makespan}_\text{j}\} \tag{11}$$

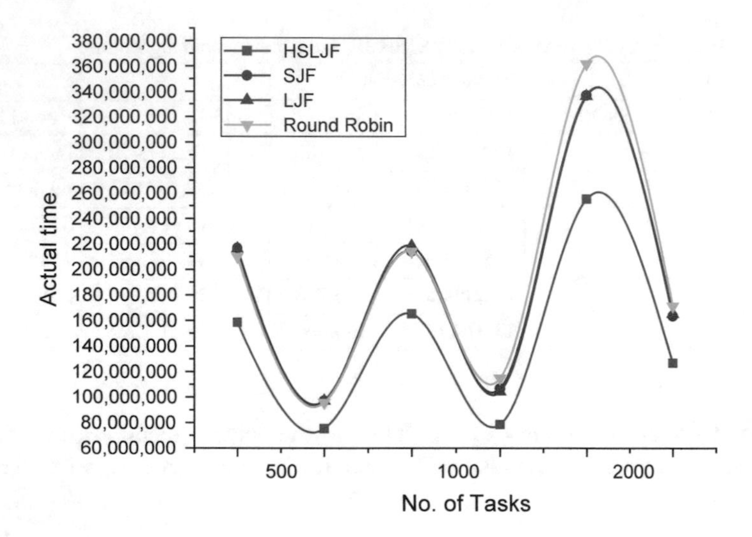

Fig. 4 Comparison of actual execution time

Table 5 Resource utilization of HSLJF, SJF, LJF, and Round Robin algorithms

Experiments		Resource utilization			
No. of tasks	No. of VMs	**HSLJF**	SJF	LJF	Round Robin
500	4	**0.904**	0.654	0.659	0.705
500	8	**0.889**	0.577	0.577	0.651
1000	8	**0.959**	0.614	0.507	0.647
1000	16	**0.641**	0.495	0.485	0.411
2000	16	**0.840**	0.522	0.516	0.490
2000	32	**0.622**	0.515	0.528	0.406

The cloud provider focuses on maximizing the efficiency of resource utilization while achieving acceptable QoS for their users. The efficiency result of resource utilization is illustrated in Table 5 for all algorithms. Increasing the tasks number on VMs would maximize the resource utilization in HSLJF algorithm when compared to the resources utilization for the same number of tasks executed on VMs with the other algorithms.

In Fig. 5, it is clear that the proposed HSLJF algorithm increases resource utilization when compared to the existing algorithms.

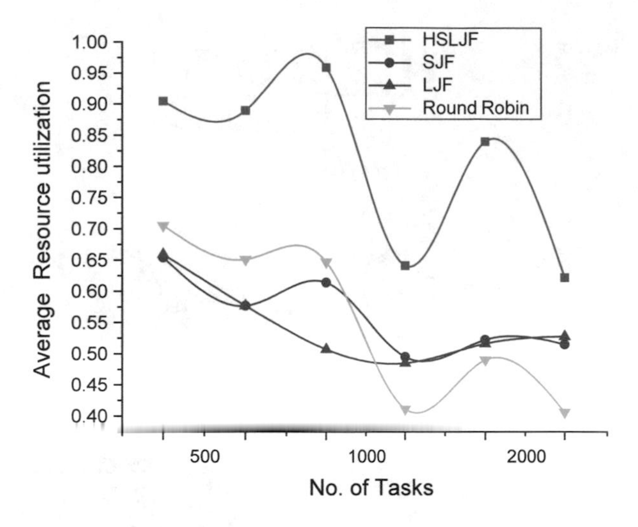

Fig. 5 Comparison of resource utilization

Table 6 Throughput of HSLJF, SJF, LJF, and Round Robin algorithms

Experiments		Throughput			
No. of tasks	No. of VMs	**HSLJF**	SJF	LJF	Round Robin
500	4	**0.000114637803868783**	0.0000814771263122101	0.0000817777774435736	0.0000863554577167409
500	8	**0.000121055906589084**	0.0000914671342072524	0.0000915529975389443	0.0000956269417050513
1000	8	**0.000142659823940658**	0.000138085881220941	0.000135420636677692	0.000142659823940658
1000	16	**0.000190777759999324**	0.000138342165946208	0.000142341543236986	0.000133453833247833
2000	16	**0.000224848209070323**	0.000166846951679968	0.000166848573167265	0.000158048630457251
2000	32	**0.000228445940623056**	0.000172091062529457	0.00017065137920029	0.000170533295986541

Throughput (TH): It is the total number of tasks completed at per unit time. The best task scheduling algorithm should increase the throughput of the system. Throughput can be calculated as mentioned in Eq. 12 [24].

$$\mathrm{TH} = \frac{n}{Makespan} \tag{12}$$

n refers to total number of tasks.

The compared results of the proposed HSJLF algorithm with other algorithms for throughput is in Table 6, in which the results proved that HSJLF returned better performance when the number of tasks and VMs' number are increased as shown in Fig. 6.

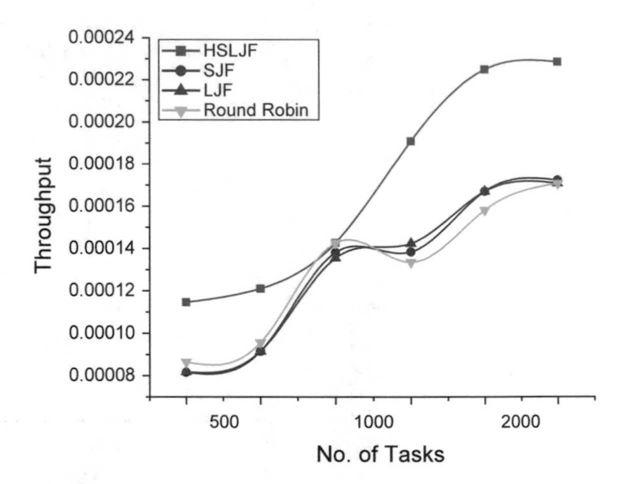

Fig. 6 Comparison of throughput

7 Conclusion

In this paper, we proposed a hybrid shortest–longest scheduling algorithm which tries to handle the starvation problem, which is a very sensitive issue, where most of the state-of-the-art scheduling algorithms suffer. We considered the length of the task and the capabilities of each VM to assign the tasks to the most appropriate VMs, so as to overcome the starvation problem by also considering and satisfying both the user and provider requirements. The experimental results proved that the proposed HSLJF algorithm has outperformed the existing algorithms in minimizing the average makespan, response time, and the actual execution time of tasks while maximizing resource utilization and throughput. In future work, we will concentrate on calculating the cost of resources used.

References

1. Al-Dulaimy, A., et al.: Job submission in the cloud: energy aware approaches. In: Proceedings of the World Congress on Engineering and Computer Science, vol. 1 (2016)
2. Leena, V.A., Ajeena Beegom, A.S., Rajasree, M.S.: Genetic algorithm based bi-objective task scheduling in hybrid cloud platform. Int. J. Comput. Theory Eng. **8**(1), 7 (2016)
3. Devi, D.C., Rhymend Uthariaraj, V.: Load balancing in cloud computing environment using improved weighted round robin algorithm for nonpreemptive dependent tasks. Sci. World J. **2016** (2016)
4. Panda, S.K., Gupta, I., Jana, P.K.: Task scheduling algorithms for multi-cloud systems: allocation-aware approach. Inf. Syst. Front. 1–19 (2017)
5. Thaman, J., Singh, M.: Current perspective in task scheduling techniques in cloud computing: a review. Int. J. Found. Comput. Sci. Technol. **6**(1), 65–85 (2016)
6. Alworafi, M.A., et al.: Task-scheduling in cloud computing environment: cost priority approach. In: Proceedings of International Conference on Cognition and Recognition. Springer, Singapore (2018)

7. Shoja, H., Nahid, H., Azizi, R.: A comparative survey on load balancing algorithms in cloud computing. In: 2014 International Conference on Computing, Communication and Networking Technologies (ICCCNT), pp. 1–5. IEEE (2014)
8. Kruekaew, B., Kimpan, W.: Virtual machine scheduling management on cloud computing using artificial bee colony. In: Proceedings of the International Multi Conference of Engineers and Computer Scientists, vol. 1 (2014)
9. Elmougy, S., Sarhan, S., Joundy, M.: A novel hybrid of shortest job first and round robin with dynamic variable quantum time task scheduling technique. J. Cloud Comput. **6**(1), 12 (2017
10. Yeboah, T., Odabi, I., Hiran, K.K.: An integration of round robin with shortest job first algorithm for cloud computing environment. In: International Conference on Management, Communication and Technology, vol. 3
11. Tiwari, D., Tiwari, D.: An efficient hybrid SJF and priority based scheduling of jobs in cloud computing. Int. J. Mod. Eng. Manag. Res. **2**(4), 26 (2014)
12. Suri, P.K., Rani, S.: Design of task scheduling model for cloud applications in multi cloud environment. In: International Conference on Information, Communication and Computing Technology. Springer, Singapore (2017)
13. Ibrahim, E., El-Bahnasawy, N.A., Omara, F.A.: Job scheduling based on harmonization between the requested and available processing power in the cloud computing environment. Int. J. Comput. Appl. **125**(13) (2015)
14. Al-maamari, A., Omara, F.A.: Task scheduling using hybrid algorithm in cloud computing environments. J. Comput. Eng. (IOSR-JCE) **17**(3), 96–106 (2015)
15. Srinivasan, R.K.I., Suma, V., Nedu, V.: An enhanced load balancing technique for efficient load distribution in cloud-based IT industries. In: Intelligent Informatics, pp. 479–485. Springer, Berlin, Heidelberg (2013)
16. Ru, J., Keung, J.: An empirical investigation on the simulation of priority and shortest-job-first scheduling for cloud-based software systems. In: 2013 22nd Australian Software Engineering Conference (ASWEC). IEEE (2013)
17. Banerjee, S., et al.: Development and analysis of a new cloudlet allocation strategy for QoS improvement in cloud. Arab. J. Sci. Eng. (Springer Science & Business Media BV) **40**(5) (2015)
18. Aldulaimy, A., et al.: Job classification in cloud computing: the classification effects on energy efficiency. In: 2015 IEEE/ACM 8th International Conference on Utility and Cloud Computing (UCC). IEEE (2015)
19. Singh, P., Dutta, M., Aggarwal, N.: A review of task scheduling based on meta-heuristics approach in cloud computing. Knowl. Inf. Syst. 1–51 (2017)
20. Alworafi, M.A., Mallappa, S.: An enhanced task scheduling in cloud computing based on deadline-aware model. Int. J. Grid High Perform. Comput. (IJGHPC) **10**(1), 31–53 (2018)
21. Dutta, M., Aggarwal, N.: Meta-heuristics based approach for workflow scheduling in cloud computing: a survey. In: Artificial Intelligence and Evolutionary Computations in Engineering Systems, pp. 1331–1345. Springer, New Delhi (2016)
22. Kalra, M., Singh, S.: A review of metaheuristic scheduling techniques in cloud computing. Egypt. Inform. J. **16**(3), 275–295 (2015)
23. Dhari, A., Arif, K.I.: An efficient load balancing scheme for cloud computing. Indian J. Sci. Technol. **10**(11) (2017)
24. Panda, S.K., Gupta, I., Jana, P.K.: Task scheduling algorithms for multi-cloud systems: allocation-aware approach. Inf. Syst. Front. 1–19 (2017)

A Moment-Based Representation for Online Telugu Handwritten Character Recognition

Krishna Chaithanya Movva and Viswanath Pulabaigari

Abstract Characters from various languages like Telugu consist of many strokes along with spatial relations among them. The representation for these strokes while being written on a mobile screen or any other personal computers needs to possess invariance to scaling, translation and if necessary, rotational aspects. An old technique for representing shapes has been explored and modified slightly to produce better results. The current work proposes that geometric moment-based features could be adapted to represent a stroke which possesses the needed invariance properties and the usage of neural networks to recognize the corresponding characters from the stroke combinations and the positional information of the strokes.

Keywords Statistical moments · Handwritten character recognition · Telugu
Stroke representation · K-nearest neighbors · Artificial neural networks
HP-Labs-India's Telugu character dataset

1 Introduction

With the advent of portable computers like smartphones and personal digital assistants in the field of technology, the physical keyboards are slowly being replaced by the virtual keyboards. The virtual keyboards could also prove cumbersome when dealing with languages like Japanese, Chinese, Indic languages, etc. When it comes to other languages particularly those which are syllabic, says Telugu, it becomes difficult for the amateur users to type using the virtual keyboard. Handwritten input and Machine Translation come handy and user-friendly in such scenario. Handwritten input recognition is not new at this point. Early research in handwriting recognition dates back to the late nineteenth century. There have been a lot of advancements in

K. C. Movva (✉) · V. Pulabaigari
Indian Institute of Information Technology, Sri City, Chittoor 517646, Andhra Pradesh, India
e-mail: chaitanya.m14@iiits.in

V. Pulabaigari
e-mail: viswanath.p14@iiits.in

P. Nagabhushan et al. (eds.), *Data Analytics and Learning*,
Lecture Notes in Networks and Systems 43,
https://doi.org/10.1007/978-981-13-2514-4_3

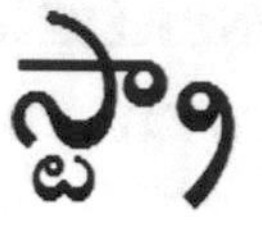

Fig. 1 Four characters showing few combinations of vottus and maatras

this technology till date along with the development of the machine learning and pattern recognition fields. Recent advancements include Google's Handwritten Input tool that incorporates large-scale neural networks and approximate nearest neighbor search and many other sequence and image classification models. Most of the models extract or try to learn the features from the given data. It is rather wise to extract features and perform classification than to make the model learn features by itself when the computation power could not be afforded. In the current work, a moment-based representation technique has been described that becomes a part of the feature extraction process while classifying a handwritten character. To test the features, a dataset consisting of characters from Telugu has been used which is discussed in Sect. 2. Telugu is a language that has more than 800 million users across the globe and is one of the most prominent languages spoken in India and Asia as well. One of the reasons why handwriting recognition is relatively difficult for Telugu scripts when compared with other languages is the complexity of the orthography in Telugu. Telugu is a syllabic language and contains 41 consonants, 16 vowels, and 3 vowel-modifiers whose combinations form more than 10,000 syllabic orthographic units. Telugu characters consist of rounded shapes and very few vertical strokes. The orthography is composed with various vowel sound syllables (*maatras*) modifying the basic consonants. *Vottus* are symbolized pure consonant sounds and can be combined with other consonant or vowel-modified consonant symbols. Vottus and maatras can be positioned at locations surrounding the base character as shown in Fig. 1. Thus, Telugu proves to be one of the challenging scripts for handwriting recognition.

There are many shape recognition techniques that are widely popular since the advent of pattern recognition field. The current work proposes that usage of distributions of few properties of stroke points like local variance and local moments [1] can be used to represent a stroke and also suggests some features to improve the shape recognition scheme in [1]. A stroke could be viewed as a sequence of points from pen-up to pen-down. Each Telugu character is comprised of some strokes and thus individual stroke identification and classification can lead us to individual character classification.

2 Data Collection

There are only a few Telugu stroke datasets available online. HP-Labs-India's Isolated Handwritten Telugu Character dataset used in [2] is one such dataset and consists of approximately 270 samples of each of the 166 Telugu characters written by native

(a) (b)

Fig. 2 Telugu vowel /Ru/ written as a combination of **a** three strokes, **b** one stroke

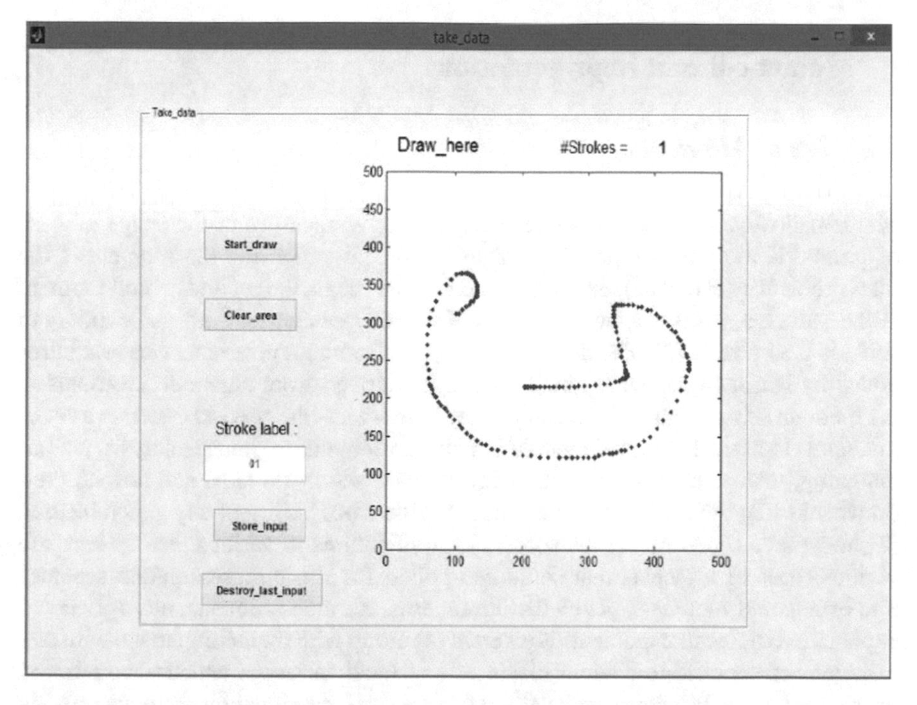

Fig. 3 A MATLAB program showing the generation of stroke data for a single-stroke Telugu character '/ā/' (first vowel in Telugu)

Telugu writers and was collected using AceCad's Digimemo electric clipboard devices and a Digimemo-DCT application. Since the current work also requires stroke samples, a simple MATLAB program has been used initially for creating a dataset of strokes comprising the Telugu vowels. In [3], data from three persons were collected that contains 525 training samples and 105 test samples and this data has been used to create an application that recognizes vowels in Telugu literature. A sample of the data collection process could be witnessed in Fig. 3. Basically, a sequence of x-y coordinates is generated from pen-down to pen-up for every stroke. One considerable point to make here is that there could be many ways of drawing the same stroke. For example, the Telugu vowel "/Ru/", as shown in Fig. 2 can be written in more than one ways.

To obtain good classification results, all such possibilities have to be included in the dataset. Data collection also counts on various application-based aspects such as ease of segmentation of syllabic units, the possibility of sharing stable shapes across syllabic units, etc., and not just based on linguistic criteria [4]. In the end, to check the authenticity of the approach, a dataset of strokes comprising the Telugu vowels is generated which is a subset of the HP-Labs-India's Isolated Handwritten Telugu Character Dataset [2]. The outcome is discussed further in the results section.

3 Moment-Based Representation

3.1 Local Moments

Mathematically, a moment is a specific quantitative measure of the shape of a set of points. If the points represent probability density, then the 0th moment is the total probability (i.e., one), the first moment is the mean, the second central moment is the variance, third is skewness, and the fourth moment(with normalization and shift) is the kurtosis [5]. The direct description of moment invariants was first introduced by Hu in [6] showing how they can be derived from algebraic invariants in his theorem of moment invariants. Geometric moments are used to generate a set of invariants that were then widely used in pattern recognition, image analysis, pattern matching, texture analysis, etc. Hu's invariants became classical and despite their drawbacks like the theory itself does not provide a possibility of any generalization [7], they have found numerous successful applications in various areas. There are various types of moments that could be applied for shape representation schemes like orthogonal moments, geometrical moments, Zernike moments, etc. A detailed explanation of Zernike moments is given in [1] along with their comparison with distributions of centroid-to-contour distances and local variances which were referred to as local moments in terms of invariant features and classification accuracy proving that the local moments are more efficient than the Zernike moments owing to handwritten symbol recognition. Similar symbol representation schemes are adapted in the current work for individual stroke representation. Apparently, an efficient way for representing handwritten symbols would be the one that is invariant to shifting or translation, scaling and if necessary, rotation.

3.2 Preprocessing

The local moments in the current paper have all invariance properties for a given shape as in [1]. Using invariant features rather than the raw data could actually help the model train faster. Preprocessing is an important step to train the model efficiently. Perhaps, sometimes the preprocessing steps are included in the feature extraction steps itself. Some of the preprocessing steps that are relevant to the current problem include smoothing, wild-point correction, boundary normalization, interpolation, etc. A stroke could be piece-wise linear as any program generated stroke data is not perfectly continuous. A Gaussian filter can be used to smooth the strokes. A common smoothing technique averages a point with only its neighboring points [8]. The smoothing achieved depends on the size of the window function and the smoothing kernel parameters used, as in [9]. Most of the Telugu characters are rounded and hence smoothing is required for good recognition accuracy of the system.

3.3 Stroke Representation and Features

The strokes are recognized individually from pen-down to pen-up where each stroke is a set of positive real points. Let 'S' be a stroke which is represented by the sequence of *n* (*n*>*0*) points $P_1, P_2, \ldots, P_n$ where P_i is an x-y coordinate (x_i, y_i). The centroid is stored as $C = \frac{1}{n}\sum_{i=1}^{n} P_i$. For a point P_i, the Eucledian distance to the centroid is denoted as $l(P_i, C)$ which is hereby called as the Centroid-Point-length. To reduce large variations among these lengths, they are normalized by the maximum length, i.e.,

$$\bar{l}(P_i, C) = \frac{l(P_{i,}, C)}{\max\{l(P_1, C), \ldots, l(P_n, C)\}}$$

For a given finite set of points S and a neighborhood radius $r = u * \max\{l(P_1, C), \ldots, l(P_n, C)\}$, we define three random variables instead of two (as in [1]). $A{:}S \rightarrow \mathbf{R}^+$, $V{:}S \rightarrow \mathbf{R}^+$ and $R{:}S \rightarrow \mathbf{R}^+$ where *u* is a hyper-parameter which is determined empirically. Similarly, we define distributions F_A, F_V, F_R and probabilities P_A, P_V, and P_R as follows:

$$F_A = P_A(A \leq t) = \frac{|\{A(P_i) \leq t | P_i \in S)\}|}{|\{A(P_i) | P_i \in S\}|}$$
$$F_V = P_V(V \leq t) = \frac{|\{V(P_i) \leq t | P_i \in S)\}|}{|\{V(P_i) | P_i \in S\}|}$$
$$F_R = P_R(R \leq t) = \frac{|\{R(P_i) \leq t | P_i \in S)\}|}{|\{R(P_i) | P_i \in S\}|},$$

where the functions $A(P_i)$ and $V(P_i)$ are the average and variance of the normalized centroid-point lengths within the neighborhood of radius r respectively. The function $R(P_i)$ corresponds to the normalized centroid-point angles made by the line segment $\overleftrightarrow{P_i C}$ with the horizontal, say $\theta\left(\overleftrightarrow{P_i C}\right)$ and is given by $R(P_i) = \frac{\theta\left(\overleftrightarrow{P_i C}\right)}{2\pi}$. The distributions F_A and F_V alone were shown to possess various invariant properties in [1] like scaling, translation, rotation, and mirror-reflection. Now that the language strokes or symbols have been involved, certain strokes like */kommu/* and */ra/*-vottu as in Fig. 4 could be confusing when the features are invariant to rotation and flipping (mirror-reflection). So, another distribution F_R has been added to the current set of features.

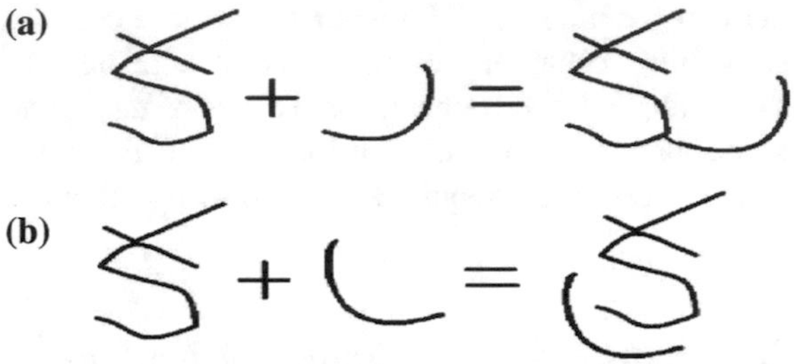

Fig. 4 **a** A stroke called /kommu/ being combined with the letter /ka/ to form /ku/. **b** A stroke called /ra-vottu/ being combined with the letter /ka/ to form /kra/

It could be shown that the distribution F_R, varies over rotation and flipping unlike F_A and F_V.

Proof For a given stroke S, let there be a point P ∈ S, C be the centroid of S and let P′ be P's corresponding point in S′, which is a transformation of S. If the new stroke S′ is formed by rotating S anticlockwise by an angle α ($0<\alpha\leq 2\pi$), then the line $\overleftrightarrow{PC}$ is also rotated by an angle α and let the new line be $\overleftrightarrow{P'C}$. We can say that $\theta\left(\overleftrightarrow{P'C}\right)=\theta\left(\overleftrightarrow{PC}\right)+\alpha$ (and $\theta\left(\overleftrightarrow{P'C}\right)=\theta\left(\overleftrightarrow{PC}\right)-\alpha$ if S was rotated clockwise). As $0<\alpha\leq 2\pi$, $\theta\left(\overleftrightarrow{P'C}\right)\neq\theta\left(\overleftrightarrow{PC}\right)$ and hence the stroke is being treated as a different stroke when it is rotated. In other words, the rotational variance for a stroke is being preserved in the features.

4 Stroke Classification and Character Recognition

The derived features can be used to train a model to recognize a given test character. Usually, in stroke-based models each stroke is labeled with the hypothesized character lists that the stroke may represent. There is a wide range of techniques that could be applied at this stage. In [10], a segmental two layered Hidden Markov Model was used where the strokes were recognized first and later the relationship among the strokes. Considerable results were achieved with only a few training samples per symbol. Although Telugu scripts were not discussed in [1, 10], shape representation schemes proposed in them can seemingly be applied to Telugu characters. In the current work, similar techniques shall be used. The derived features are given to a k-Nearest Neighbor model that computes on symmetric Kullback–Leibler distance [1] and then the stroke combinations are identified by a simple artificial neural network model.

4.1 K-Nearest Neighbor (KNN) Model

The k-Nearest-Neighbor classification algorithm essentially involves finding the similarity between the test sample and every sample in the training set. Then, based on the similarity measures, k closest neighbors to the test pattern are identified and the frequently occurring class among the k neighbors is the classification result. Choosing the hyper-parameter k and the distance metric is crucial for the working of this model. Efficient value of k can be determined using cross-fold validation over a range of values for k. For large datasets, k can be larger to reduce the error. In the current character recognition scheme, the individual strokes are identified using the k Nearest Neighbor model and then a vector that encodes the individual stroke classes together with their relative positions is constructed. For a stroke S, let the probability functions P_A, P_V, and P_R be denoted as $P_{A,S}$, $P_{V,S}$, and $P_{R,S}$, respectively. The symmetric Kullback–Leibler distance which is a popular way of measuring distance between probability distributions will be the distance metric used for the current problem. From a stroke S_i to a stroke S_j, the symmetric Kullback–Leibler distance over P_A is given by

$$\begin{aligned} D_{KL}\left(P_{A,S_i}, P_{A,S_j}\right) &= KL\left(P_{A,S_i}, P_{A,S_j}\right) + KL\left(P_{A,S_j}, P_{A,S_i}\right) \\ &= D_{KL}\left(P_{A,S_j}, P_{A,S_i}\right) \end{aligned}$$

$$\text{where } KL\left(P_{A,S_i}, P_{A,S_j}\right) = \sum_{i=1}^{p} P_{A,S_i}(a_i).\ \log \frac{P_{A,S_i}(a_i)}{P_{A,S_j}(a_i)}$$

and where the p intervals $a_1, a_2, \ldots, a_p$ are the result of quantizing the range of A. As the random variable A deals with the averages of the normalized centroid-point lengths, the intervals a_i lies in between 0 and 1, $\forall 1 \leq i \leq p$. Also, if a_i = (m,n] for $0 \leq m < n \leq 1$, $P_{A,S_i}(a_i) = F_A(n) - F_A(m)$. Similarly, let V and R also be quantized into intervals $v_1, v_2, \ldots, v_q$ and $r_1, r_2, \ldots, r_w$ respectively (q, w > 0). Then,

$$\begin{aligned} D_{KL}\left(P_{V,S_i}, P_{V,S_j}\right) &= KL\left(P_{V,S_i}, P_{V,S_j}\right) + KL\left(P_{V,S_j}, P_{V,S_i}\right) \\ D_{KL}\left(P_{R,S_i}, P_{R,S_j}\right) &= KL\left(P_{R,S_i}, P_{R,S_j}\right) + KL\left(P_{R,S_j}, P_{R,S_i}\right) \end{aligned}$$

Finally, the measure of divergence between S_i and S_j that involves all the distributions is given by

$$D(S_i, S_i) = D_{KL}\left(P_{A,S_i}, P_{A,S_j}\right) + D_{KL}\left(P_{V,S_i}, P_{V,S_j}\right) + D_{KL}\left(P_{A,S_i}, P_{A,S_j}\right)$$

Then, the relative positions of the strokes with respect to the base stroke in a character is calculated by determining which bin (Fig. 5) the centroid of the stroke goes into and a fixed length vector is made out of this information and the stroke label obtained from the k-Nearest-Neighbor model. If multiple strokes fall into the same

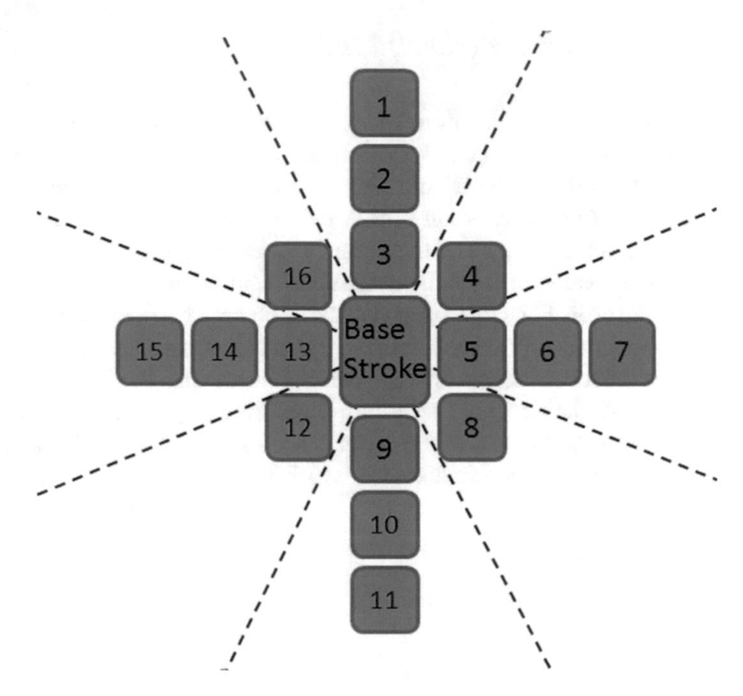

Fig. 5 Grid to determine the position of the strokes with respect to the base (first) stroke

quadrant, then the second stroke goes into a second bin in the respective quadrant and so on. The positioning of the bins has been considered primarily based on the structure of the Telugu characters. Normally, 16 bins as shown in Fig. 5 would be enough to cover most of the characters in Telugu. The fixed length vector is then fed to an artificial neural network model for character classification.

4.2 Artificial Neural Network (ANN) Model

An artificial neural network is a computational model based on the structure and functioning of the biological neural networks. They can provide the necessary mapping between a given set of inputs and outputs which is why they have been used here. The vector from the k-Nearest-Neighbor model contains the information about the strokes that are recorded till now and is helpful in classifying what character could be formed from the available set of stokes. A simple three-layer neural network model with 37 input neurons, 50 hidden neurons, and 16 output neurons has been trained with the *Levenberg–Marquardt* training algorithm and mean square error performance function to identify a character. The input size was 37 that includes the output from the k-Nearest-Neighbor model in its one hot encoded form which is of length 21 and the positional information from the grid (Fig. 5) which is of 16 in size. The 16 neurons

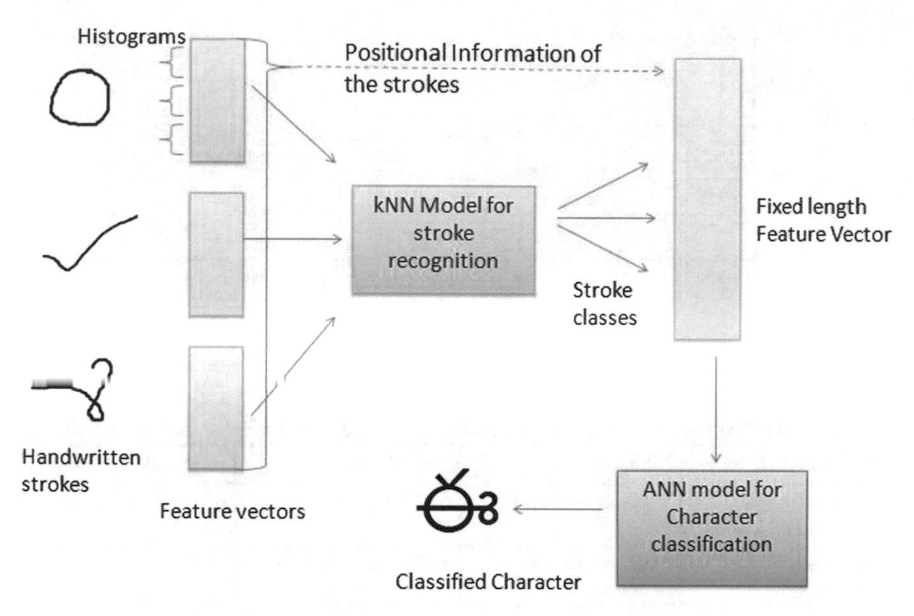

Fig. 6 Block diagram of the current character recognition model

in the output correspond to the 16 vowels in the Telugu language. The entire model could be witnessed in Fig. 6. Normally, extracting features such as those described in the above sections are often skipped in the artificial neural network models. It is because the artificial neural networks are assumed to learn the classifying functions through the set of weights by continuously updating the weights while minimizing the loss in the process of training. Still, the feature extraction steps help to improve the accuracy of the model entirely. It is always better to tell the model what features to train on rather than giving the raw data and train on some random set of features to get good accuracies.

5 Results

The random variables A, V, and R are equally quantized in the range of [0, 1] with interval lengths 0.1, 0.05, and 0.1, respectively. In the above-discussed procedure (in Sects. 3.3 and 4.1), the parameter 'r' which is the neighborhood radius, is replaced with a parameter 'v' corresponding to the number of nearest neighbors from the given point P and is applied over the dataset in [3] that consists of strokes comprising the vowels in Telugu literature. This replacement does not affect the procedure much, as we have to deal with parameters in both cases (as r varies with u in the first case) that are to be determined empirically. It yielded an accuracy of about 83.81% over 630 samples of which 83% (525 samples) was considered as training data and the

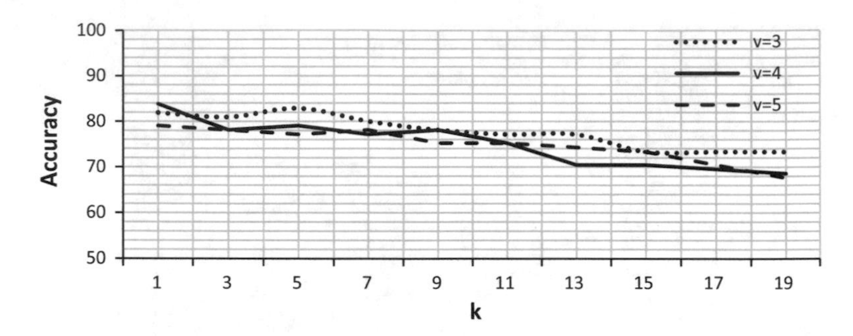

Fig. 7 k versus classification accuracy graph for the Matlab application with v = 3, 4 and 5

Table 1 Table showing the accuracies on the dataset [3] used for the kNN and ANN models

Classification type	Model	Dataset size	Accuracy (test) %
Individual stroke classification	k-nearest neighbor	630 samples	83.81 (at k = 1 and v = 4)
Character classification	Artificial neural networks	800 samples	91.5

remaining 17% (105 samples) as test data. Figure 7 shows the k Vs accuracy graph for strokes comprising Telugu vowels with the features proposed in this work. Further, the artificial neural network model has been employed to identify the character from the vector generated by the k-Nearest-Neighbor model. It yielded a result of 91.5% accuracy on a dataset [3] that consists of 800 samples of Telugu Vowels. Though the dataset is small, the point to consider here is the features used for this model. They form an extended version of features in [1], where the problem is chosen was hand-drawn shape classification. The current set of features has been considered based on the current problem which is handwritten character classification (Table 1).

To compare the current set of features with the shape representation scheme in [1], a subset of the HP-Labs-India's Isolated Handwritten Telugu Character Dataset from [2] that consists of various strokes comprising of the Telugu Vowels, has been generated. This new dataset consists of 23 classes that are derived from the strokes that form the Telugu Vowels and in total it has 5081 samples after eliminating the improper samples. Feature sets from both the schemes are fed to the k-Nearest-Neighbor model (Sect. 4.1) and the results are shown in Figs. 8 and 9 for the features from [1] and in Figs. 10 and 11 for the features described in the above sections respectively. In Figs. 8 and 9, the accuracy was high for small values of k and it went decreasing until it gets stabilized at some point further. On comparing these results with the plots in Figs. 10 and 11, it is clear that the new set of features, i.e., those which also preserve the rotation information stabilize at higher accuracies and are better than the features from [1] for handwritten character classification.

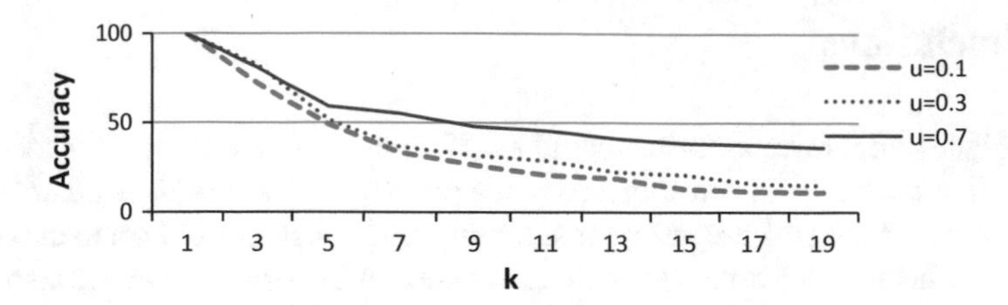

Fig. 8 k versus accuracy (kNN) graph for the first Telugu vowel with hyper-parameter u = 0.1, 0.3 and 0.7 and features described in [1]

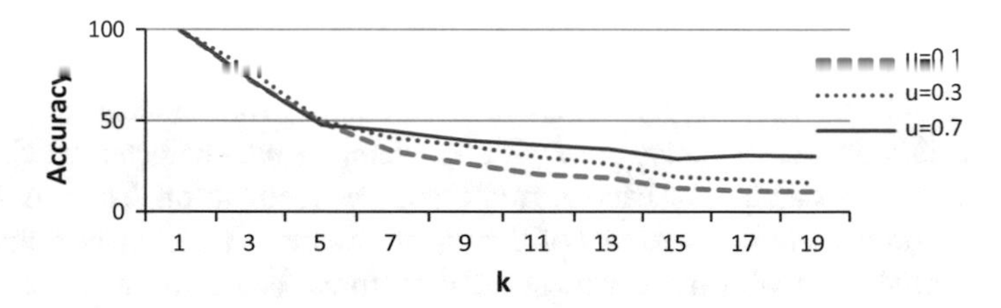

Fig. 9 k versus accuracy (kNN) graph for the second Telugu vowel with hyper-parameter u = 0.1, 0.3 and 0.7 and features described in [1]

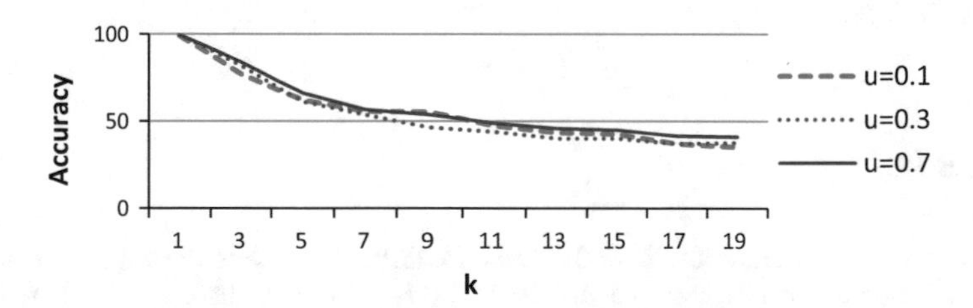

Fig. 10 k versus accuracy (kNN) graph for the first Telugu vowel with hyper-parameter u = 0.1, 0.3 and 0.7 and features in the current work

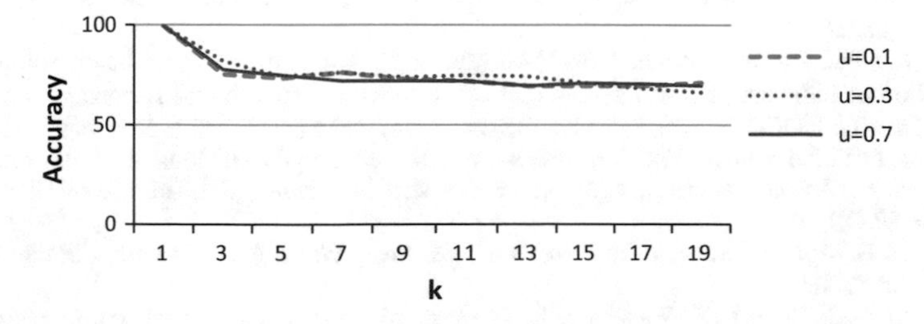

Fig. 11 k versus accuracy (kNN) plot for the second Telugu vowel with hyper-parameter u = 0.1, 0.3 and 0.7 and features in the current work

6 Conclusions

A technique and a feature-set that could help in developing a recognizer for Telugu handwritten scripts have been proposed based on two other works [1, 10]. The features given in the current work are invariant to scaling and translation but not to the rotation as there is a need to preserve the rotational aspects of the stroke. They are also shown to yield better results (Figs. 10 and 11) than those in [1] (Figs. 8 and 9) using the data samples from HPL Telugu character dataset. These results also show that with properly fine-tuned features, classification models can achieve good results. Though most of the current state-of-the-art methods use image data to train their recognizers, pen data could also prove to achieve good results as does in the current work and also takes less storage. The artificial neural network models are proven to give good classification accuracy even with few samples per character which makes them efficient and suitable for handwritten character recognition. There is a lot of scope for improvement in the results of the current work as the data available itself was very limited and with more concentrated features, better results as of models trained on image data could be achieved.

Acknowledgements The work presented in this paper is partly funded by a DIC project with reference: DIC_Proj14_PV.

References

1. Viswanath, P., Gokaramaiah, T., Prabhakar Rao, G.V.: A Shape Representation Scheme for Hand-Drawn Symbol Representation, MIWAI 2011, LNAI, vol. 7080, pp. 213–224 (2011)
2. Prasanth, L., Babu, V., Sharma, R., Rao, G., Dinesh, M.: Elastic matching of online handwritten Tamil and Telugu scripts using local features. In: Ninth International Conference on Document Analysis and Recognition, 2007, ICDAR 2007, vol. 2, pp. 1028–1032. IEEE (2007)
3. Chaitanya, M.K.: Telugu vowels stroke data. https://github.com/krishr2d2/Telugu_Vowels_stroke_data (2017)
4. Agrawal, M., Bhaskarabhatla, A.S., Madhvanath, S.: Data collection for handwriting corpus creation in Indic scripts. In: International Conference on Speech and Language Technology and Oriental COCOSDA (ICSLT-COCOSDA 2004), New Delhi, India, November 2004
5. Moment (mathematics), Wikipedia. https://en.wikipedia.org/wiki/Moment_(mathematics)"
6. Hu, M.-K.: Visual pattern recognition by moment invariants. IRE Trans. Inf. Theory **8**(2), 179–187 (1962)
7. Flusser, J.: Moment invariants in image analysis. Proc. World Acad. Sci. Eng. Technol. **11**(2), 196–201 (2006)
8. Tappert, C.C., Suen, C.Y., Wakahara, T.: The state of the art in online handwriting recognition. IEEE Trans. Pattern Anal. Mach. Intell. **12**(8), 787–808 (1990)
9. Swethalakshmi, H., Jayaraman, A., Chakravarthy, V.S., Sekhar, C.C.: Online handwritten character recognition of Devanagari and Telugu characters using support vector machines. In: Tenth International Workshop on Frontiers in Handwriting Recognition. Suvisoft (2006)
10. Artieres, T., Marukatat, S., Gallinari, P.: Online handwritten shape recognition using segmental hidden markov models. IEEE Trans. Pattern Anal. Mach. Intell. **29**(2) (2007)

A Switch-Prioritized Load-Balancing Technique in SDN

K. A. Vani, J. Prathima Mabel and K. N. Rama Mohan Babu

Abstract Software defined networks is an emerging paradigm that offers a split architecture by separating the functionalities of control plane and Data plane. Both these planes work independently, i.e., control plane and data plane are decoupled. SDN aggregates the intelligence of entire network elements into the control plane. Various load balancing methods have been proposed in traditional networks but these approaches cannot satisfy the requirement of load balancing in all scenarios. This paper apprehends the different load balancing techniques available in SDN. The key idea of load balancing using switch prioritization via flow prioritization is discussed with simple simulation scenario. In comparison to other techniques the proposed idea is new and helps in achieving better load balancing in SDN.

1 Introduction

With the upward growth of network applications, it is becoming hard to scale and manage networks using traditional methods. This is because of the tight coupling between the control plane and the data planes in traditional networks. The networking devices are not programmable in traditional networks. To address these issues a new network paradigm called Software-defined networking (SDN) was introduced. The control plane and data planes are decoupled in SDN which makes it possible to program the networking devices. Any communications between the decoupled planes are carried out by an OpenFlow API. The Ethernet protocol at the data link layer makes use of a OpenFlow switch. The OpenFlow switches fetch the data from the

K. A. Vani (✉) · J. Prathima Mabel · K. N. Rama Mohan Babu
Department of Information Science & Engineering, Dayananda Sagar College of Engineering, Bengaluru, India
e-mail: vaniram.reddy@gmail.com

J. Prathima Mabel
e-mail: prathimamabel@gmail.com

K. N. Rama Mohan Babu
e-mail: rams_babu@hotmail.com

P. Nagabhushan et al. (eds.), *Data Analytics and Learning*,
Lecture Notes in Networks and Systems 43,
https://doi.org/10.1007/978-981-13-2514-4_4

controllers. The controller is mainly responsible for building the flow entries in the flow tables. The controller acts as a brain of SDN since it is capable of acquiring the global view of the network. The controllers provide this global view to the OpenFlow switches via which it is made possible to implement flexible and efficient networks. Minimizing network latency and improving the network throughput are the important aspects of any network. To tackle this issue load balancing techniques can be adopted. Load balancing techniques help in better utilization of the resources and can provide better bandwidth utilization in the network. However providing load balancing in traditional network is intricate for the network administrator as the devices here are proprietary and it is required to configure each device to run a network protocol. The solution to this crisis is SDN.SDN is a sovereign that brings in innovation in configuring and managing the networks [1]. The paper is organized as: Section 2 provides an overview of SDN. Section 3 brings in the load balancing technologies in SDN. Section 4 describes the performance issues of each technique and final the paper concludes in Sect. 5.

2 The Overview of Software Defined Network

The concept of SDN came into light at Stanford University with the clean slate project. SDN offers a greater flexibility in managing the networks as it comprises of two planes the control and data planes which are decoupled from each other The control plane acts as a brain of SDN since every decision on each flow is decided by the controller. The data plane act as a forwarding device as it just has to follow the instructions of the controller. The communication between Control plane and Data plane is taken care by a special API called OpenFlow. The layered architecture of SDN is as depicted in Fig. 1. There are three different layers infrastructure layer, control layer, and application layer. The communication among the networking devices are established at the infrastructure layer. The control plane or the controller helps in logically managing the network also it provides global view of entire network. The next layer is application layer which deals with deployment of various network applications. Some of the applications may be load balancing [2], IDS [3], scheduling [4], routing [5].

3 Load Balancing Techniques

High performance and distributing loads among multiple servers in any network is achieved by incorporating an efficient load balancing technique. Load balancing techniques are classified into static and dynamic [6, 7].

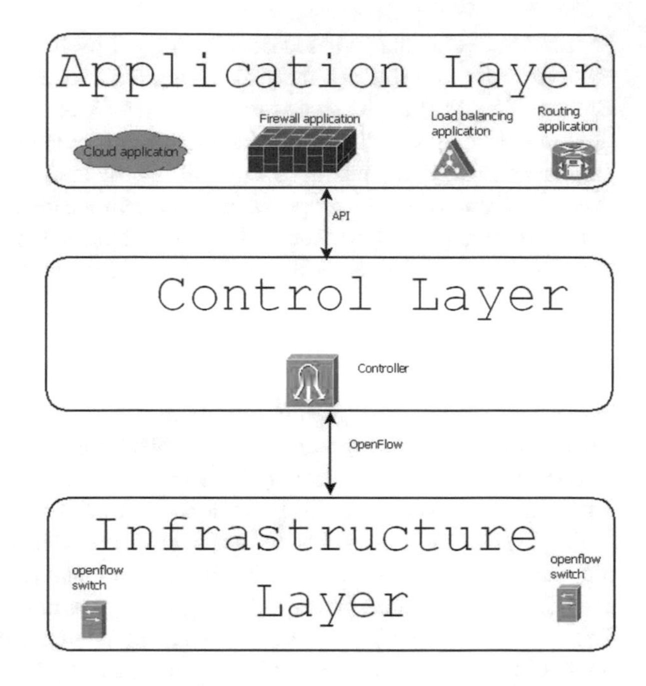

Fig. 1 SDN architecture

Static load balancing:
This technique makes use of average behavior of the system rather than current state of a system. The methods that are used in this technique are non-deterrent, i.e., once the static loads balancing algorithms are round robin [8], central manager [9], threshold algorithm [10], and randomized algorithm [11]. The aim of this technique is to minimize the execution time and delays. The major lacuna of this technique is it ignores the current system state while making the current allocation.

Dynamic load balancing:
This technique tracks changes in the system and load distribution is done as per the requirement of the given system. The methods or policies defined here are dynamic, i.e., they react to according to state of the system. Some of the dynamic loads balancing algorithms are least connection [12], token routing [13], Central queue [9], and Local queue [14].

Associated work in SDN:
Load balancing can be realized in many ways. In SDN load balancing is rendered either on the control plane or on the data plane. Apart from these several other techniques are based on various factors like server load, controller load, flow-based load distribution, traffic distribution to manage load across different servers, policy-based load distribution, load distribution during security attacks, etc.

Load balancing techniques on control plane:
As described in [15] the load on the controller is balanced based on genetic algorithm. In this method to assign switches to the controllers randomly it makes use of

population coding that is based on mapping solution. This algorithm distributes the load across the controllers if the load reaches beyond given threshold. The other work Balance flow [16] is a designed for wide area OpenFlow networks that distributes flow request dynamically among the controllers for achieving fast response also the controller utilization is maximized by shifting the load from high loaded controller to less loaded controller. As proposed in [17] to reduce the latency and to maximize the cluster throughput a dynamically distributed load balancing algorithms for control plane is discussed. This algorithm makes use of a concept called wardrop equilibrium. The main objective of this algorithm is to make controller to learn about different flow rates from each switch and to have a convergence to stabilize the policies. So that the traffic is balanced among the cluster of controllers.

Load balancing techniques on data plane:
Most of the load balancing techniques is implemented on control plane. More recently in HULA [18] the load balancing is adopted in data plane. The authors present an algorithm that overcomes the limitations of traditional load balancing techniques scalability at the edge switches and use of customized hardware. As discussed in [19] the control tasks to switches are delegated using stateful dataplane SDN that could provide multi-path load balancing along with gain in performance and scalability. The other load balancing techniques on data plane are discussed in [20] that is based on distributing the loads across different routes using hashing techniques.

Server load balancing:
It was hard to gain efficient load balancing using traditional networks. To overcome this lacuna and to balance the server load many techniques were used. In SDN to achieve load balancing on the server side an efficient server load balancing scheme based on ant colony optimization technique may be adopted [21]. In this method the server and routes leading to server are balanced. Here to find the best server and best path for server it collects server load and statistics of a network.

Path load balancing:
To increase the network reliability and optimizing link utilization path load balancing [22] may be employed. In this method fuzzy synthetic evaluation method [FSEM] is used. Here the flow handling rules are defined in the controller. FSEM has a global view of the entire network that makes it possible for it to control the paths that change dynamically.

Load balancing based on routing:
In [23] Long et al. proposes LABERIO a load balancing algorithm based on path switching for balancing transmissions dynamically. The authors devised two algorithms one is single hop LABERIO algorithm. This algorithm finds out the biggest flow on the link and assigns it as object flow. The main objective of this algorithm is to allocate alternate path for the flow which consumes more bandwidth. The other is multi-hop LABERIO algorithm. This algorithm on each interval it picks the overloaded hops and sets it as object flow. The main aim of this algorithm is to provide path switching which performs better in link overloaded scenarios.

Switch-controller load balancing:
A new approach to balance load based on switch-controller mapping is presented in [TNOVA] [24]. The project aims at managing the load across multiple controllers and switches. This is based on maintaining the connection between switch and controllers that estimates current controllers load.

Load balancing during security attacks:
DDoS is common and important security issue that needs to be taken care of in any network. Achieving load balancing during security attacks such as DDoS is pretty much interesting. This is addressed in [25]. Here it follows two solutions based on DDoS Mitigation. One is attacking traffics are filtered and in the other survival techniques are used to effectively balance the load among the end points.

Traffic Engineering based load management:
Traditionally to achieve performance optimization traffic engineering techniques were widely used in ATM and IP/MPLS networks. SDN provides a global view of entire network that aids in managing the traffic flows among different controllers. As discussed in [26] load balancing can be implemented both at control and data planes. However this paper highlights that considering traffic engineering for flow management and providing consistent view of entire network adds a greater advantage of improving scalability, performance, and high availability and reliability. The comparative study of these load balancing technique are summarized in Table 1.

4 Proposed Idea

In the above-cited works most of them have implemented load balancing using different methods. Some of them are suitable for different scenario's handling different kinds of traffic and balancing the load. Apart from these works our proposal mainly aims at provisioning Load balancing based on prioritizing the flows in the switches by the controller. The key idea of this proposal is to provide scalability, availability and elasticity for applications that are distributed geographically. As shown in Fig. 2 the load balancing application help access the data plane forwarding devices like switches based on the priorities. The priorities are given to switches in two ways

(i) Based on the type of traffic the cluster is handling
(ii) Based on their current load status.

We first discuss the first case where the load distribution takes place based on the type of traffic the cluster is handling. For example, as depicted in the diagram below. The controller defines the flow rules and sets the priority for each of the flows. The switches that handle TCP traffic are grouped as one cluster, i.e., SC_1. Likewise, the switches that handle UDP traffic are grouped as one cluster, i.e., SC_2. Suppose the traffic generated is TCP the cluster of switches SC_1 is given first priority. Since the TCP flows require guaranteed service. If the flow generated is UDP it is sent to the cluster of switches SC_2. Since the UDP flow does not care for guaranteed service.

Table 1 Comparison of load balancing techniques

Associated work	Technique used	Implementation level	Advantage	Disadvantage
Load balancing techniques on control plane	Genetic algorithm	Control plane	Provides capability of handling arbitrary constraints. It also provides optimization in deploying switches to controller	The accuracy of different fitness functions used in the algorithm are not discussed in detail
	Balance flow using cross controller communication		Reduction in Propagation delays in whole network	Detection of imbalance between the controllers is not addressed properly
	Dynamically distributed load balancing algorithms		Increase in cluster throughput	It is not clear that how the network attains equilibrium at first place
Load balancing techniques on data plane	HULA	Data plane	It is very effective in adapting to the volatile conditions of network in addition to it provides more stability	Since the distance vector method is used for distribution of network utilization information it creates more traffic on the switch
	Stateful dataplane		It helps in increasing the performance and scalability by delegating the control tasks to switches	Many control packets are exchanged between the switch and controller which creates extra traffic on the link
Server load balancing	Ant colony optimization	Server	Increased performance by optimal resource utilization	The load distribution changes in each iteration

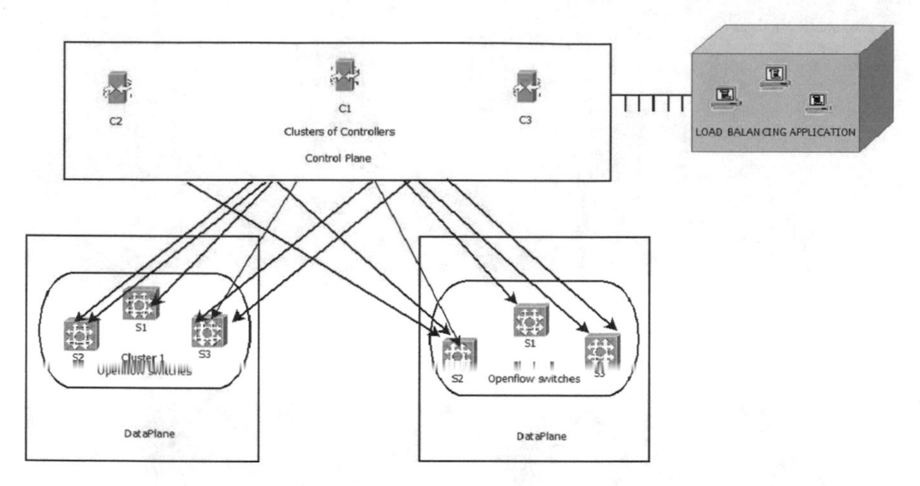

Fig. 2 Proposed load balancing architecture

Based on the kind of traffic generated the flows will be handled by the respective clusters. This method will increase the performance of the network since the flows are handled by dedicated clusters of switches. The second method is load distribution based on the load handled by the switches. Since the controller in SDN will have the overview of the entire network and the load status of all the dataplane elements it is possible for the controller to define the flow to a particular switch dynamically. For example the switches are connected to the cluster of controllers. When the flow arrives the switch follows rules defined in the flow table. The controller will have the complete load status of the current network and the load handled by each end device. Based on the workload each switch is handling the flows can be distributed among the cluster of switches. That is, if the switch with less load on it will be having highest priority to serve the current request. If the controller finds if one of the switch is free to handle the request it will be sent to the switch with highest priority. The load balancing can be achieved by considering the relative difference between each switch in a cluster. The performance will be better if the difference is less. The flows can be defined in terms of relative difference. In case of $\$_f$, FC_i represents the flow count assigned by the controller to each switch.

$$\$_f = \left\{ \frac{|f1 - f2|}{|f1 + f2|} \ \ldots \text{s.t. } 0 \leq \$_f \leq 1 \right.$$

If the flow count reaches the threshold level [80%] then controller assigns another switch to handle the flow. The load among different cluster of switches can be balanced by prioritizing the flows.

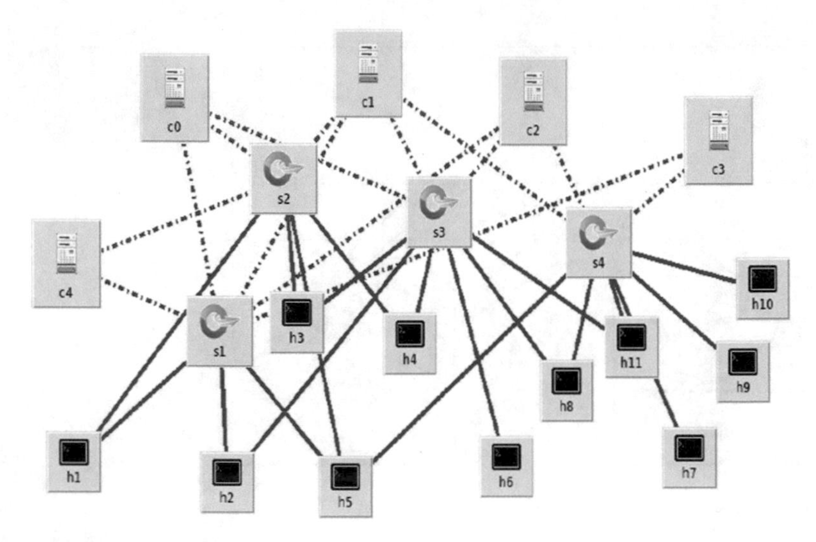

Fig. 3 Experimental setup

Switch Load Balancing Algorithm

Require: Priflow
1:for each(pflow){
//dynamically controller chooses the least loaded switch
2:dst = getLeastLoadedSwitch(pflow);
// dynamic load distribution algorithm select the least loaded switch//
3:load = apply theload statistics of network elements
4:for each(switch in cluster){
5:installFlowEntry(switch);
6:}
7:perform local update after each iteration
8:}

Experimental Results

As per the proposed idea the results are simulated using mininet 2.3.0.0. As shown in the setup in Fig. 3 we considered five openflow Controllers [c0 to c4], four switches from s1 to s4 and 11 hosts h1–h11. The result shows that switch s4 is heavily loaded. Now host h7 sends a request to s4. The switch s4 fails to respond to h7. At this scenario the controller c2 distributes the load of s4–s1 which was having less load. The experimental results and setup files are as shown in Figs. 4, 5, 6 and 7.

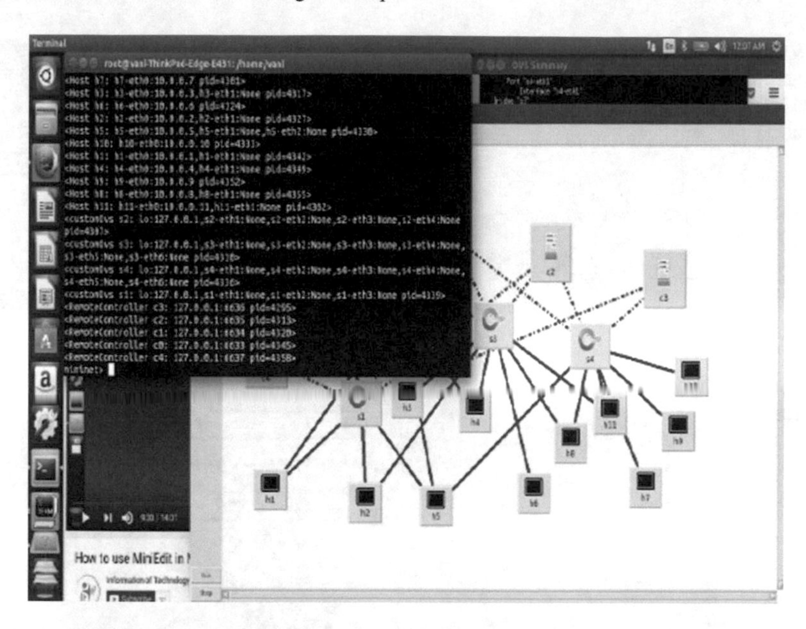

Fig. 4 TCP dump flows

Fig. 5 OVS flows

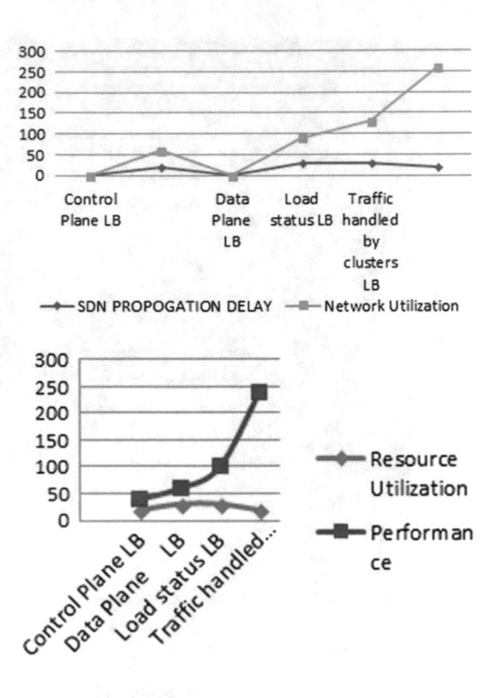

Fig. 6 Reduction in propagation delay and increase in network utilization

Fig. 7 Increase in performance with less resources

5 Conclusion

In this proposed architecture and from the given algorithm better load balancing scheme can be achieved based on the concept of switch prioritization. In comparison with other architectures the load among all the data plane elements can be balanced more efficiently. This algorithm prioritizes the load based on the predictors decision as given in the above algorithm. Hence this method yields better results.

References

1. Feamster, N., Rexford, J., Zegura, E.: The Road to SDN: An Intellectual History of Programmable Networks (2013)
2. Zou, T., Xie, H., Yin, H.: Supporting software defined networking with application layer traffic optimization. Google Patent 13 801 850 (2013)
3. Chung, C.J., Khatkar, P., Xing, T., Lee, J., Huang, D.: NICE: network intrusion detection and countermeasure selection in virtual network systems. IEEE Trans. Depend. Secure Comput. **10**(4), 198–211 (2013)
4. Li, L.E., Mao, Z.M., Rexford, J.: Toward software-defined cellular networks. In: Proceedings of the European Workshop on Software-Defined Networking (EWSDN), Darmstadt, Germany, pp. 7–12 (2012)
5. Kotronis, V., Dimitropoulos, X., Ager, B.: Outsourcing the routing controll logic: better Internet routing based on SDN principles. In: Proceedings of the 11th ACM Workshop Hot Topics Networks, Redmond, WA, USA, pp. 55–60 (2012)

6. Tkachova, O., Chinaobi, U., Yahya, A.R.: A load balancing algorithm for SDN Sch. J. Eng. Technol. (SJET) **4**(11), 527–533 (2016). ISSN 2321-435X
7. Chen-xiao, C., Ya-bin, X.: Research on load balance method in SDN. Int. J. Grid Distrib. Comput. **9**(1), 25–36 (2016)
8. Mitzenmacher, M., Upfal, E. Probability and Computing: Randomized Algorithms and Probabilistic Analysis, p. 110. Cambridge University Press (2005)
9. Ren, X., Lin, R., Zou, H.: A dynamic load balancing strategy for cloud computing platform based on exponential smoothing forecast. In: Proceedings of the International Conference on Cloud Computing and Intelligent Systems (CCIS), pp. 220–224. IEEE (2011)
10. Sharma, S., Singh, S., Sharma, M.: Performance analysis of load balancing algorithms. Proc. World Acad. Sci. Eng. Technol. **28**, 269–272 (2008)
11. Motwani, R., Raghavan: Randomized algorithms. ACM Comput. Surv. **28**, 33–37 (1996)
12. Shaoliang, T., Ming, Z., Shaowei, W.: An improved load balancing algorithm based on dynamic [illegible] Comput. Eng. Des. **28**, 572–573 (2007)
13. Hiranwal, S., Roy, K.C.: Adaptive round robin scheduling using shortest burst approach based on smart time slice. Int. J. Comput. Sci. Commun. **2**(2), 319–323 (2010)
14. Leinberger, W., Karypis, G., Kumar, V.: Load Balancing Across Near-Homogeneous Multiresource Servers, 0-7695-0556-2/00. IEEE (2000)
15. Kang, S.B., Kwon, G.I.: Load balancing strategy of SDN controller based on genetic algorithm. Adv. Sci. Technol. Lett. **129**, 219–222 (2016)
16. Hu, Y., Wang, W., Gong, X., Que, X., Cheng, S.: Balanceflow: controller load balancing for openflow networks. In: Proceedings of IEEE 2nd International Conference on Cloud Computing and Intelligent Systems, CCIS'12, vol. 2, pp. 780–785, October 2012
17. Cimorelli, F., Priscoli, F.D., Pietrabissa, A., Celsi, L.R., Suraci, V., Zuccaro, L.: A distributed load balancing algorithm for the control plane in software defined networking. In: 24th Mediterranean Conference on Control and Automation (MED) 21–24 June 2016
18. Katta, N., Hira, M., Kim, C., Sivaraman, A., Rexford, J.: HULA: Scalable load balancing using programmable data planes. In: 2nd ACM SIGCOMM Symposium on Software Defined Networking Research (SOSR) (2016)
19. Cascone, C., Pollini, L., Sanvito, D., Capone, A.: Traffic management applications for stateful sdn data plane. In: 2015 Fourth European Workshop on Software Defined Networks (EWSDN), pp. 85–90, September 2015
20. Akyildiz, I.F., et al.: A Roadmap for Traffic Engineering in Software Defined Networks. Comput. Netw. **71**, 1–30 (2014)
21. Sathyanarayana, S., Moh, M.: Joint route-server load balancing in software defined networks using ant colony optimization. In: Proceedings of the International Conference on High Performance Computing and Simulation (HPCS), Innsbruck, Austria, July 2016
22. Li, J., Chang, X., Ren, Y., Zhang, Z., Wang, G.: An effective path load balancing mechanism based on SDN (2014)
23. Long, H., Shen, Y., Guo, M., Tang, F.: LABERIO: dynamic load-balanced routing in OpenFlow-enabled networks. In: Proceeding of the 27th International Conference on Advanced Information Networking and Applications, pp. 290–197, 25–28 March 2013
24. T-NOVA|Deliverable D4.21 SDN Control Plane 7th framework programme, NAAS over virtualized infrastructures. Grant agreement no. 619520
25. Belyaev, M., Gaivoronski, S.: Towards load balancing in SDN-networks during DDoS-attacks. In: First International Science and Technology Conference (Modern Networking Technologies) (MoNeTeC), pp. 1–6 (2014)
26. Akyildiz, I.F., Lee, A., Wang, P., Luo, M., Chou, W.: Research challenges for traffic engineering in software defined networks. IEEE Netw. **30**(3), 52–58 (2016)

A Robust Human Gait Recognition Approach Using Multi-interval Features

V. G. Manjunatha Guru and V. N. Kamalesh

Abstract This paper has demonstrated a simple and robust interval type features based gait recognition approach. For each an individual, first gait energy image (GEI) is generated. Next, gradient matrices are obtained based on the neighborhood gradient computational procedures (i.e., horizontal/vertical, opposite, and diagonal pixels). Then, multi-interval features are extracted from these matrices in order to achieve the gait covariate conditions. For the gallery and probe interval matching, a new dissimilarity measure is proposed in this work by counting the minimum number of arithmetic operations required to transform one interval into the other. Lastly, a simple KNN classifier is incorporated in the classification procedure. Three standard datasets (Chinese Academy of sciences B and C) are used for the experimental procedures and satisfactory results are obtained. The effective comparative analysis with the current state-of-the-art algorithms has shown that the proposed approach is robust to changes in appearance and different walking speed conditions.

Keywords Human gait · Interval features · Symbolic

1 Introduction

Gait recognition is the identification of a person or persons by their gait or features of gait, usually from closed circuit television (CCTV) footage and comparison to footage of a known individual. The major use of gait recognition is in the visual monitoring and forensics applications [1–9].

V. G. Manjunatha Guru (✉)
GFGC, Honnāli, India
e-mail: vgmanjunathji@gmail.com

V. G. Manjunatha Guru · V. N. Kamalesh
VTU RRC, Belagavi, Karnataka, India

V. N. Kamalesh
TJIT, Bengaluru, India

P. Nagabhushan et al. (eds.), *Data Analytics and Learning*,
Lecture Notes in Networks and Systems 43,
https://doi.org/10.1007/978-981-13-2514-4_5

Hadid et al. [1] commented that biometric traits can spoof in certain cases. For example, 3D mask can spoof the facial recognition system. However, gait is potentially difficult to spoof as it is behavioral and encompasses the whole body. This means that maximal effort (i.e., available time and subject's knowledge) is required to imitate someone's walking styles.

Gait recognition is a very challenging problem in real time scenarios. Since, it is dependent on the psychological status of the walker which means the same person's gait varies at different time intervals. For example, the person may walk fast when he/she is in a hurry/busy/stress mood. The same person may walk slowly when he/she is in relaxing mood. Hence, the use of single-valued features may not be able to capture the variation of gait information properly. So that, these factors should be addressed in order to build the real-time gait recognition system. However, this paper has addressed the interval-valued type features-based gait recognition approach in order to make insensitive to gait covariate conditions.

The concept of symbolic data (i.e., interval, ratio, and multi-valued, etc.) provides a more natural and realistic analysis to the problems. These unconventional features have proved that they outperform the conventional features in terms of performance [10–13].

Bock and Diday [14] have explained the theoretical information about symbolic data analysis, and also the use of symbolic data in recognition systems. With the backdrop of symbolic literatures [2–7, 10–13], it is found that the concept of symbolic data analysis has been well studied in the field of object recognition and biometric applications.

Due to the rich merits of symbolic data approach, the few symbolic approach based attempts have recently made on gait recognition. Mohan and Nagendraswamy [2] have used the simple features like height, width, ratio of height to width and distance between two feet in the form intervals. The same authors have proposed a fusion based method in another work [3]. In [3], Authors have extracted the features like height, width, step length and axis of least inertia from the each silhouette.

Mohan and Nagendraswamy [4] have split the GEI (Gait Energy Image) into four equal blocks. Further, they have applied the LBP (Local Binary Pattern) technique on each block. The extracted features were represented in the form of intervals.

Mohan and Nagendraswamy [5] have explored the change energy images (CEI) as the gait representation method. They have used the Radon transformation for feature extraction procedure and a simple symbolic based similarity measure for classification procedure. The same author's another work [6] has utilized the LBP technique to GEI and to the region bounded by legs (RBL) of GEI. The extracted features were represented in the form of intervals.

2 Proposed Method

The proposed work is categorized into three major phases such as interval feature extraction, intervals matching measurement and classification procedure. For this implementation, the publicly freely available datasets [8, 15] are used. These datasets provide the gait sequences in silhouette format which means the background is black and human region is white. Since, silhouettes are insensitive to the color and texture of cloth.

2.1 Preprocessing

The datasets (i.e., Chinese Academy of Sciences B and C) have noisy images due to the lighting conditions and poor illumination. Hence, morphological operators are used in order to remove the isolated noise, to fill the small gaps and to fill the hole in the silhouette. Then, bounding rectangle is used to extract the silhouette region. The same procedure is applied to all silhouettes in a sequence. Further, all silhouettes align to the same center by considering the centroid of the upper portion of the human body.

2.2 *Gradient Matrices*

This paper utilizes the popular gait representation method (i.e., GEI) [16]. Since, it overcomes the practical challenges such as noise, poor illumination, distortion, lack of frames and different lighting conditions, etc. For each an individual, GEI is obtained as indicated in the formula (1) where "g" represents the gait energy image, "I" indicates the silhouette and "Nf" indicates the number of frames. Further, it is cropped and resized to the fixed size (i.e. 128 × 88).

$$G = \frac{1}{N_f} \sum_{i=1}^{N_f} I_i \tag{1}$$

In GEI, each pixel is modified by the neighborhood gradient information. The small neighborhoods are insensitive to the gait covariate conditions and have discriminative capability. Hence, neighborhood of size 3 × 3 is selected. Since, it is small in size and also gives the central pixel. The sample neighborhood structure is depicted in Fig. 2 where Gi,j represents the central pixel.

As indicated in Fig. 2, Gi,j represents the each location of GEI and it is modified by the three kinds of neighborhood computational procedures in order to capture the rate of change at each pixel in different directions. These are 4—horizontal/vertical pixels (Ghv), 4—diagonal pixels (Gdiag) and 4—opposite pixels (Gopp).

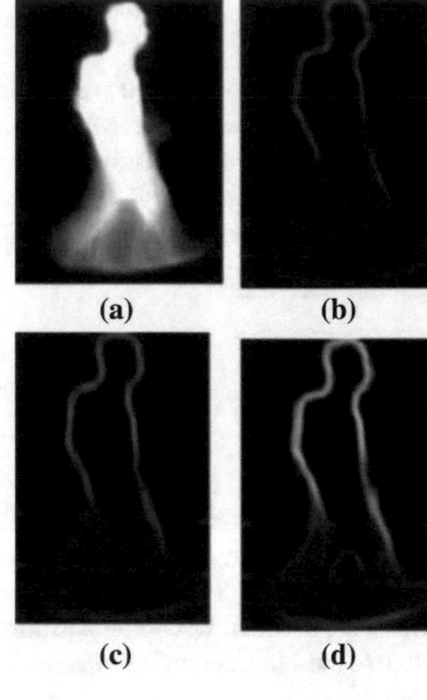

Fig. 1 GEI and gradient matrices **a** GEI **b** G_{hv} **c** G_{diag} **d** G_{opp}

Fig. 2 3 × 3 neighborhood

$$\begin{pmatrix} G_{i-1,j-1} & G_{i-1,j} & G_{i-1,j+1} \\ G_{i,j-1} & G_{i,j} & G_{i,j+1} \\ G_{i+1,j-1} & G_{i+1,j} & G_{i+1,j+1} \end{pmatrix}$$

In the loop over the "M" rows and "N" columns of GEI, the three procedures (i.e., Ghv, Gdiag, and Gopp) perform the subtraction and give the mean of the absolute subtractions. Subtraction in a homogeneous region produces zero and indicates an absence of dynamic characteristics. Each gradient procedure will generate the separate matrix. Hence, three gradient matrices (i.e. Ghv, Gd, and Gopp) are obtained for the individual GEI.

For Ghv computation, subtract the each of the horizontal/vertical pixels which are next to the center of a 3 × 3 area from the center pixel. In Gdiag, subtract the each of the diagonal pixels next to the center of a 3 × 3 area from the center pixel. Whereas, Gopp subtract the pixels in opposite directions, i.e., the upper left minus lower right, upper right minus lower left, left minus right, and top minus bottom. Figure 1 clearly depicts the GEI and the three gradient images based on these gradient procedures (Fig. 2).

2.3 Multi Interval Features

For each person's gait, the feature matrix "f" is generated from the three gradient matrices (Ghv, Gopp, and Gdiag) of size M × N. Each row of "f" indicates the concatenated list of row-wise- and column-wise variance values of the particular gradient matrix. However, the size of "f" is 3 × (M+N).

$$f = \begin{bmatrix} (rowwise\ \sigma^2(G_{hv})), (columnwise\ \sigma^2(G_{hv})) \\ (rowwise\ \sigma^2(G_{opp})), (columnwise\ \sigma^2(G_{opp})) \\ (rowwise\ \sigma^2(G_{diag})), (columnwise\ \sigma^2(G_{diag})) \end{bmatrix}$$

For each column of "f", mean (μ) and standard deviation (σ) values are computed. Then, the interval feature is obtained by performing the arithmetic operations on μ and σ. However, the upper bound (Max) of the interval is $\mu+\sigma$ and lower bound (Min) is $\mu-\sigma$. The same procedure applies to all columns of "f". Hence, the final feature vector (Ii) is generated as indicated in (2). In which, the variable "i" represents the column of "f" and its range from 1 to (M+N).

$$I_i = [Min_i - Max_i], \tag{2}$$

where $Max_i = \mu_i + \sigma_i$ and $Min_i = \mu_i - \sigma_i$.

From the observations of the above paragraphs, multi interval features are obtained for each person's gait which means the both probe and gallery samples are represented by multi interval features. Therefore, finding the measure between the intervals is more essential for the classification task.

2.4 Symbolic Dissimilarity Measure and K-Nearest Neighbor (KNN) Classifier

In this paper, we explore a new symbolic dissimilarity measure for the interval data. This is inspired by the traditional edit distance measure which works with strings. The proposed measure works with numerical features. Our symbolic edit distance measure is a way of quantifying how dissimilar two intervals are to one another by counting the minimum number of arithmetic operations required to transform one interval into the other. The below Algorithm 1 and 2 implements the proposed methodology.

Algorithm 1 Dissimilarity Measure and KNN (X, Y)

Input: The both probe sample (X) and gallery samples (Y) are represented by the multi interval features. Each sample is consisting of M+N intervals.

Output: Resultant class of probe sample.

Description: The symbol "D_k" is the dissimilarity value between two *k*th intervals of gallery and probe sample. The value of variable "k" ranges from 1 to M+N.

The symbol "T(X, Y_i)" is the sum of all dissimilarity measures between the intervals of probe sample (X) and gallery (Y_i). The value of variable "i" ranges from 1 to n where "n" indicates the number of samples in gallery set(Y).

Step 1: For each probe sample(X), calculate T(X, Y_i) for each i = 1 … n

$$T(X, Y_i) = \sum_{k=1}^{M+N} D_k \tag{3}$$

$$D_k = |X_k| + |Y_{ik}| - 2|X_k \cap Y_{ik}|, \tag{4}$$

where $|X_k| = Min_p - Max_p$ and $|Y_{ik}| = Min_q - Max_q$.

The value of $| X_k \cap Y_{ik} |$ can be determined by the Algorithm 2.

Step 2: Arrange the "T" in ascending order

Step 3: Let "k" be positive integer. Taking the first "k" values from the above sorted list

Step 4: Let "k_i" denotes the number of points belonging to the *i*th class among "k" points (k_j); If $k_i > k_j$, then put "X" in class "i"

Step 5: End of the Algorithm 1.

Algorithm 2 Dissimilarity Measure Between Two Intervals (X_k,Y_{ik})

Input: X_k i.e. [$Min_p - Max_p$] and Y_{ik} i.e. [$Min_q - Max_q$] are k-th intervals of probe(X) and gallery sample(Y) respectively

Output: $|X_k \cap Y_{ik}|$ is the dissimilarity measure

Description: The symbol "Φ" is null i.e. zero

Step 1:

if ($Min_q > Max_p$)
$|X_k \cap Y_{ik}| = \Phi$
else
if ($Max_q < Min_p$)
$|X_k \cap Y_{ik}| = \Phi$
else
if ($Min_q == Min_p$ and $Max_q == Max_p$)
$|X_k \cap Y_{ik}| = Max_p - Min_p$
else
if ($Min_q < Min_p$ and $Max_q > Max_p$)
$|X_k \cap Y_{ik}| = Max_p - Min_p$
else
if ($Min_q > Min_p$ and $Max_q < Max_p$)
$|X_k \cap Y_{ik}| = Max_q - Min_q$
else
if ($Min_q >= Min_p$ and $Max_q >= Max_p$)
$|X_k \cap Y_{ik}| = Max_p - Min_q$

else
if ($Min_q <= Min_p$ and $Max_q <= Max_p$)
$|X_k \cap Y_{ik}| = Max_q - Min_p$
else
Step 2: End of an Algorithm 2.

3 Experiments

This paper utilizes three considerably largest, publicly freely available standard datasets [8, 15, 17]. Among the three, CASIA A consists of multi view sequences. CASIA B consists of multi view, carrying and wearing condition sequences. CASIA C consist of backpack and different walking speed variation sequences. Hence, these datasets address the prominent gait covariates conditions which are commonly encounter in human day life activities. However, the below experiments are conducted on these datasets with different kind of experimental settings and procedures. The three types of experimental procedures such as leaving one out hold out and K-fold procedures are efficiently incorporated for the effective experimentation purpose.

3.1 CASIA A Dataset

CASIA A is one of the publicly available and small multi-view dataset. This dataset consists of three prominent viewing angle sequences with normal walking condition (i.e., lateral, oblique, and frontal views). For each viewing angle, there are four sequences. Hence, each subject consists of 12 sequences. However, the whole dataset consist of 240 sequences (i.e., 20 subjects × 12 sequences).

Experiment 1. In order to effectively evaluate the proposed approach with multi-view conditions, each view is considered as a particular class. However, 60 classes (i.e., 3 views per person × 20 people) are considered for this experiment rather than considering 20 subject's problems.

This experiment has incorporated hold out type experimental settings. For each view, two sequences are used for training and the remaining two sequences are used for testing. However, total 120 sequences are used as the gallery set and remaining the same number of sequences are used as the probe set. For this experimental setting, the proposed method achieved the 98.75% and 100% respectively, for K = 1 and K = 3.

Experiment 2. This section also used 60 classes for the experiment. In order to efficiently evaluate the proposed approach with large number samples, this experiment has incorporated leaving one out type experimental settings. In each turn, 1 sequence is used for testing and the remaining is used for training.

In each time, 1 sequence is used as a test set and the remaining 239 sequences are used as a training set. This kind of the experimental setup is repeated for 240 times. For this experimental setting, the proposed method achieved the 98.75% and 100% respectively, for K = 1 and K = 3.

3.2 CASIA B Dataset

CASIA B is one the popular, publicly available, considerably largest and widely used multi-view dataset. This dataset consists of 124 person's gait sequences. In this, each person walked in 11 view directions in front of the recording device, i.e., 00, 180, 360, 540, 720, 900, 1080, 1260, 1440, 1620, and 1800. Each view has 10 sequences, i.e., two carrying bag (CB), two wearing cloth (WC), and six normal walk sequences (NW). However, the dataset consist of total 13,640 gait sequences (i.e. 11 views × 10 sequences per view × 124 people).

Experiment 1. In order to effectively evaluate the proposed approach with multi-view conditions, each view is considered as a particular class. However, 1,364 classes (i.e., 11 views per person × 124 people) are considered for this experiment rather than considering 124 subject's problems.

This experiment has incorporated hold out type experimental setting. For each viewing angle, five sequences are used in the training (i.e., 1 CB + 1 WC + 3 NW sequences) and the remaining five sequences are used for the testing. However, total 6,820 sequences are used as a training set and remaining the same number of sequences are used as a test set. For this experimental setting, the proposed method achieved the 95.30% and 100% respectively, for K = 1 and K = 5.

Experiment 2. This section also used 1,364 classes for the experiment. In order to verify the proposed approach with gait covariate conditions, CB and WC sequences are used as a test set and NW sequences are used as a training set.

This experiment has incorporated hold out type experimental setting. For each viewing angle, four sequences, i.e., two carrying bag (CB), two wearing cloth (WC) are used for the testing, and six normal walk (NW) sequences are used in the training process. However, total 5,456 sequences (i.e., 4 sequences per view × 11 views per person × 124 person's) are used as a test set and the remaining 8,184 sequences (i.e., 6 sequences per view × 11 views per person × 124 person's) are used as a training set. For this experimental setting, the proposed method achieved the 94.11% and 100% respectively, for K = 1 and K = 5.

Experiment 3. This section also used 1,364 classes for the experiment. In order to efficiently evaluate the proposed approach with large number samples with gait covariate conditions, this experiment has incorporated leaving one out type experimental settings. In each turn, 1 sequence is used for testing and the remaining is used for training.

Table 1 Results on gait covariate conditions

Covariates	CCR	
	K = 1 (%)	K = 5 (%)
Carrying BAG	93.47	98.97
Cloth	95.30	99.04
Normal walk	95.44	99.04

In each time, 1 sequence is used as a test set and the remaining 13,639 sequences are used as a training set. This kind of experimental setup is repeated for 13,640 times. For this experimental settings, the proposed method achieved the 96% and 100% respectively for K = 1 and K = 10.

Experiment 4. This section has incorporated the K-fold type cross-validation procedure in order to verify the robustness of the proposed method with respect to the covariate conditions. CASIA B dataset consist of three gait covariates conditions (i.e., carrying, wearing coat and normal walk). However, the whole dataset is split into three sets (i.e. K = 3) by considering each covariate as a separate class. In each turn, one of the "K" subsets is used for testing and remaining the K − 1 subsets are used for training. The same is repeated for three times. Table 1 shows the CCR rates for the prominent covariate conditions.

Experiment 5. For this experimentation, each subject is considered as the three classes with respect to the views (i.e., lateral, oblique, and frontal). Hence, 372 classes (i.e., 124 Subjects × 3 classes) are considered for this experiment instead of 124 subjects in order to verify the proposed approach.

This experimental setup is incorporated k-fold cross-validation procedure in order to effectively show the cross-view validation. However, the whole dataset is partitioned into 11 subsets with respect to the viewing angles (i.e., 0°, 18°, 36°, 54°, 72°, 90°, 108°, 126°, 144°, 162°, and 180°). Each subset consists of particular view sequences. Each time, one of the "k" subsets is used as a probe set and the remaining k − 1 subsets are combined together to form a gallery set. In each turn, total 12,400 sequences (i.e., 1240 sequences per view × 10 views) are used as the training set and the rest 1,240 sequences (i.e. 1240 sequences × 1 view) are used as the testing set. For this experimental setting, the proposed method achieved the 87.97% and 95.30% of average CCR respectively, for K = 1 and K = 10.

3.3 CASIA C Dataset

CASIA C is the publicly available and speed variation dataset. This dataset consists of 153 person's gait sequences. This dataset has total 1,530 lateral view sequences with backpack and different walking speed conditions, i.e., slow, medium, and fast walk. Each subject consists of 10 sequences, i.e., two backpack (BP), four normal walk (NW), two slow walk (SW), and two fast walk (FW).

Table 2 Results on gait covariate conditions

Covariates	CCR	
	K = 1 (%)	K = 3 (%)
Backpack	94.77	98.03
Slow walk	96.40	100
Fast walk	97.38	100
Normal walk	98.36	100

Experiment 1. This section has incorporated the hold out type experimental setup. For each person, 5 sequences (i.e., 1 BP + 2 NW + 1 SW + 1 FW) are used for the training and the remaining the same number of sequences are used for the testing. However, total 765 sequences are used as a training set (i.e., 5 sequences per person × 153 persons) and the same number as a test set. For this experimental setting, the proposed method achieved the 94.77% and 100% respectively, for K = 1 and K = 5.

Experiment 2. In order to verify the efficiency of the proposed method with speed variation and backpack conditions, this section also has incorporated the hold out type experimental setup. For each person, 6 sequences (i.e., 2 BP + 2 SW + 2 FW) are used for the testing and the remaining four sequences (i.e., 4 NW) are used for the training. However, total 918 sequences are used as a test set (i.e., 6 sequences per person × 153 persons) and 612 sequences (i.e., 4 sequences per person × 153 persons) are used as a training set. For this experimental setting, the proposed method is achieved the 93.68% and 100% respectively, for K = 1 and K = 97.93%.

Experiment 3. This section has incorporated the K-fold type cross-validation procedure in order to verify the effectiveness of the proposed method with respect to the covariate conditions. CASIA C dataset consist of four gait covariates conditions (i.e., backpack, slow, fast, and normal walk). However, the whole dataset is split into four sets (i.e., K = 4) by considering each covariate as a separate class. In each turn, one class is used for testing and remaining the three classes are used for training. The same is repeated for four times. Table 2 shows the CCR rates for this experiment.

Experiment 4. This section has incorporated leaving one out type experimental setting for total 1,530 sequences. In each time, 1 sequence is used as a test set and the remaining 1,529 sequences are used as a training set. This kind of the experimental setup is repeated for 1,530 times. For this experimental setting, the proposed method achieved the 97.97% and 100% respectively, for K = 1 and K = 7.

4 Comparative Results

The proposed work has been tested on two considerably largest datasets, i.e., CASIA (B and C). The multi-view scenario with the other covariates has effectively addressed in this work. The extensive experiments have conducted on these datasets with different experimental procedures and settings. The promising results have shown that the proposed attempt outperforms the current symbolic approach based gait attempts.

The prominent comparisons which are made on the current symbolic approach based literatures are described in below paragraphs.

The paper has explored the multi-interval features based gait recognition algorithm. However, this work has been compared with the current symbolic approaches and other related works. For the effective comparisons with the recent literatures, the same datasets are used. Also, the same kind of experimental procedure and settings are used. Mohan Kumar and Nagendraswamy [3] have conducted the experiments on CASIA B dataset. They have achieved the 79.30% of average CCR for 900 view angle sequences. In their work, four normal walking sequences per subject were used as the gallery and the six sequences per subject were used as the probe (i.e., two carrying bag + two wearing coat + two normal walking sequences). Further, another work of Mohan Kumar and Nagendraswamy [3] reported the 79.01% of average CCR for the same experimental setting which is used in the above literature. For the same above mentioned experimental setting, the proposed work achieved the 91.73% of average CCR and shown that it is better than the recent interval features based attempts.

In real-time scenario, the person can walk in any direction in front of the camera. Hence, the multi-views (i.e., side view, frontal, and oblique) are one of the prominent gait challenges and are effectively addressed in this paper. In comparison with the symbolic approach based literatures, attempts [2, 3] have used only side view sequences (0°) in their experiments. However, there is no experimental proof about other views in their works.

With the help literature study, it is found that the recent symbolic approach based works have considered interval valued type features as the training patterns and crisp-valued type features as the testing patterns. The crisp-valued type features sometimes may fail to describe the gait information properly. Since, the same person's gait information varies at different time intervals, i.e., different walking speed, carrying, wearing and backpack conditions, etc. Hence, these conditions can be effectively addressed by the interval features in order to build the robust gait recognition system. With this backdrop, the proposed work has considered the interval features for the both training and testing phases.

5 Conclusion

This paper highlighted the benefits of the interval features in the prominent gait covariate conditions. In real-time scenario, view and walking speed conditions are more prominent in the gait recognition system compared to other gait covariates, since a person can walk in any direction in front of the recording device. Also, he/she can walk at any speed depending on the psychological parameters such as stress, relax, and hurry, etc. Hence, the proposed approach is effectively tested for these pragmatic conditions by utilizing the standard datasets (i.e., CASIA B and C). CASIA B is one of the popular and widely used dataset for the multi-view experiments. CASIA C is also one of the largest, publicly available and widely used dataset for the walking speed experiments. The extensive experiments are conducted on these datasets by including the different kind of experimental settings and procedures. The experiments have given the promising results for the scenarios such as multi-view, speed variation, backpack, carrying bag, and clothing conditions. The effective comparison with the current interval features based approaches shown that the proposed approach is better in the context of multi views, speed transition and robust to changes in appearance. Also, the proposed approach has some merits compared to the current attempts in literature such as it is easy to implement, understand, takes less time and space requirement due to its simple data structures utilization and computation.

Acknowledgements The authors would like to thank the creators of CASIA A, B and C datasets for providing the publicly available gait datasets.

References

1. Hadid, A., Ghahramani., M., Kellokumpu, V., Pietikainen, M.: Can gait biometrics be spoofed. In: Proceedings of the International Conference on Pattern Recognition, ICPR12 (2012)
2. Mohan Kumar, H.P., Nagendraswamy, H.S.: Gait recognition based on symbolic representation. Int. J. Mach. Intell. (Bioinfo) **3**(4), 295–301 (2011). ISSN: 0975-2927 and E-ISSN: 0975-9166
3. Mohan Kumar, Nagendraswamy: Fusion of silhouette based gait features for gait recognition. Int. J. Eng. Tech. Res. (IJETR) **2**(8) (2014). ISSN: 2321-0869
4. Mohan Kumar, Nagendraswamy: Symbolic representation and recognition of gait: an approach based on LBP of split gait energy images. Signal Image Process. Int. J. (SIPIJ) **5**(4) (2014)
5. Mohan Kumar, H.P., Nagendraswamy, H.S.: Change energy image for gait recognition: an approach based on symbolic representation. Int. J. Image Gr. Signal Process. (MECS) (2014). https://doi.org/10.5815/ijigsp.2014.04.01
6. Mohan Kumar, H.P., Nagendraswamy, H.S.: LBP for gait recognition: a symbolic approach based on GEI plus RBL of GEI. In: International Conference on Electronics and Communication Systems (ICECS), Coimbatore, India. IEEE (2014)
7. Hiremath, P.S., Prabhakar, C.J.: Extraction and recognition of nonlinear interval-type features using symbolic KDA algorithm with application to face recognition. Res. Lett. Signal Process. (Hindawi Publishing Corporation) **2008**, Article ID 486247, 5 pp. (2008). https://doi.org/10.1155/2008/486247
8. Yu, S., Tan, D., Tan, T.: A framework for evaluating the effect of view angle, clothing and carrying condition on gait recognition. In: Proceedings of the 18th International Conference on Pattern Recognition (ICPR), Hong Kong (2006)
9. Manjunatha Guru, V.G., Kamalesh, V.N.: Vision based human gait recognition system: observations, pragmatic conditions and datasets. Indian J. Sci. Technol. **8**(15) (2015). ISSN (Print): 0974-6846. ISSN (Online): 0974-5645
10. Guru, D.S., Kiranagi, B.B.: Multi valued type dissimilarity measure and concept of mutual dissimilarity value useful for clustering symbolic patterns. Pattern Recognit. **38**(1), 151–156 (2006)
11. Guru, D.S., Nagendraswamy, H.S.: Symbolic representation of two-dimensional shapes. Pattern Recognit. Lett. 144–155 (2007)
12. Gowda, K.C., Diday, E.: Symbolic clustering using a new dissimilarity measure. Pattern Recognit. **24**(6), 567–578 (1991)
13. Gowda, K.C., Diday, E.: Symbolic clustering using a new similarity measure. IEEE Trans. SMC **22**(2), 368–378 (1992)

14. Bock, H.H., Diday, E. (eds.): Analysis of Symbolic Data. Springer, Heidelberg, Germany (2000)
15. Tan, D., Huang, K., Yu, S., Tan, T.: Efficient night gait recognition based on template matching. In: Proceedings of the 18th International Conference on Pattern Recognition (ICPR06), Hong Kong (2006)
16. Han, J., Bhanu, B.: Individual recognition using gait energy image. IEEE Trans. Pattern Anal. Mach. Intell. **28**(2) (2006)
17. Wang, L., Tan, T., Ning, H., Hu, W.: Silhouette analysis based gait recognition for human identification. IEEE Trans. Pattern Anal. Mach. Intell. (PAMI) **25**(12), 1505–1518 (2003)

Keyword Spotting in Historical Devanagari Manuscripts by Word Matching

B. Sharada, S. N. Sushma and Bharathlal

Abstract Huge quantities of ancient manuscripts are there in various National archives. Digitization of these manuscripts and historical documents is very important for making its access efficiently. Handwritten keyword spotting is very challenging task for Devanagari documents due to large variations in writing styles. Keyword spotting of unconstrained offline handwritten documents is performed based on matching scheme of word images. This paper presents an efficient keyword spotting approach for handwritten Devanagari documents. Experiments are conducted on historical datasets consisting of manuscripts by Oriental Research Institute @ Mysuru.

Keywords Handwritten documents · Keyword spotting · Segmentation
Feature extraction · Keypoints

1 Introduction

Current researches in the area of document image analysis require designing numerous systems for digitization of historical manuscripts. George Washington's papers at the Library of Congress, Isaac Newton's papers at the University of Cambridge Library are some instances of historical manuscripts [1]. In the analysis and processing of historical documents, there are many confronts such as noise removal, enhancement of tainted documents, artifact removal and skew elimination due to poor quality of documents [2]. Word spotting has received significant attention in recent years and various systems are designed for English, Chinese, Arabic and Latin.

B. Sharada · S. N. Sushma (✉) · Bharathlal
Department of Studies in Computer Science, University of Mysore, Mysuru, India
e-mail: sushisn2007@gmail.com

B. Sharada
e-mail: sharadab21@gmail.com

Bharathlal
e-mail: bharathlal14@gmail.com

P. Nagabhushan et al. (eds.), *Data Analytics and Learning*,
Lecture Notes in Networks and Systems 43,
https://doi.org/10.1007/978-981-13-2514-4_6

But for Indic scripts especially Devanagari, information retrieval is challenging due to huge vocabulary and divergent writing styles [3].

Keyword spotting is designed as an alternative method of an absolute transcription of documents. It is a confined task of retrieving keywords from document images. Keyword spotting is to spot similar words in documents. It is usually achieved by calculating a similarity measure between the query word and a segmented candidate words present in documents [4]. The vital application of keyword spotting is indexing a handwritten document which leads to digitization and exploration of valuable old manuscripts in order to safeguard world's cultural heritage. The efficiency of word spotting depends on the holistic matching scheme designed for comparing a query word and the candidate words stored in the database [5]. Usage of suitable preprocessing, segmentation methods, and extraction of unique features is the main criteria for improving the performance of the keyword spotting systems.

Experiments are conducted on historical handwritten Devanagari manuscript of the Oriental Research Institute @ Mysuru. The handwritten Devanagari document is scanned which is used for testing and validation. The paper is arranged as follows: Sect. 2 elaborates the related works of keyword spotting. Section 3 illustrates the proposed System overview. Section 4 discusses on the performance and evaluation of proposed scheme. Section 5 focuses on the conclusion and future research activities.

2 Related Work

There are two categories of keyword spotting based on segmentation based or segmentation-free method. For word spotting using character shape coding, the word images are further segmented into smaller units and corresponding set of predefined codes are generated which is sensitive to error in segmentation of character. This method is called analytical approach which involves segmentation of word images into characters. In word shape coding, the whole word image is used for matching [6]. For the process of matching of words, cohesive elastic distance is used and Gradient angles are chosen as local features. This is called holistic approach which involves image-by-image matching technique. Line base segmentation method is used based on DTW which spots the word automatically in [7]. For detecting similar words, BLSTM Neural Networks and CTC Token Passing algorithm are also been employed [8].

Query-by-string approaches are like OCR-based handwriting recognition systems which usually require training process followed by testing. The word images are ranked based on the similarity measures in query-by-example methods. For the word images unique features like upper/lower word profiles, binary gradient structural features, ink transition profiles, Profile features, convexity features, discrete cosine transformation are used for representation of words [9–13]. In order to improve the word image representation, multiple features were investigated and combined which is called as Compound features method. Various distance metrics are used for computing similarity between words. For compressed version of each document image, sliding window method, and codebook of shapes are used for spotting the words in documents for segmentation-free approach [14–17].

Majority of the word spotting systems are applied to English, Latin, Arabic and Chinese scripts, while fewer attentions has been given to Indic scripts. Devanagari is the basic script of many languages in India, such as Hindi, Sanskrit, Nepali, and Marathi. Devanagari is half syllabic in nature. Devanagari script contains 13 vowels and 36 consonants. Devanagari word has three zones such as Lower, Middle and Upper Zone. The Upper Zone contains the modifiers, and Lower Zone contains lower modifiers. The Upper Zone and middle Zone are always separated by the header line called shirorekha. The writing style of Devanagari is from left to right along a horizontal line [18]. Devanagari has got the position of national vernacular, as Hindi is the Indian national language.

3 System Overview

In a keyword spotting scenario, we have dataset of handwritten devanagari document images and the query word is compared with all words in the database to detect similar word images. The proposed keyword spotting approach is explained in the following sections. The proposed system has four important stages such as Preprocessing, Segmentation, Feature extraction and Matching. In Fig. 1, the Overview of the statistical method for keyword spotting system is graphically depicted.

3.1 Preprocessing

Handwritten documents usually need preprocessing to overcome from problems such as slant and touching letters, overlapping, low quality, and low resolution document images due to the presence of noise. So before actual process the quality of the document image is improved using preprocessing method. For eliminating the noise present in the document image, Gaussian filter is applied. Based on threshold value obtained by Otsu's method, the image is binarized. Using horizontal projection profile Skew angle is detected and Skew angle correction is performed [19]. The width of a line like object from many pixels wide is reduced to just single pixel using thinning process. Normalization process is performed to eradicate all sort of differences occurred while writing in the document.

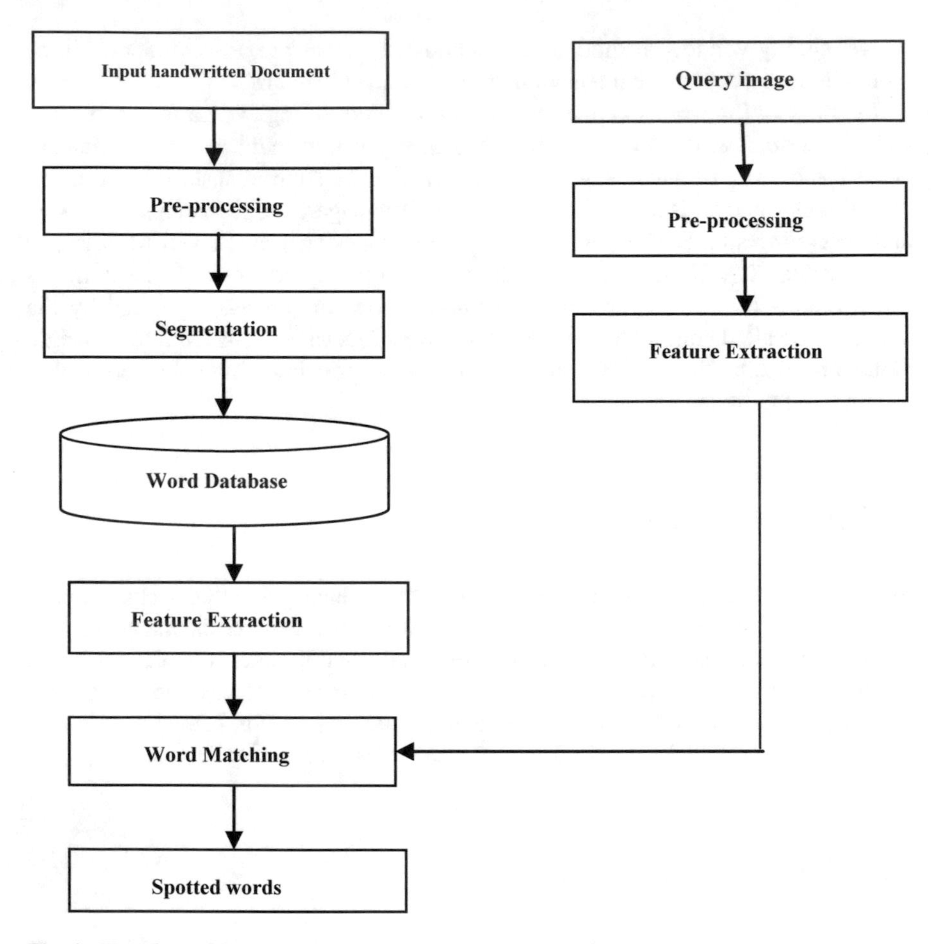

Fig. 1 Flowchart of the proposed approach

3.2 Segmentation

For the segmentation phase, preprocessed word image is given as input. In this stage the binary image is separated into lines and words. Segmentation is performed using projection profile method. In line segmentation, by locating the valleys of the projection profile, the text lines are identified by computing by a row-wise sum of black pixels. The vertical projection profiles are used for segmenting the text line image into words. The zero valley peaks is identified as the profile which is represented the word space. Segmented line from first stage is taken as input for second stage that is word segmentation [20]. Line is then segmented into word in this stage.

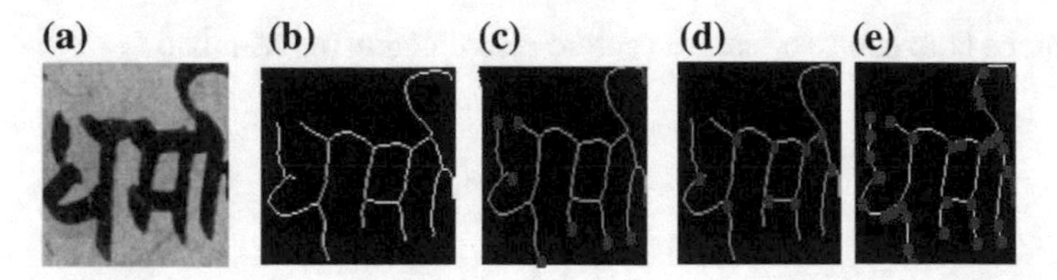

Fig. 2 Illustration of feature extraction with an example. **a** The original word image **b** skeletonization **c** end-points **d** branch-points **e** orientation points

3.3 Feature Extraction

Feature extraction technique is dimensionality reduction procedure where the given input information is represented by group of distinctive features [21]. This representation is used in further image matching and retrieval operations. For the representation of the handwritten word image, four important statistical features are identified in the proposed system.

From the thinned segmented word images end-points, branch-points, orientation points and number of connected components of word image are considered as feature set. The beginning/end of a word are the end-points, intersection point are the branch-points and the loops of word where the point direction of orientation changes are the orientation points. In Fig. 2, some of the examples are illustrated.

3.4 Keyword Spotting by Matching Word Images

Word image matching is the main phase of the keyword spotting system. The matching between query word image and candidate images in database is computed using three different similarity measures: Euclidean distance, Manhattan Distance and Cosine Similarity Metric. For the representation of query word images and all the candidate word images in the database, a standard Vector Space Model is applied [22, 23]. The distance between two feature vectors of query word image and word images in database is determined using similarity metrics for matching word images.

Euclidean Distance between two feature vectors is calculated using:

$$d_e = \sqrt{\sum_{i=1}^{n} (x_i - yi)^2} \tag{1}$$

Distance between two feature vectors calculated using Manhattan

$$d_m = \sqrt{\sum_{i=1}^{n} |x_i - yi|} \tag{2}$$

Distance between two feature vectors calculated using Cosine Similarity

$$d_c = \sum_{i=1}^{n} \frac{x_i y_i}{\sqrt{\sum_{i=1}^{n} x_i^2 \sum_{i=1}^{n} yi^2}} \tag{3}$$

where x and y are the query image and the candidate image. de, dm, and dc are similarity measures using Euclidean Distance, Manhattan Distance and Cosine Similarity Metric respectively. For ranking of the candidate words, these similarity distances are used.

4 Experiments and Results

The performance of the proposed approach is evaluated to a huge collection of historical Devanagari handwritten manuscripts and the keyword spotting results are compared with three similarity measures. As a standard handwritten Devanagari word dataset is not available. For our experiments we used two sets of documents. One set from a handwritten historical Devanagari manuscript and the other from handwritten Devanagari document. We have collected data from 20 native Hindi speakers. The

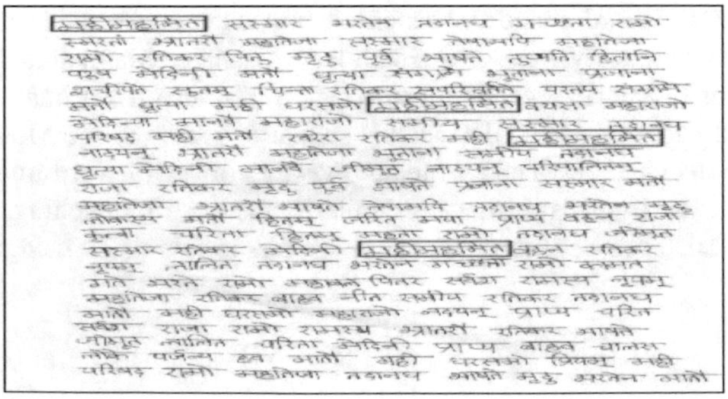

Fig. 3 Keyword spotting results for the query word "mahemahamita"

Devanagari documents contains a total of 1057 handwritten word images. Based on the word image matching criteria using similarity measures, the candidate word images are ranked. For the query word image, the top-ranked candidate word images are considered as the spotted words. Euclidean distance, Manhattan Distance, and Cosine Similarity distance are computed between the query word and all the candidate words in database. Precision is the percentage of the spotted words that exactly match the query word. Recall is the percentage of the words, same as query word that are successfully retrieved from the word database. The qualitative results of keyword spotting are shown in Fig. 3.

The values of Precision, Recall, and F-measure are shown in Tables 1, 2 and 3 respectively. The subsequent graphical representation is shown in Figs. 4, 5 and 6. The Comparison of performance of keyword spotting systems using different similarity measures is shown in Fig. 7.

Table 1 Keyword spotting results for 10 devanagari words in terms of recall and precision as well as F-measure using Euclidean distance

Image	Precision	Recall	F-measure
आरामदायक	84.6154	64.7059	73.3333
हिन्दी	60.4651	86.6667	71.2329
जानवर	98.4455	98.3456	97.7765
महेश	78.6667	70.8333	79.9613
आशीर्वाद	84.5161	80.6719	71.4286
अंग्रेजी	76.4706	96.2963	85.2459
कर्नाटक	83.3333	83.3537	83.2319
समर्थ	98.9865	77.4286	63.4146
कृष्णा	83.9422	92.8571	96.2963
विकास	96.6667	76.3158	85.2941
Average	84.61081	82.74749	80.72154

Table 2 Keyword spotting results for 10 devanagari words in terms of recall and precision as well as F-measure using Manhattan Distance

Image	Precision	Recall	F-measure
आरामदायक	87.4541	71.2841	73.7427
हिन्दी	72.2889	85.3611	73.1802
जानवर	98.2413	98.2212	97.5287
महेशा	78.6667	73.6479	77.6122
आशीर्वाद	87. 1987	84.1949	79.571
आंग्रेजी	76.8209	97.8197	87.0911
कर्नाटक	85.4558	85.3651	83.8893
समय	98.7622	75.9581	68.3971
कृष्णा	86.6355	89.4718	94.3844
विकास	97.6121	79.9441	81.2951
Average	86.88194	84.1268	81.66918

Table 3 Keyword spotting results for 10 devanagari words in terms of recall and precision as well as F-measure using Cosine Similarity Metric

Image	Precision	Recall	F-measure
आरामदायक	83.3282	62.3241	70.3143
हिन्दी	62.3824	84.7696	69.4997
जानवर	95.1128	88. 6153	91.881
महेशा	75.4733	71.5157	69.0471
आशीर्वाद	75.5022	76.4791	67.0912
आंग्रेजी	74.0477	88.4933	86.1592
कर्नाटक	80.1133	81. 3327	82.2723
समय	91.6517	85.4286	83.493
कृष्णा	80. 2732	90. 7041	89.3519
विकास	89.2719	77.0619	82.5813
Average	80.76483	78.01033	79.1691

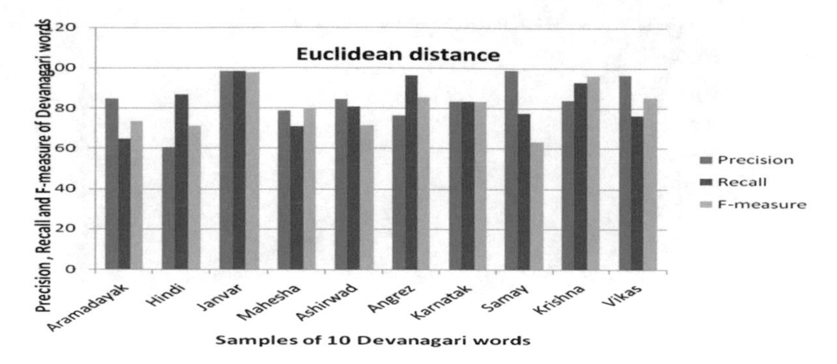

Fig. 4 Graphical representation of precision, recall and F-measure using Euclidean distance

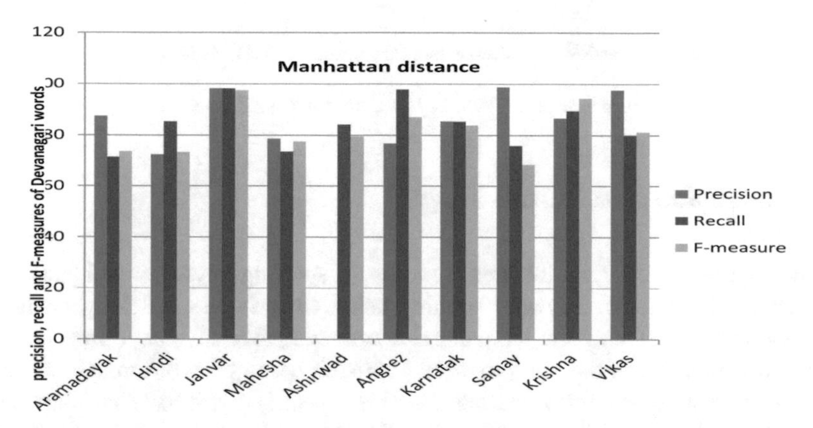

Fig. 5 Graphical representation of precision, recall and F-measure using Manhattan distance

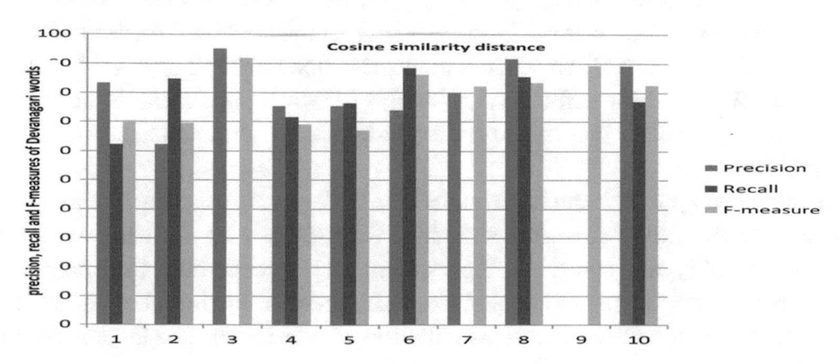

Fig. 6 Graphical representation of precision, recall and F-measure using Cosine similarity distance

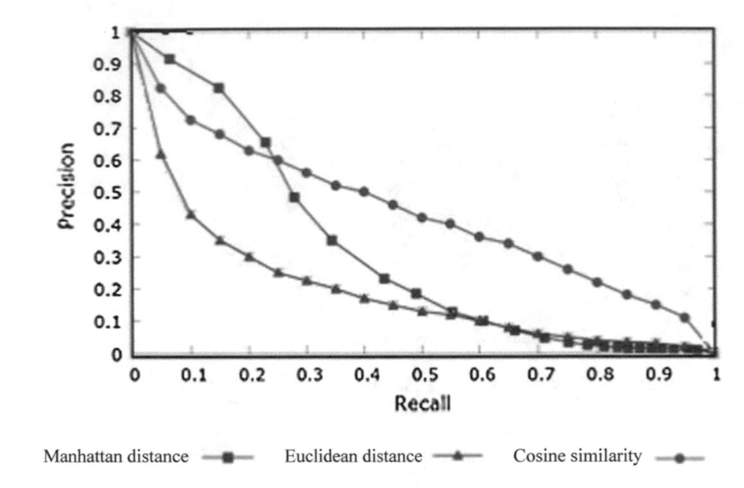

Fig. 7 Comparing performance of different keyword spotting systems

5 Conclusion and Future Scope

In this approach, statistical features are used for spotting keywords in handwritten Devanagari documents. Keyword spotting using word image matching method is a robust approach. Using the Devanagari documents, extensive experimentation is achieved which shows that the proposed approach based on word image matching can considerably outperform existing methods. Specifically, it has been shown that on a set of 23 pages. When compared to Euclidean distance and Cosine similarity metrics, the average matching accuracy of Manhattan distance is 84% which shows significant accuracy. It can be concluded that the proposed method effectively achieves spotting of similar words in handwritten documents using word matching. Finally, considering the enormous complexity of Devanagari script, the contribution of the present approach may be considered significant with satisfactory matching performances.

The proposed approach can be extended for the keyword spotting in various Indic languages. In the future research, structural features can be used for word image representation. Keyword spotting using segmentation-free methods from handwritten documents and adopting a complete indexing method for historical manuscripts can be the future research. The exact representation of word images is the key factor for overall accuracy of the keyword spotting system.

References

1. Rath, T.M., Manmatha, R.: Word spotting for historical documents. Int. J. Doc. Anal. Recognit. (IJDAR) **9**, 139–152 (2007)
2. Messaoud, I.B., Amiri, H., Abed, H.E., Margner, V.: Document preprocessing system—automatic selection of binarization. In: IAPR International Workshop on Document Analysis Systems (2012)
3. Srihari, S.N., Srinivasan, H., Huang, C., Shetty, S.: Spotting words in Latin, Devanagari and Arabic scripts. Indian J. Artif. Intell. **16**, 2–9 (2006)
4. Mozaffari, S., Faez, K., Märgner, V., Abed, H.E.: Two-stage lexicon reduction for offline Arabic handwritten word recognition. Int. J. Pattern Recognit. Artif. Intell. **22**, 1323 (2008)
5. Rath, T.M., Manmatha, R.: Word image matching using dynamic time warping. In: CVPR, vol. 2, p. 521 (2003)
6. Liu, C.L., Kim, J., Kim, J.H.: Model-based stroke extraction and matching for handwritten Chinese character recognition. Pattern Recognit. **34**, 2339–2352 (2001)
7. Kim, S., Park, S., Jeong, C., Kim, J., Park, H., Lee, G.: Keyword spotting on korean document images by matching the keyword image. Digit. Libr. **3815**, 158–166 (2005)
8. Chiang, Jung-Hsien: A hybrid neural network model in handwritten word recognition. Neural Netw. **11**, 337–346 (1998)
9. Adamek, T., Connor, N.O.: Efficient contour-based shape representation and matching. In: ACM SIGMM International Workshop on Multimedia Information Retrieval (2003)
10. Novikova, T., Barinova, O., Kohli, P., Lempitsky: Large-lexicon attribute consistent text recognition in natural images. In: Computer Vision—ECCV (2012)
11. Konidaris, T., Gatos, B., Ntzios, K., Pratikakis, I., Theodoridis, S., Perantonis, S.J.: Keyword guided word spotting in historical printed documents using synthetic data and user feedback. Int. J. Doc. Anal. Recognit. **9**, 167–177 (2007)
12. Andreev, A., Kirov, N.: Some variants of Hausdorff distance for word matching. Rev. Natl. Center Dig. **12**, 3–8 (2008)
13. Cha, S.H., Tappert, C.C., Srihari, S.N.: Optimizing binary feature vector similarity measure using genetic algorithm and handwritten character recognition. Int. Conf. Doc. Anal. Recognit. **2**, 662 (2003)
14. Lowe, D.G.: Distinctive image features from scale-invariant keypoints. Int. J. Comput. Vis. **60**, 91–110 (2004)
15. Abirami, S., Manjula, D.: Profile based information retrieval from printed document images. Int. Conf. Comput. Gr. Imaging Vis. **3**, 268–272 (2013)
16. Cho, N.I., Mitra, S.K.: Warped discrete cosine transform and its application in image compression. IEEE Trans. Circuits Syst. Video Technol. **10**, 1364–1373 (2000)
17. Marti, U.V., Bunke, H.: Using a statistical language model to improve the performance of an HMM-based cursive handwriting recognition system. J. Pattern Recognit. Artif. Intell. **15**, 65–90 (2001)
18. Shrivastava, A., Malisiewicz, T., Gupta, A., Efros, A.A.: Datadriven visual similarity for cross-domain image matching. In: ACM TOG, vol. 30, pp. 154 (2011)
19. Ma, H., Doermann, D.: Adaptive Devnagari OCR using generalized Hausdorff image comparison. ACM Trans. Asian Lang. Inf. Process. **2**, 193–218 (2011)
20. Papandreou, A., Gatos, B., Louloudis, G., Stamatopoulos, N.: Document image skew estimation contest. In: ICDAR, pp. 1444–1448 (2013)
21. Serrano, J.R., Perronnin, F.: Local gradient histogram features for word spotting in unconstrained handwritten documents. In: Proceedings of the International Conference on Frontiers in Handwriting Recognition (2008)

22. Rothfeder, J.L., Feng, S., Rath, T.M.: Using corner feature correspondences to rank word images similarity. In: Computer Vision and Pattern Recognition Workshop, pp. 30–35 (2003)
23. Almazan, J., Gordo, A., Fornes, A., Valveny, E.: Word spotting and recognition with embedded attributes. IEEE TPAMI **36**, 2552–2566 (2014)
24. Plamondon, R., Srihari, S.: Online and off-line handwriting recognition: a comprehensive survey. IEEE Trans. Pattern Anal. Mach. Intell. **22**, 63–84 (2000)

A Parallel Programming Approach for Estimation of Depth in World Coordinate System Using Single Camera

C. Rashmi and G. Hemantha Kumar

Abstract A novel parallel processing approach for computation of Z coordinate based on epipolar geometry of a scene captured with a single camera (monovision) is presented in this paper. The algorithm uses single camera images, before it is used to capture data, the camera is first calibrated. The calibration procedure calculates internal parameters such as distortion coefficient, focal length, and principal point. After the camera's internal parameters have been defined and set, the image is rectified. The main approach is to retrieve depth information from two or more images of one and the same object or scene. A feature point from one image is mapped to the corresponding feature points on another image. As this process is time-consuming, it is parallelized using OpenMP. A Fundamental matrix is estimated from the corresponding feature points and is calculated based on the epipolar constraint property between two views. Hence intrinsic parameters of camera and matched feature points in space are used to yield Z coordinate by triangulation.

Keywords Monovision · OpenMP · Fundamental matrix · Epipolar geometry Triangulation

1 Introduction

Distance measurement of an object is one of the main tasks for robotic applications of real world. Typically multiple cameras are used for distance measurement. Due to burden of computation, higher complexity and usage of multiple cameras which are expensive in stereo vision, a monovision technique using single fixed camera is utilized for the depth estimation. For some applications, a monovision approach is

C. Rashmi (✉) · G. Hemantha Kumar
High Performance Computing Project, DoS in CS,
University of Mysore, Mysuru 570006, India
e-mail: rashmi.hpc@gmail.com

G. Hemantha Kumar
e-mail: ghk.2007@yahoo.com

P. Nagabhushan et al. (eds.), *Data Analytics and Learning*,
Lecture Notes in Networks and Systems 43,
https://doi.org/10.1007/978-981-13-2514-4_7

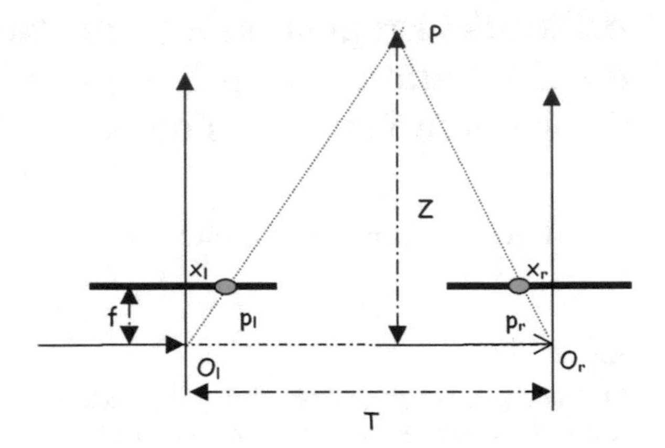

Fig. 1 Triangulation

used due to lower complexity, lower computational burden and cost of single camera is inexpensive compared to multiple cameras. Accuracy of the measurement of Z coordinate can be improvised by the combination of other sensors with the single camera. Using the size of given object and camera's focal length, object distance from camera is measured with the consideration of motion which is rigid but not with the general due to articulated and deformable scenes based on single camera. Structure from motion [1] and optical flow [2] are two common techniques for depth measurement when the camera is in motion. Structure from motion comprises of disparity measurement from correspondence feature points between frames and feature flow based image gradient is used in optical flow. Algorithm relies on matching the feature points of previous frame with current frame of video and distance of an object from camera is extracted at each frame.

BRISK [3] is one of the methods for detection of features in depth estimation. This paper presents an optimized algorithm that produces 3D information, in the form of a depth map, from an image sequence provided by a single camera. This problem is often called structure from motion. Depth perception main mechanism is the disparity measurement to recover the distance of target scene captured with slightly different viewpoints. By the usage of triangulation method [4], Z coordinate is measured from the correspondence keypoints in the pair of images. Coplanar-based triangulation approach is illustrated in Fig. 1 for the retrieval of depth information.

Left and Right image projections of 3D point P are P_l and Pr. Based on the relationship obtained from the correspondence keypoints of image pair left and right as in Fig. 1, depth value z of the point P can be obtained as

$$z = F\frac{T}{D}, \tag{1}$$

where D is the measure of displacement between the corresponding image pair keypoints as $D = Xr - X_l$. Hence, finding the disparity map is essential for the triangulation. Matching the correspondence feature points in image pair is the most time consuming aspect. Stereo correspondence is also known as stereo matching. Given

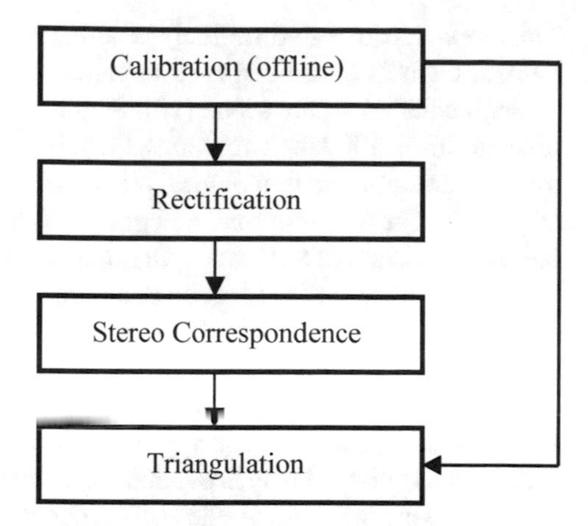

Fig. 2 Stereo vision

an image point on the left image, the correspondence points is found out in right image and matched using brute force approach. Hence this work can be overviewed as in Fig. 2 depicting the stereo vision sequence where calibration calculation is done offline for internal parameters such as focal length and distortion coefficient. Later the feature points are detected and matched known as rectification, using triangulation approach depth is estimated.

This paper is organized as literature survey is depicted in Sect. 2, depth estimation methodologies are illustrated in Sect. 3 and Sect. 4 illustrates the details of implementation considering checkerboard for testing followed by conclusion.

2 Literature Survey

Light-Based Depth Estimation [5] is initial approach for depth estimation. Active and passive are the two methods to obtain the distance of a point. Markov random Field (MRF) [6] learning algorithm is used for the depth estimation that holds for variety of indoor and outdoor environments [7]. Color information is used in stereo vision for depth estimation. Multi-scale stereo correspondence algorithm is presented in their approach. In order to improve both quality of results and response time multiresolution technique is used. Motion segmentation approach [8] is used for a single camera captured dynamic scene for estimating depth. A set of moving objects which are not dependent produces depth map from geometric principles of complex dynamic scenes is obtained from an optical flow based dynamic scene segmentation. Coded aperture system [9] based method is employed for depth estimation. A sequence of monocular video comprising of deformable and moving objects used for estimation introducing spatial temporal domain based on nonlocal-means (NLM) filtering for the reconstruction is applied. The method presented in this paper is applied for regions having similar texture with its neighboring frames belonging to the same surface.

This approach is tested on real scenarios increasing the computational efficiency. Semantic labels have been used to guide 3D reconstruction for performing semantic segmentation of the scene [10]. Semantic class of knowledge is incorporated in their approach allowing two things. First, by conditioning on semantic labels, a better model depth as a function of local pixel appearance and second, making use of geometry priors, for example road and grass textures supporting ground planes for other objects, sky is at the farthest depth possible. A global optimization scheme called as bundle optimization [11] is used to assign unique depth values for pixels that fall in different depth layers producing temporal and sharp consistent boundaries of object among different frames. Outliers, image noise, and occlusions are handled by bundle optimization framework for stereo reconstruction. Photo-consistency constraint is not only imposed in their approach but also geometric coherence with multiple frames in a statistical way is explicitly associated [12]. Here it is presented a method for depth learning from single monocular images, by combining deep convolution neural network and continuous conditional random field in a unified CNN framework. A deep structured learning scheme is proposed learning an unary and pairwise potentials of continuous CRF in a unified deep CNN framework. Their proposed methodology does not exploit any geometric cues [13]. Modeled a compact catadioptric stereo system for perception of depth. This system uses two sets of planar mirrors made of aluminum creating two virtual cameras which has resulted in an inexpensive for depth perception using single camera.

From the literature survey, it is noted that depth is estimated easily using two cameras for rigid object and also for moving object in a video sequence and also it is noticed that the parallel implementation of distance measurement of an object from a single camera is not much carried out as it is a challenging work.

3 Depth Estimation Using Single Camera

Extraction of geometric structures from the images using single camera usually called a monocular approach is presented in this section. Here camera calibration method of Zhang is performed as the initial step for the calculation of extrinsic and intrinsic characteristics. An extrinsic characteristics are position and orientation of camera. Internal characteristic such as principal point, distortion coefficient, focal length are the camera's intrinsic parameters. Initially first step for 3D computer vision is estimation of intrinsic parameters as it removes the lens distortion which degrades the accuracy. Chessboard pattern is used for calibrating camera as it results in accurate measurement of parameters from its corner points. Estimation of depth usually comprises of two steps as illustrated in architecture in Fig. 3. First feature points are detected and their correspondences are found and 3D world coordinate is estimated.

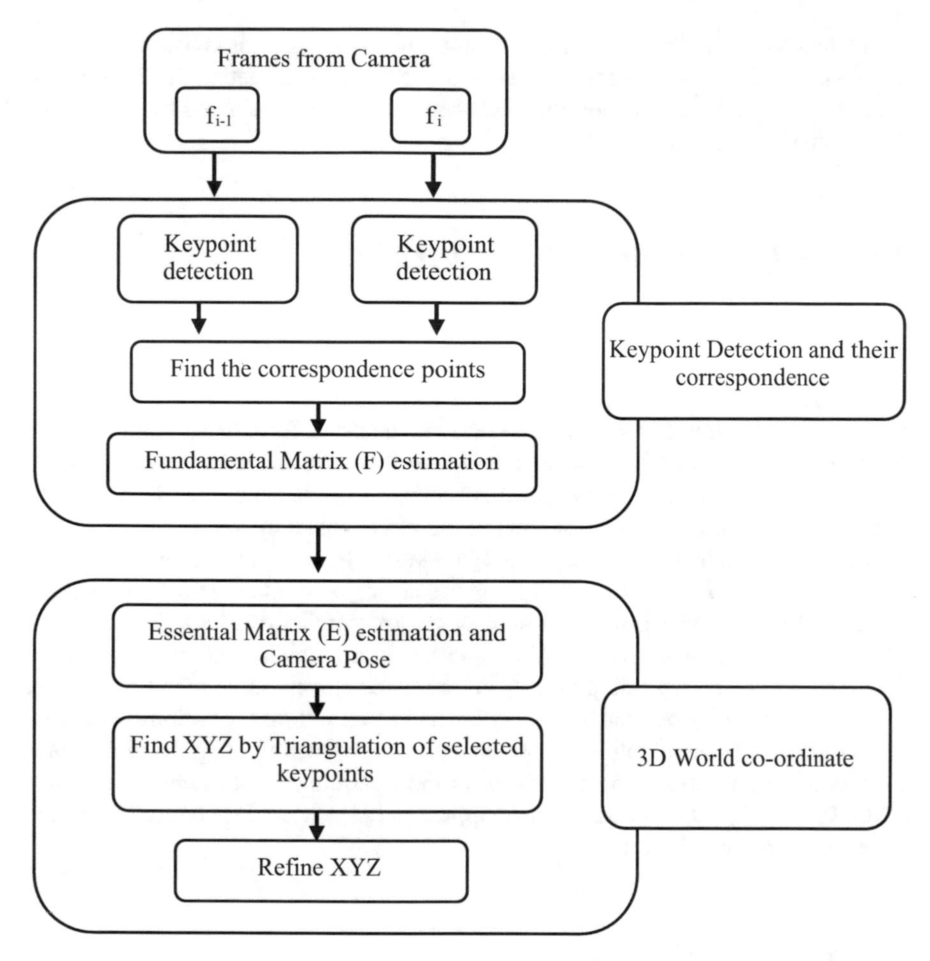

Fig. 3 Architecture

3.1 Corner Feature Point Detection

Vision tasks such as motion, monocular, and stereo estimation require finding the corresponding features across two or more views. Corner detection is one of the feature detection methods in extracting certain kinds of features and infer the contents of an image. There are many algorithms for corner detection such as SUSAN, FAST, Harris operator, Shi and Tomasi, Hessian feature strength measured and level curve curvature. Harris and Stephens (Plessey) based detection of corner is the most commonly and popular approach for the detection of corner points [14]. It is based on the autocorrelation function of the signal. The point is marked as corner point, if there is gradient variation in horizontal and vertical directions. A score is created from the determinant and trace of second moment matrix in turn determining corner points in a local window. If the score is large then that region is considered as corner. Harris detector is most informative and repetitive. This method is illumination change, rotation and translation invariant and hence it is considered for the corner detection of chessboard.

3.2 The Fundamental Matrix with RANSAC

Epipolar geometry is obtained by the estimation of Fundamental Matrix F. This is done by the two images captured from slightly different viewpoints of same scene, perfect point correspondence estimation relies on the fundamental matrix estimation, for the detection of outliers in the correspondences. Both linear and nonlinear methods are incorporated for the estimation of fundamental matrix. F can be estimated directly using only seven point matches, as fundamental matrix has seven degrees of freedom. A method for solving the fundamental matrix using eight point matches, if more than seven point matches are available. A minimization technique is implemented, if the points in both images are usually subjected to noise. RANSAC is a robust method which removes spurious corresponding matches. This method removes these spurious corresponding matches in the fundamental matrix estimation. The set of corresponding point pairs contains errors. Some pixels have been tracked completely wrong and some have errors in their sub-pixel position estimation. Therefore, it is essential not to use erroneous points when running the eight-point algorithm. One solution to this problem is the Random Sample Consensus Algorithm. RANSAC is a non-deterministic approach but will produce better results the more iterations are allowed.

3.3 The Essential Matrix

The Essential matrix E maps a 3D point defined in the first and second image reference through an Epipolar line. It gives us the relative rotation and translation between two views. Equation 2 is the representation for calculation of essential matrix.

$$E = K^T F K \tag{2}$$

Given that a 3D world coordinate projected in one image lies on epipolar line l1, the projection of the same coordinate in the other image can be found along the corresponding epipolar line l2. The essential matrix E, describes this relation in homogeneous normalized image coordinates pi, as $p^{T2}\ Ep^1 = 0$. One observation is that $E = RS$ where S is in Eq. 3.

$$S = \begin{pmatrix} 0 & -T_z & T_y \\ T_z & 0 & -T_x \\ -T_y & T_x & 0 \end{pmatrix} \tag{3}$$

Further, R and T can be determined from E through a singular value decomposition to $E = UDV^T$. The rotation is then given by R = UWV T or $R = UW^T V^T$ where W is of Eq. 4 and T = u3 or T = −u3 with u3 as the last column of U. Note that T has an unknown scale as it lie in the null space of E, i.e., ET = 0.

$$W = \begin{pmatrix} 0 & -1 & 0 \\ 1 & 0 & 0 \\ 0 & 0 & 1 \end{pmatrix} \tag{4}$$

3.4 The Normalized Eight-Point Algorithm

Fundamental matrix is calculated by the simplest algorithm eight-point from correspondence points of an image pair involving a set of linear equations with no information on extrinsic and intrinsic characteristics of camera. The essential matrices can be computed from five or more point correspondences in the frames. The five degrees of freedom relate to the three rotation angles and the two degrees of freedom for a normalized translation vector. A simpler equation can be derived if eight-point correspondences are used instead. The eight-point algorithm uses Eq. 5.

$$p_{1x}p_{2x}e_{11} + p_{1y}p_{2x}e_{12} + p_{2x}e_{13} + p_{1x}p_{2y}e_{21} + p_{1y}p_{2y}e_{22} + p_{2y}e_{23} + p_{1x}e_{31} + p_{1y}e_{32} + e_{33} = 0 \tag{5}$$

The equation system can be constructed and solved using eight not degenerated point pairs. This is because E has nine elements but has an unknown scale. The calculation has numerical problems because the range of the pixel coordinates is from zero to a thousand while the third homogeneous coordinate is usually one. The solution to this problem is to translate the coordinate system to the center of mass and scale to get an average distance of p2 to the origin.

3.5 Triangulation

Computation of depth X from correspondence image pair $x \leftrightarrow x^1$ is obtained by the triangulation method and is denoted as τ. Depth X can be computed as X = τ (x, x^1, P, P^1). It is assumed that there are no errors in projection matrices P, P^1 but occurs only in the measured image coordinates. Triangulation method is invariant to transformation of an appropriate class for the depth estimation. It is desirable to use an invariant affine triangulation method for computation of X coordinate if

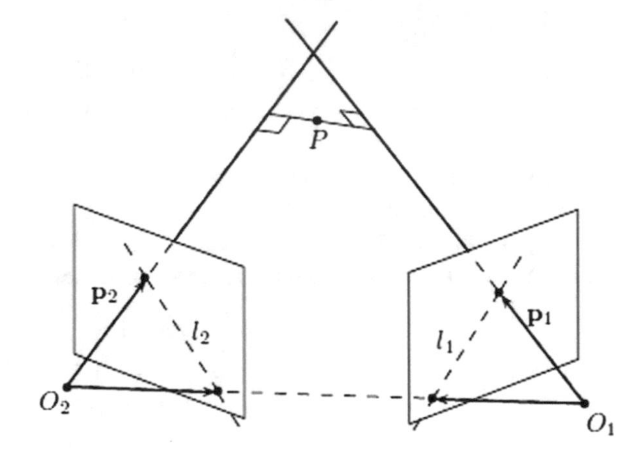

Fig. 4 P is the observed point of epipolar plane with O1 and O2 as the two centers of projection

the matrices of camera are known to an affine transformation. Here a triangulation method that is projective invariant is preserved is presented. The key idea is the 3D point X estimation, which exactly satisfies the supplied camera geometry, hence it projects as $\ddot{x} = P\ddot{X}$ $\ddot{x}' = P'\ddot{X}$ and aim is to estimate X from the image correspondence x and x^1 between left and right images. Such a method of triangulation is projective-invariant because only image distances are minimized, and the points $\ddot{x}$ and $\ddot{x}'$ which are the projections of $\ddot{x}$ do not depend on the projective frame in which $\ddot{x}$ is defined, i.e., a different projective reconstruction will project to the same points.

The method linear triangulation is an analogue to Direct Linear Transform (DLT) method. Equations $x = PX$, $x^1 = PX^1$ of correspondence image pair left and right are combined into a form $AX = 0$ which is linear in X. Initially cross product eliminated the homogeneous scale factor giving three equations for each point of an image, of which two are linear not dependent. If the translation vector, rotation matrix and camera matrix is known for a corresponding point pair, the 3D point P can be triangulated. There is always an estimation error in the pixel positions, which means that the rays from the optical center through the pixel will not intersect. One way of estimating the 3D point is the mid-point method. The method chooses the point on the middle of the shortest line segment between the two rays as shown in Fig. 4. If the translation T and rotation matrix R are known, the end points of the segments in the camera centered coordinate system of the first camera can be written as ap_1 and $T + bR^T p_2$. The direction of the line segment is $p_1 \times R^T p_2$, which is orthogonal to both rays. By solving the equation system $ap_1 - bR^T p_2 + c(p_1 \times R^T p_2) = T$ for a, b and c and hence the 3D point P can be computed as $P = f(T/x - x^1)$. Where $P =$ Depth, $f =$ Focal length, $d =$ Disparity (correspondence points), $T =$ Baseline.

4 Implementation Details

In this section, experiment is performed on checkerboard as it comprises of corner points, considering it as a feature points from an image sequence. Three Checkerboard video file of 20 frames per second which is moved at a distance of 24, 28 and 35 cm from camera is considered for experimentation. Following subsections describe the methodology for the depth estimation of checkerboard along with the illustration of performance evaluation.

4.1 Parallelization of Corner Detection and Finding the Correspondence Points

Harris-Plessey [12] methodology is used for finding the corner points by considering points of high curvature. Harris-Plessey is based on Moravec corner detector [13]. Harris corner detector can be expressed by an autocorrelation function, yields the corner points when thresholded by looking for variation in gradient for horizontal and vertical directions respectively in some local window. An algorithm developed by Jean-Yves Bouguet [15] is implemented to refine the corners estimated by the Harris-Plessey detectors to sub-pixel accuracy. Sub-pixel corner is detected accurately by the dot product. Many equations are formed from the corner point Q and several points like P around it equating to zero in turn solving, corner is detected with higher precision as in Fig. 5. Shared memory programming OpenMP [16–18] is utilized for the corner detection of checkerboard as the corners of checkerboard are independent to each other. After the corners are extracted correspondence points are found. Finding the correspondence points between image pairs and aligning a matching into two corresponding sets of 2D points is a time consuming process, hence it is parallelized using shared memory programming paradigm known as OpenMP. A multithreaded parallelization consists of compiler directives, runtime libraries and environment variables. A wide range of application from loop level to functional level parallelism is exhibited by openMP directives and hence an extended parallel programming model. A set of parallelism creation is provided by openMP directives. Loop-level parallelism is exhibited in finding the correspondence matches from first image to second image. Individual loops are parallelized in fine-grained parallelism called as loop-level parallelism, where multiple threads are spawned in a parallel loop and after execution of parallel, the single thread resumes its execution as illustrated in Fig. 6. Figure 7a, b depicts the process of aligning the two images called as stereo rectification in serial and parallel thus resulting in reduction of time consumption. Brute force matcher based on hamming distance is used to estimate the similarity of feature points between two images. The following code section depicts the parallel alignment of matching into two corresponding sets of 2D points. Thus in optimizing the feature point detection of corner points and finding the correspondence pair respectively and hence used for finding the fundamental matrix.

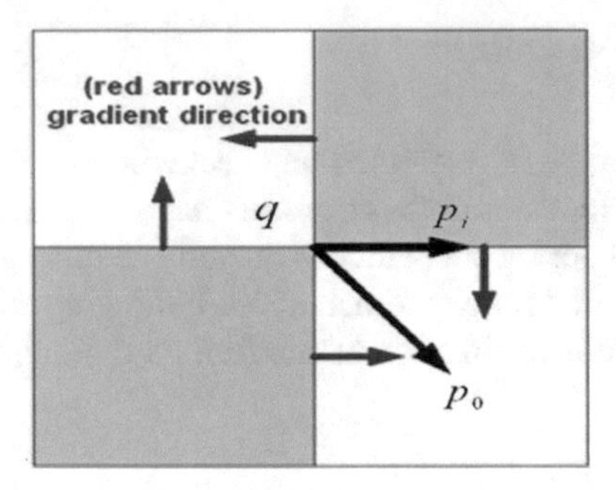

Fig. 5 Sub-corner detection

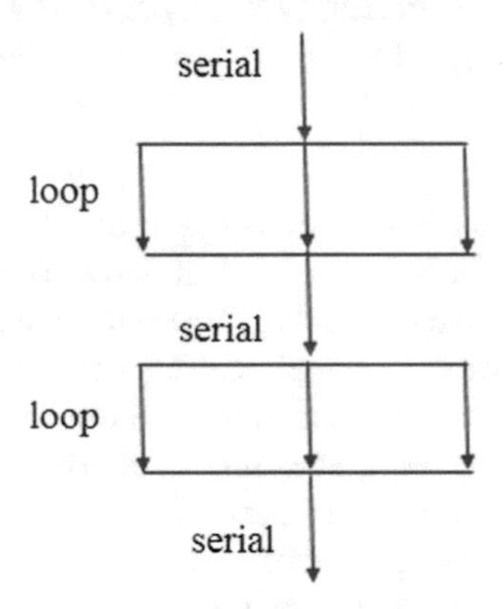

Fig. 6 Loop level parallelism

```
Vector<point_2f>imgpts_1,imgpts_2;
#pragma omp parallel for
for (a=0; a<matches.size(); a++)
{
//queryidx is the 'left' image
Imgpts_1.push_back(keypts_1[matches[a].queryidx].pt);
//trainidx is the 'right' image
Imgpts_2.push_back(keypts_2[matches[a].trainidx].pt);
}
```

4.2 Estimation of Fundamental Matrix F

This computes the fundamental matrix, F that shows the correspondence relation between feature points of two images and also checks the epipolar singularity by using RANSAC (Random Sample Consensus) algorithm with eight-point approach. It also filters all outliers from correspondence feature points and keeps only the inliers.

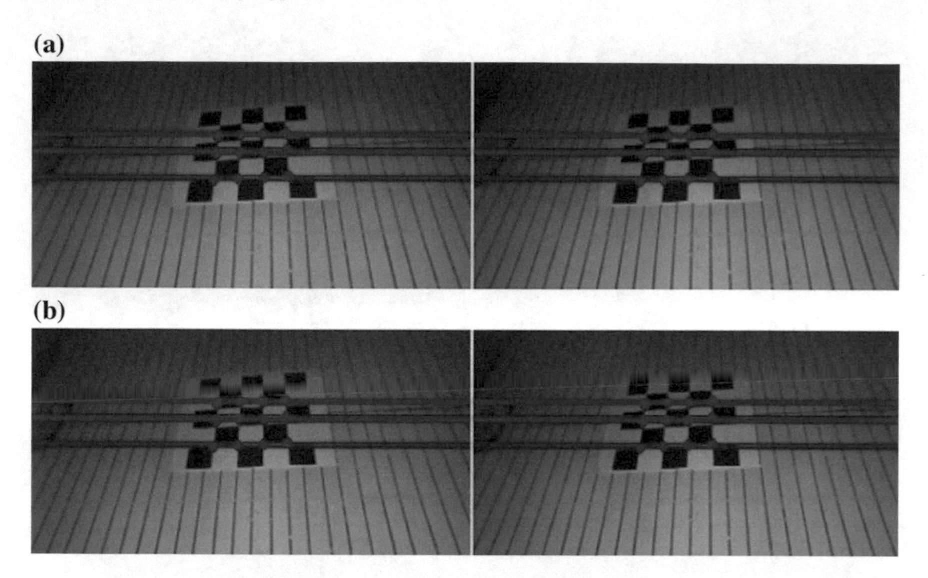

Fig. 7 **a** Image alignment by feature point correspondence (serial). **b** Image alignment by feature point correspondence (parallel)

4.3 *Find XYZ by Triangulation*

This calculates the projection matrices P1 and P2 for both images, and then sets P1 = [I|0], previous image is treated as base image, so rotation matrix is an identity matrix and translation vector = 0, and P2 = [R|T], here R and T are the selected camera pose of current frame. In the next step, 3D points using linear triangulation can be estimated as follows, build matrix A for homogenous equation system Ax = 0. When camera projection matrices are known and the image coordinates of a 3D point are known in both images the position of the 3D point Xi can be computed. This is called triangulation. Using correspondence points pair (x_i and x_{0i}) and camera projection matrices (P1 and P2), we estimate the 3D coordinates of the matched image points. Since the projection of X is known in the both images we can write as x = P1X and x0 = P2X. Then using the same reasoning as when dealing with the DLT (Direct linear transform) method.

$$X_i^*(PX_i) = \begin{pmatrix} yp^{3T} - wp^{1T} \\ wp^{1T} - xp^{3T} \\ wp^{2T} - yp^{1T} \end{pmatrix} Xi = 0, \quad (6)$$

where p^{jT} are the rows of P and X = (x, y, w). The same holds for P1 and x0. Only two of the rows are linearly independent, so one can be removed. Assuming that w = 1 and reordering the terms and stacking the two equations from each image on top of each other gets us to Eq. 7.

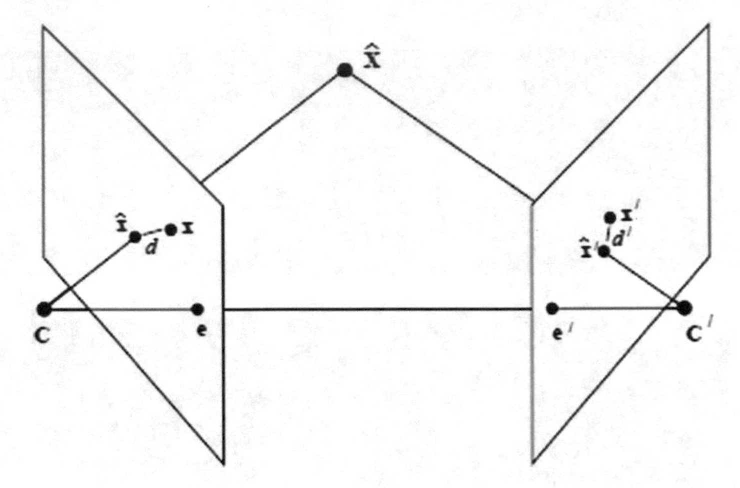

Fig. 8 Minimization of geometric error

$$AX = \begin{pmatrix} xp^{3T} - p^{1T} \\ yp^{3T} - p^{2T} \\ x^{1}p^{13T} - p^{1T} \\ y^{1}p^{13T} - p^{12T} \end{pmatrix} X = 0 \tag{7}$$

These are four equations in four unknowns, so in the general case there is a unique solution. But the X vector is a homogeneous vector so the equation system is redundant, three equations would have been enough. Using SVD Singular Value Decomposition solution can be found. Let $A = UDV^T$, then the solution X is the last column of V. Finally, linear triangulation is calculated iteratively for the minimization of reprojection error of reconstructed point to the original image coordinate.

4.4 Geometric Error Cost Function

An observation comprises of correspondence spurious feature point pair $x \leftrightarrow x^1$ not satisfying epipolar constraint. Correct values of corresponding feature points of an image pair $\ddot{x} \leftrightarrow \ddot{x}'$ lying close to the points of measurement $x \leftrightarrow x^1$ satisfying epipolar constraint $\ddot{x}'F\ddot{x} = 0$. The points $\ddot{x}$ and $\ddot{x}'$ that reduces the function $c(x, x^1) = d(x, \ddot{x})^2 + d(x^1, \ddot{x}^1)^2$ subjecting to $\ddot{x}^{1T}F\ddot{x} = 0$ where an euclidean distance between the points is represented as d(*,*), which corresponds for the reduction of reprojection error for a point $\ddot{x}$ mapping to $\ddot{x}$ and $\ddot{x}'$ by projection matrices consistent with F as in Fig. 8.

The estimated 3 space $\ddot{x}$ projects to two images at $\ddot{x}'$, $\ddot{x}$ satisfying the epipolar constraint unlike the measured points x and x^1. Minimization of reprojection error $d^2 + d'^2$ is done by choosing the point $\ddot{x}$.

4.5 Cluster Z

Cluster of estimated Z points is calculated using five-class k-means method and then computes 3D world coordinate (X, Y, Z) (cluster center) from a Z points cluster that has highest number points. For the cluster index for finding the z coordinate Single Program Multiple Data (SIMD) based methodology is utilized for calculation of an average Z coordinate.

4.6 Evaluation

The results of proposed parallel processing approach has been analyzed on an Intel Xeon E3-1220 V2 3.50 GHz CPU, 8.00 GB RAM, considering the camera with focal length of 2.3 mm. The proposed approach is implemented using an open source image processing library opencv2.3 in Microsoft Visual Studio 2010 Environment. Algorithm developed is tested for video files of checkerboard of 20 fps, where object is moving at a distance of 24, 28 and 35 cm depth from camera with pixel resolution 640×400, pixel size of 4.2 μm. The estimated result for the object which is moving at 35 cm is 34.9 with the deviation of −1.73%. The estimated result for object which is 24 cm away from camera is 23.14 cm with −3.57% deviation. Table 1 illustrate estimated results of parallel execution using four cores. Table 2 illustrates the Comparison results of serial and parallel and Fig. 9 represents the graphical representation of comparison results.

Table 1 Estimated result

Size (MB)	Actual distance (cm)			Estimated result (cm)			Deviation (%)
	X	Y	Z	X	Y	Z	
13.1	3	0	24	3	0	23.14	−3.57
15.9	3	0	28	3.26	0	24.98	−10.80
18.3	3	0	35	2.1	0	34.9	−1.73

Table 2 Comparison results

Size (MB)	Time (s)		
	Serial	Parallel	Speedup
13.1	4.604	3.994	1.152
15.9	5.427	4.201	1.291
18.3	7.372	5.218	1.412

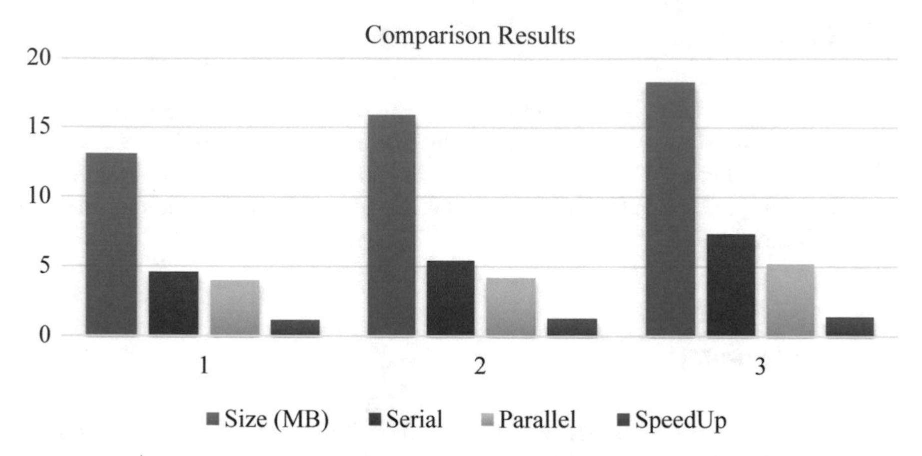

Fig. 9 Comparison results for file of size 13.1, 15.9 and 18.3 MB

5 Conclusion and Future Enhancement

A novel parallel approach for depth estimation of scene captured using single camera is presented in this paper. Two or more cameras will be used usually to record 3D geometry, when the usage of multiple cameras is either impractical or too expensive, however there is a need to record 3D geometry. The parallel algorithm which is developed uses video captured using single fixed camera. Data level based parallelism is exhibited on detection of feature points and requires in addition that each feature point on previous image is matched with the corresponding feature points of current image of a video using the Brute force approach and stereo-rectified. A RANSAC approach is used to remove the outliers and an optimal linear triangulation methodology is applied for the depth estimation. Hence, from the experimental results it is estimated that the Speedup is 1.28$\times$ on standalone systems with 95% accuracy. 3D model of an image can be reconstructed using single camera. As this is an advanced technology in computer vision can be used for real-time approaches.

Acknowledgements The work by Rashmi C. was supported by High Performance Computing Project lab, University of Mysore, Mysuru.

References

1. Recker, S., Gribble, C., Shashkov, M.M., Yepez, M., Hessflores, M., Joy, K.I.: Depth data assisted structure-from-motion parameter optimization and feature track correction. In: Applied Imagery Pattern Recognition Workshop (AIPR). IEEE Xplore (2015). https://doi.org/10.1109/aipr.2014.7041930(2014)
2. Honegger, D., Greisen, P., Meier, L., Tanskanen, P., Pollefeys, M.: Real-time velocity estimation based on optical flow and disparity matching. Intell. Robots Syst. (IROS) (2012). https://doi.org/10.1109/iros.2012.6385530

3. Leutenegger, S., Chli, M., Siegwart, R.Y.: BRISK: binary robust invariant scalable keypoints. In: ICCV'11 Proceedings of the 2011 International Conference on Computer Vision, pp. 2548–2555 (2011). https://doi.org/10.1109/iccv.2011.6126542
4. Sonka, M., Hlavac, V.: Roger Boyle Image Processing, Analysis and Machine Vision. PWS Publishing Company (1999)
5. Dansereau, D., Bruton, L.: Gradient-based depth estimation from 4D light fields. In: International Symposium on Circuits and Systems (2004). https://doi.org/10.1109/iscas.2004.1328805
6. Saxena, A., Schulte, J., Ng, A.Y.: Depth estimation using monocular and stereo cues. In: IJCAI'07 Proceedings of the 20th International Joint Conference on Artificial Intelligence, pp. 2197–2203 (2007)
7. Compañ, P., Satorre Cuerda, R., Rizo Aldeguer, R., Molina-Carmona, R.: Improving depth estimation using colour information in stereo vision. In: Proceedings of the Fifth International Conference on Visualization, Imaging and Image Processing (2005)
8. Ranftl, R., Vineet, V., Chen, Q., Koltun, V.: Dense monocular depth estimation in complex dynamic scenes. Comput. Vis. Pattern Recognit. (2016)
9. Martinello, M., Favaro, P.: Depth estimation from a video sequence with moving and deformable objects. In: IET Conference on Image Processing (IPR), pp. 1–6. IEEE Xplore (2012). https://doi.org/10.1049/cp.2012.0425
10. Liu, B., Gould, S., Koller, D.: Single image depth estimation from predicted semantic labels. In: IEEE Conference on Computer Vision and Pattern Recognition (CVPR) (2010). https://doi.org/10.1109/cvpr.2010.5539823
11. Zhang, G., Jia, J., Wong, T.-T.: Consistent depth maps recovery from a video sequence. IEEE Trans. Pattern Anal. Mach. Intell. **31**(6), 974–988 (2009). https://doi.org/10.1109/tpami.2009.52
12. Liu, F., Shen, C., Lin, G., Reid, I.: Learning depth from single monocular images using deep convolutional neural fields. IEEE Trans. Pattern Anal. Mach. Intell. **38**(10), 2024–2039 (2016)
13. Seal, J.R., Bailey, D.G., Gupta, G.S.: Depth perception with a single camera. In: 1st International Conference on Sensing Technology Palmerston North, New Zealand, pp. 96–101 (2005)
14. Hartley, R., Zisserman, A.: Multiple View Geometry in Computer Vision, 2nd ed. United States of America by Cambridge University Press, New York (2003)
15. Harris, C., Stephens, M.: A combined corner and edge detector. In: Proceedings of Fourth Alvey Vision Conference, pp. 147–151 (1988)
16. Moravec, H.: Obstacle Avoidance and Navigation in the Real World by a Seeing Robot Rover. Technical Report CMU-RI-TR-3, Robotics Institute, Carnegie-Mellon University (1980)
17. Bradski, G., Kaehler, A.: Learning OpenCV. O'Reilly Media, Inc., 1005 Gravenstein Highway North, CA (2008)
18. Quinn, M.J.: Parallel Programming in C with MPI and openMP. McGraw-Hill (2004)

An Alternate Voltage-Controlled Current Source for Electrical Impedance Tomography Applications

Venkatratnam Chitturi and Nagi Farrukh

Abstract The desirable features of any current source in electrical impedance tomography (EIT) are high output impedance and a constant current for varying loads. Howland current sources are the most popular current sources for EIT applications. Alternately, a simple non-inverting amplifier circuit can be used as a voltage-controlled current source using IC LM7171 op-amp. A constant current of 5 mA (PP) is designed and tested over a wide range of frequencies. The frequency response indicates a stable current up to 1 MHz. These results are validated for an impedance of 1 kΩ which is usually considered for EIT applications. A buffer circuit is added to this current circuit to meet high output impedance matching. A minor discrepancy is visible between the theoretical value and measured value of the output current due to the tolerance band of the resistors used in the non-inverting amplifier circuit.

Keywords Voltage-controlled current source · Howland circuit · Slew rate

1 Introduction

Electrical Impedance Tomography (EIT) is relatively a new technology evolved in 1980s and rapidly developed in the 1990s. This technique is beneficial for its safe use, inexpensiveness, non-invasiveness, and portability. It is currently being used in some of the applications like monitoring of lung problems like fluid accumulation in the lungs, monitoring of heart function, monitoring of internal bleeding, imaging of breast cancer, study of pelvic fluids, study of premenstrual syndrome and so on [1].

The EIT systems generally use a certain number of electrodes which are placed on a medium with equal spacing to surround the domain that is imaged. A constant,

---

V. Chitturi (✉)
Department of Instrumentation Technology, GSSSIETW, Mysuru, India
e-mail: venkatratnamc@gsss.edu.in

N. Farrukh
Faculty of Mechanical Engineering, Universiti Tenaga Nasional,
Kajang, Selangor, Malaysia

P. Nagabhushan et al. (eds.), *Data Analytics and Learning*,
Lecture Notes in Networks and Systems 43,
https://doi.org/10.1007/978-981-13-2514-4_8

low frequency, and low magnitude alternating current (AC) is injected to a pair of electrodes and the potential differences which are the function of unspecified conductivity distribution are measured from the remaining non-current carrying electrodes. The internal conductivity or permittivity or resistivity distribution is calculated based on the set of voltage measurements corresponding to the injected current. The necessary features of the injected current source include a stable current for the varying loads, large signal-to-noise ratio (SNR) and a wide frequency bandwidth [2].

2 Literature Review

In the EIT systems, the current acts as an excitation source with the voltages being measured. Current input reduces the effects of contact impedance, produces more uniform electric field and hence can contribute towards the resolution of the reconstructed images [3]. Thus the spatial resolution of an EIT image depends on the characteristics of the current source [4]. In addition, the current source not only helps in the reduction of noise arising due to the spatial variations [5] but also ensures patient's safety [6] as compared to a voltage source. All of these require a well-designed voltage-controlled current source (VCCS) with an understanding that the measured voltage differences are too small [7].

Most of the EIT systems include a voltage-controlled oscillator (VCO) [8] followed by VCCS. The VCCS circuit converts the alternating voltage from the VCO to an alternating current to be injected into the EIT systems. An ideal current source has infinite output impedance and constant current independent of the load (Fig. 1a). However, practical current sources have a finite output impedance and current varies with the load (Fig. 1b).

An ideal current source has a load current I_L equal to the source current Is. The practical current source is characterized by an impedance which is a parallel combination of resistor 'R' and capacitor 'C' and hence the relationship between I_L and Is changes with the value of the load impedance. Normally the EIT systems using single-ended current source face the problem of high common mode voltages, as one of the two ends is grounded. This problem can be overcome using a floating current source [9].

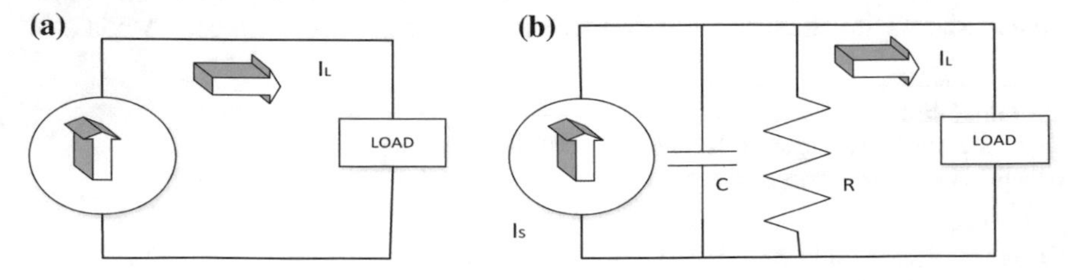

Fig. 1 **a** Ideal current source 1. **b** Practical current source

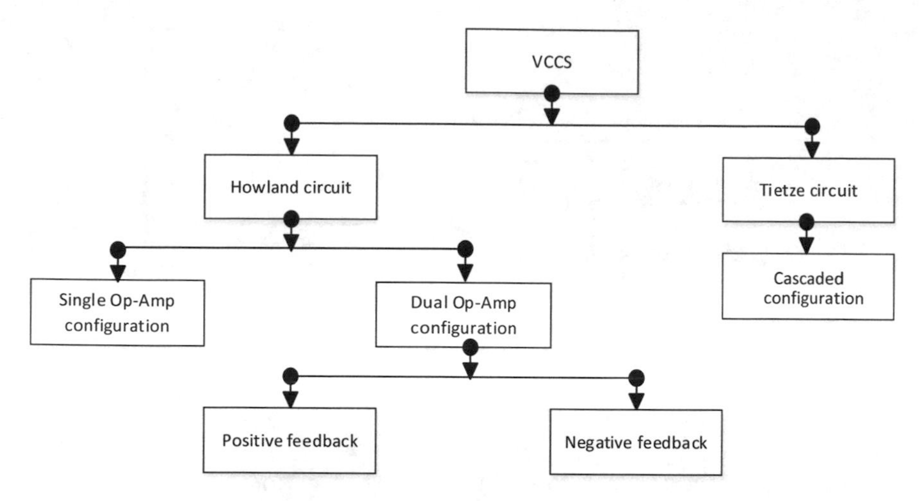

Fig. 2 Topologies of VCCS circuits [12]

The current source for an EIT system must be capable of delivering precise currents over a specified range of frequencies to the load impedances well within the expected range of values. These requirements can be analyzed in terms of frequency response, voltage compliance, and output impedance. For EIT, a sinusoidal signal with a peak current in the range of 0.1–5 mA is commonly used. Load impedance from 100 Ω to 10 kΩ is typically considered [10].

The VCCS circuit's commonly use operational amplifiers (op-amp) with high slew rates, failing which the sine wave changes to a triangular wave. Also the magnitude of the output current depends on the slew rate of an op-amp [2]. AD844 was used to design a VCCS circuit because of its high slew rate (2000 V/μs) and wide bandwidth (600 MHz). It has a high impedance at the non-inverting terminal and low impedance at the inverting terminal. The major drawback of AD844 is the transient current which is ±10 mA. The maximum output current is up to 30 mA for a load impedance of 750 kΩ in parallel with 18 pF capacitor [11].

The VCCS circuits are generally classified as shown in Fig. 2 [12].

The most popular VCCS circuit is the Howland circuit with the single op-amp configuration [2]. A steady state output current can be obtained with a tight resistor matching of $\leq$0.1%. However, larger resistor values (several MΩ) must be used to neglect the noises due to the current source [5]. Also, the Howland circuit is suitable only for kHz range of currents [2].

Hence the dual op-amp configurations with either a positive or a negative feedback circuits to overcome the stray capacitances are used. The enhanced Howland circuits generate stable currents with negative feedback even at MHz frequencies [4]. However, the output current becomes unstable with the positive feedback leading to errors [10]. Few more versions of the enhanced Howland circuits can be seen in the literature. A new enhanced Howland circuit with a general inverter circuit was

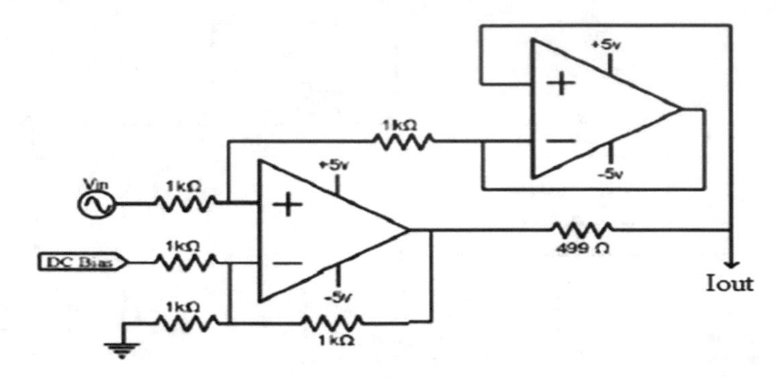

Fig. 3 Modified Howland current source circuit [14]

developed by [2] for a varying load of up to 6 kΩ. A mirrored modified Howland circuit simulation showed a stable current for frequencies beyond 100 kHz [11].

In general, the enhanced Howland current source comes with the stray capacitance effects due to the long cables. Sinking the current to the ground gives rise to high common mode voltage errors. Hence an active current sink must be used. Decoupling capacitors were used to block DC current components, which can cause unwanted redox reactions of the electrodes causing a current drift [13].

NI (National Instruments) hardware was used to develop a dual drive (voltage or current) system. The hardware included the required signal generator, timing circuit and ADC module. The voltage signal generated was converted into a current signal using a modified Howland current source circuit as shown in Fig. 3. Stray capacitance is a predominant problem in the Howland current circuits. The effects increase with an increase in the output currents in turn with the increasing frequencies. This increase would lower down the applied current to the test object. It is recommended to use shielded probes and further to use GIC (generalized impedance converter) circuits to reduce the stray capacitance effects. However, the above precautions are not suitable for different frequencies. Hence, a specific probe (or) and a specific GIC circuit must be developed for a specific operating frequency for the EIT system [14].

In general, all the enhanced Howland circuits require an additional (two or more) operational amplifier.

Tietze current sources are more stable than the enhanced Howland circuits. However, these circuits generate a current of up to only 500 μA across a load of 2 kΩ [3]. An integrated VCCS based on OTA (Operational Transconductance Amplifier) was developed by [9] with an operating frequency of 20 kHz driving a current of few μA. OTA's can overcome the resistor mismatch problems of Howland current sources. Two OTA's were used to maximize the operating frequency up to 1 MHz supporting loads up to 2 kΩ. But were capable of generating an alternating current whose magnitude was 500 μA.

A VCCS circuit was developed with additional features such as setting the offset voltage and regulating the same to make sure that maximum current is injected to the test object [15] as shown in Fig. 4.

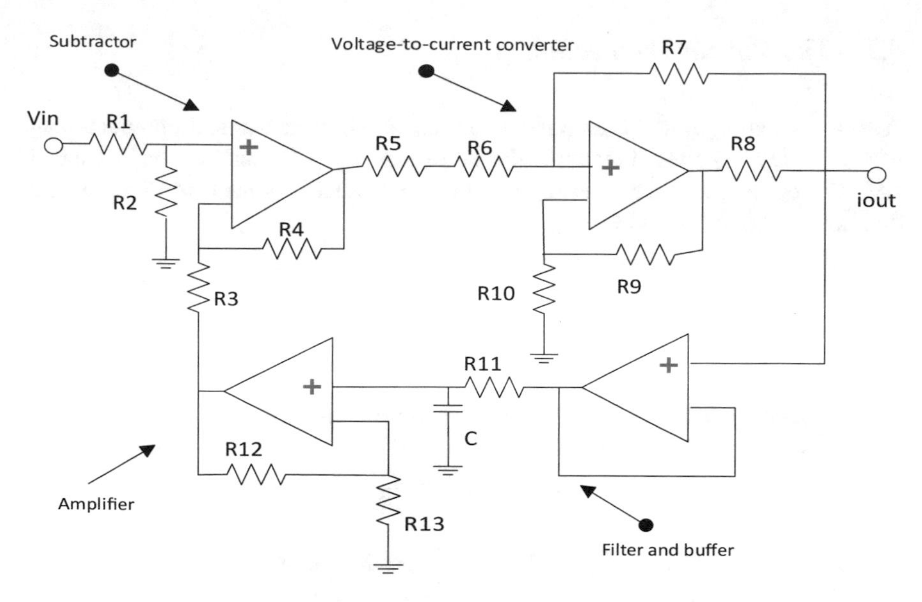

Fig. 4 Circuit diagram of the proposed stimulator circuit [15]

An input signal, Vin (externally generated) is given to a subtractor circuit. The output voltage is converted into its equivalent current iout that flows through the electrodes w.r.t. ground. Its equivalent voltage Vout is then buffered and filtered. Finally after amplification, it is subtracted from Vin in such a way that Vout, the offset voltage is controlled to zero volts. A current of 2 mA is generated for Vin = 1 V, to a maximum of 10 mA using OPA404KP IC. However the operating frequency was 22 Hz. The study was mainly carried out using stimulation devices. Hence the frequency is not suitable for EIT applications [15].

3 Methodology

Based on the comparison tests done between IC LM7171, IC THS 4061 and IC OPA604 op-amps, by applying frequencies between 1 kHz and 1 MHz for 1 kΩ load, the results showed that IC LM7171 op-amp has high output impedance with high slew rate of up to 4100 V/μs while supplying an output current up to 100 mA. However, the bandwidth and slew rate of the IC THS4061 and IC OPA604 op-amps are less [16]. Hence, IC LM7171 op-amp is chosen because of its high slew rate and a wide bandwidth of up to 200 MHz.

3.1 The Circuit Connection

A non-inverting operational amplifier configuration is used, where the output voltage of the VCO is applied directly to the non-inverting (+) input terminal of the IC LM7171 op-amp and the inverting (−) input terminal is connected to the voltage divider network of R_f and R_1.

3.2 Design of the Output Current

The closed-loop voltage gain of a non-inverting op-amp is given by Eq. (1)

$$V_{out} = V_{in}(1 + \frac{R_f}{R_1}) \quad (1)$$

An output voltage of 5 V_{PP} is designed generating an output AC current of 5 mA according to Eq. (2).

$$P = V_{rms} \times I_{rms} \times PF = \frac{V_{rms}^2}{R_L} \times PF \quad (2)$$

The load resistance, R_L is assumed as 1 kΩ which is generally the impedance range considered for EIT applications [3]. The safe let-go current for a human being is less than 6 mA [17].

3.3 The Buffer Circuit

At lower frequencies, almost all the current flows through the extracellular space of the cells of the human body and so the total impedance is largely resistive. Hence, the resulting impedance is relatively high. While, at higher frequencies, the current can cross the capacitance of the cell membrane and enters the intracellular space as well. It has access to the conductive ions in both the extra- and intracellular spaces and hence the overall impedance is lower [10]. A buffer circuit is used to overcome these impedance issues. The input impedance of the buffer is very high, however the output impedance is just few ohms. The input voltage of the buffer circuit which comes from the non-inverting op-amp is connected to the non-inverting (+) input terminal of another LM7171 op-amp and the inverting (−) input terminal of the op-amp is connected to the output terminal of the op-amp through resistor R2, which is used to remove the DC offset that is produced by the input bias current of the op-amp. The output voltage of the buffer circuit is equal to the input voltage. The circuit diagram for the non-inverting op-amp connected along with the buffer circuit is shown in Fig. 5.

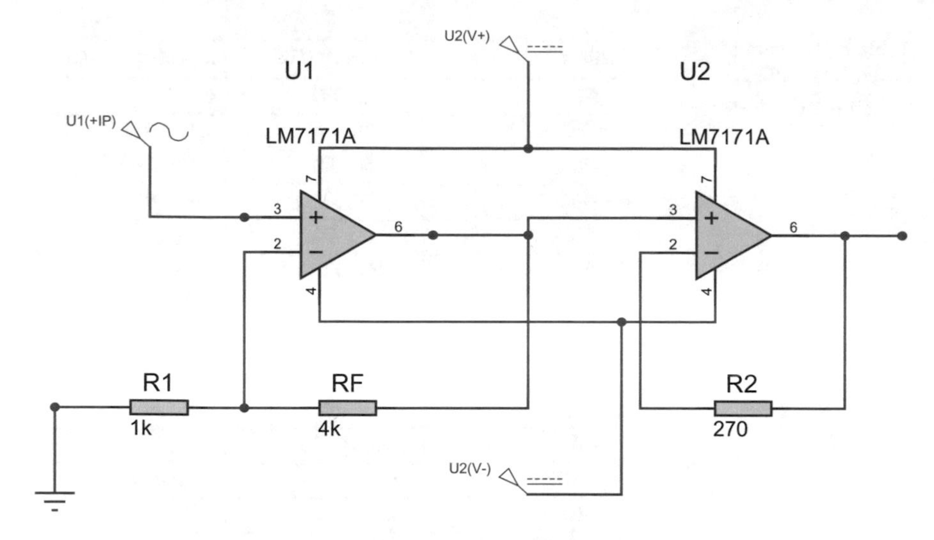

Fig. 5 Non-inverting operational amplifier connected to buffer circuit

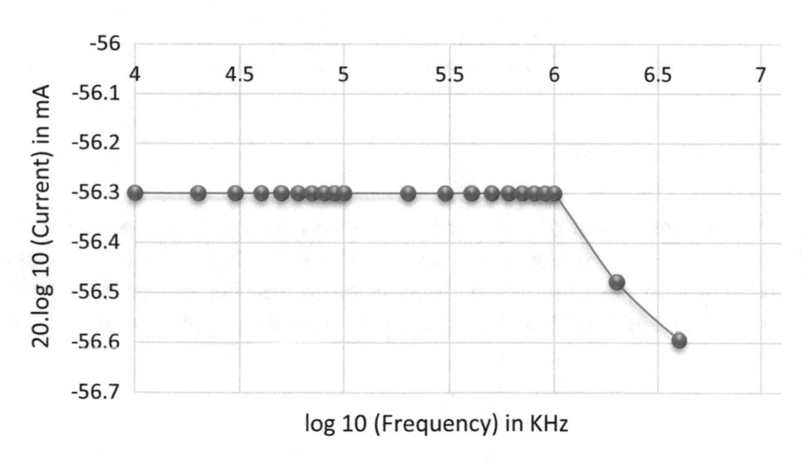

Fig. 6 Bode plot frequency response of the output current

4 Results and Discussion

4.1 Frequency Response

The output current (mA) was observed for different operating frequencies. The selected readings are tabulated in Table 1.

Figure 6 shows the frequency response of the output current in the form of a bode plot considering Table 1. The current was highly stable over a wide range of frequencies up to MHz range.

Table 1 Test current readings for different operating frequencies

20log10(Current)	Frequency (Hz)	log10(Frequency)
−56.3	10,000	4
−56.3	30,000	4.477121255
−56.3	50,000	4.698970004
−56.3	70,000	4.84509804
−56.3	90,000	4.954242509
−56.3	100,000	5
−56.3	300,000	5.477121255
−56.3	500,000	5.698970004
−56.3	700,000	5.84509804
−56.3	900,000	5.954242509
−56.3	1,000,000	6
−56.478	2,000,000	6.301029996
−56.594	4,000,000	6.602059991
−56.772	6,000,000	6.77815125
−56.893	8,000,000	6.903089987
−57.077	10,000,000	7

4.2 Simulation Result

Figure 7 shows the simulation results of the proposed study. Proteus software, developed by Labcenter electronics is a virtual system modeling and circuit simulation was used. The non-inverting op-amp connected to a buffer circuit, with an input voltage of 1 VPP for 1 kΩ load resistance generates an output current of 1.77 mA (rms).

4.3 Hardware Result

The entire experimental set up is shown in Fig. 8. Here, the non-inverting op-amp is connected to a buffer circuit, with an input voltage of 1 V_{PP} from the VCO circuit. The output current is 1.58 mA (rms) as displayed on the multimeter.

5 Conclusion

A discrepancy is visible between the theoretical value and measured value of the output current. The percent relative error is 10.734%. This can be accounted to the tolerance band of the resistors used for the non-inverting amplifier. The resistors used

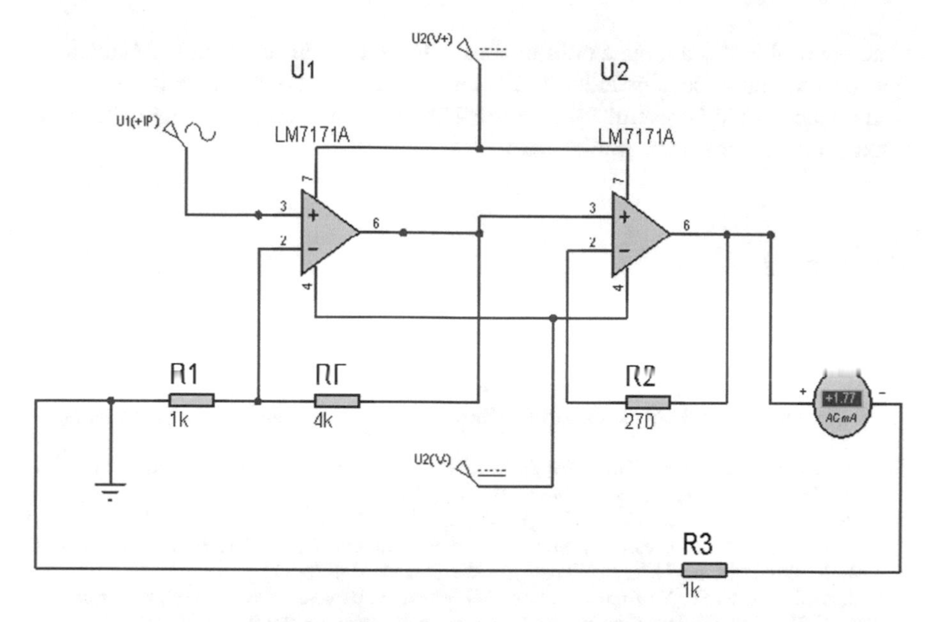

Fig. 7 Simulation results of the non-inverting operational amplifier connected to a buffer circuit

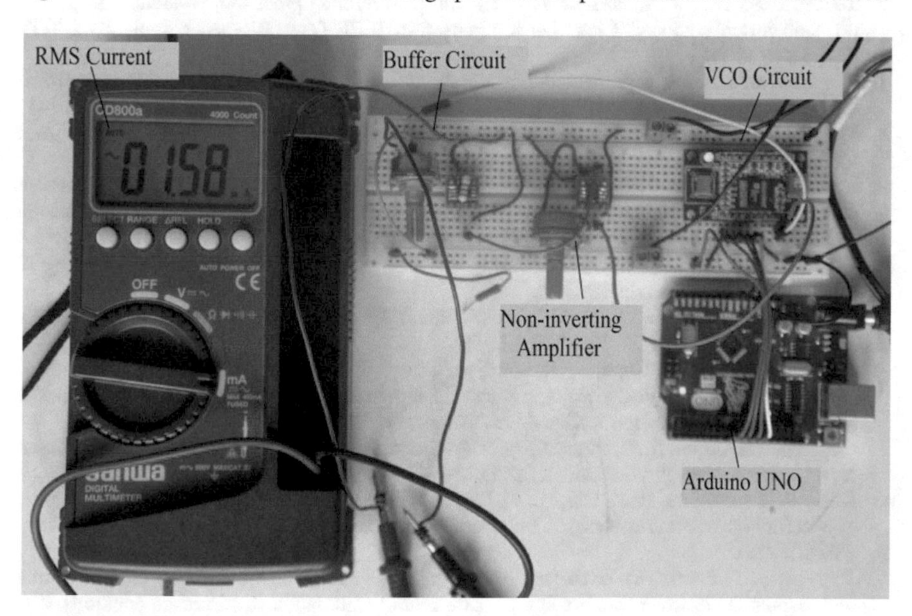

Fig. 8 Experimental set up displaying the output rms current

have a tolerance band of 5% which means that the percentage of error in the resistor's resistance can be more or less than 5% of the actual resistor's value. The next step is to improve the accuracy of the output current by selecting precision components and hence to enable this alternate current source to be used in future EIT applications. Further, the future scope includes design and development of this circuit for varying loads which could be useful for Electrical Resistance Tomography and Electrical Capacitance Tomography applications.

References

1. Saulnier, G.J., Blue, R.S., Newell, J.C., Isaacson, D., Edic, P.M.: Electrical impedance tomography. IEEE Signal Process. Mag. **18**(6), 31–43 (2001)
2. Al-Obaidi, A.A., Meribout, M.: A new enhanced Howland voltage controlled current source circuit for EIT applications. In: 2011 IEEE GCC Conference and Exhibition, pp. 327–330 (2011)
3. Alexander, S.R., Saulnier, G.J., Newell, J.C., Isaacson, D.: Current source design for electrical impedance tomography. Physiol. Meas. **24**(2) (2003)
4. Bouchaala, D., Shi, Q., Chen, X., Kanoun, O., Derbel, N.: Comparative study of voltage controlled current sources for biompedance measurements. In: 2012 IEEE 9th International Multi Conference on Systems, Signals and Devices, pp. 1–6 (2012)
5. Li, X., Dong, F., Fu, Y.: Analysis of constant-current characteristics for current sources. In: 2012 IEEE 24th Chinese Control and Decision Conference, pp. 2607–2612 (2012)
6. Tucker, A.S., Fox, R.M., Sadleir, R.J.: Biocompatible, high precision, wideband, improved howland current source with lead-lag compensation. IEEE Trans. Biomed. Circuits Syst. **7**(1), 63–70 (2013)
7. Sun, S., Xu, L., Cao, Z., Zhou, H., Yang, W.: A high-speed electrical impedance measurement circuit based on information-filtering demodulation. Meas. Sci. Technol. **25**(7), 1–10 (2014)
8. Venkatratnam, C., Nagi, F.: Development of an agilent voltage source for electrical impedance tomography applications. ARPN J. Eng. Appl. Sci. **11**(5), 3270–3275 (2016)
9. Hong, H., Rahal, M., Demosthenous, A., Bayford, R.H.: Floating voltage-controlled current sources for electrical impedance tomography. In: 2007 IEEE 18th European Conference on Circuit Theory and Design, pp. 208–211 (2007)
10. David, S.H.: Electrical Impedance Tomography—Methods, History and Applications. IOP Publishing Ltd, London (2005)
11. Wang, M., Ma, Y., Holliday, N., Dai, Y., Williams, R.A., Lucas, G.: A high-performance EIT system. IEEE Sens. J. **5**(2), 289–299 (2005)
12. Bouchaala, D., Shi, Q., Chen, X., Kanoun, O., Derbel, N.: A high accuracy voltage controlled current source for handheld bioimpedance measurement. In: 2013 IEEE 10th International Multi-conference on Systems, Signals & Devices, pp. 1–4 (2013)
13. Gaggero, P.O., Adler, A., Brunner, J., Seitz, P.: Electrical impedance tomography system based on active electrodes. Physiol. Meas. **33**(5), 831–847 (2012)
14. Khan, S., Manwaring, P., Borsic, A., Halter, R.: FPGA-based voltage and current dual drive system for high frame rate electrical impedance tomography. IEEE Trans. Med. Imaging **34**(4), 888–901 (2015)
15. Schuettler, M., Franke, M., Krueger, T.B., Stieglitz, T.: A voltage-controlled current source with regulated electrode bias-voltage for safe neural stimulation. J. Neurosci. Methods **171**, 248–252 (2008)
16. Wahab, Y.A., Rahim, R.A., Shima Mohd Fadzil, N.: A review of process tomography application in inspection system. Jurnal Teknologi **3**, 35–39 (2014)
17. Fish, R.M., Geddes, L.A.: Conduction of electrical current to and through the human body: a review. Open Access J. Plast. Surg. **9**, 407–421 (2009)

Analyzing and Comparison of Movie Rating Using Hadoop and Spark Frame Work

Akshaya Devadiga, C. V. Aravinda and H. N. Prakash

Abstract The unceasing stream of information produced by machines, sensors, vehicles, cell-phones, web-based systems of social networking, and other close ongoing resources are enticing associations to figure what they can do with this information if they possibly pick up knowledge into it [1]. In the real-time machine-driven data processing system must encounter the requirements of data-scientist, data-analysts, and data-center-operational teams without any widespread interval to third-party components [2]. RTP requires a frequent input, continual processing, and a firm output. Speed of NRTP is significant, but the processing time in minutes is also accepted in part of seconds. CEP involves combining data from multiple sources in order to detect patterns. In this paper we carry out a method of portraying an audience assessment for movie using RDD technique. This RDD is responsible for making RDDs resilient and distributed data. As we all knew this is a most challenging job, first step is that the audiences watching movie environment at dark room and contains many perspectives of persons at different scales, second step is that the continuance of a movie varies from 160 to 180-min, Finally the articulation and body-language of audience are inferred [3]. To excel these issues, the antiquity moments of a movie data was collected which capture the inferred moments of a person, and subsequently this was possessed by a category of the audience by pair-wise correlations over a period of time. Using this group representation, the work was figured out and refined the real-time ratings from set of people and evaluated the prediction of the movie.

Keywords Real-time processing · Near real-time processing
Resilient distributed data · Complete event processing

A. Devadiga · C. V. Aravinda (✉)
NMAM Institute of Technology NITTE, Karakal, Karnataka, India
e-mail: aravinda.cv@nitte.edu.in

A. Devadiga
e-mail: akshaya@nitte.edu.in

H. N. Prakash
Rajeev Institute of Technology, Hassan, Karnataka, India
e-mail: prakash.hn@yahoo.com

P. Nagabhushan et al. (eds.), *Data Analytics and Learning*,
Lecture Notes in Networks and Systems 43,
https://doi.org/10.1007/978-981-13-2514-4_9

1 Introduction

The Internet Movie Database is a superlative source to obtain thorough information about all the screenplay made. This encompasses a huge significance of data, which stores information about general drift in films. As the society is enhancing towards digital, the volume of data being produced and the storage of data is rising significantly. Analyzing of these data with traditional tools will become a challenge. Information administration designs has been progressed from the conventional data model to more complex data model in order to specify additional requirements, as batch processing and real time [2, 4].

1.1 RTP

The huge data collections from IOT and other devices gather substantial volumes of data that can clout for business purpose. A common example used for RTP is data moving, radar system and ATMs were in immediate processing plays an important role to ensure that the systems work suitably [2].

1.2 NRTP

NRTP speed is important, but the processing time is acceptable in lieu of seconds. If we consider an example of near real-time processing is the creation of operational intelligence, which is the combination of data processing and CEP. This event is involved in combining data from multiple sources in order to detect patterns (Fig. 1).

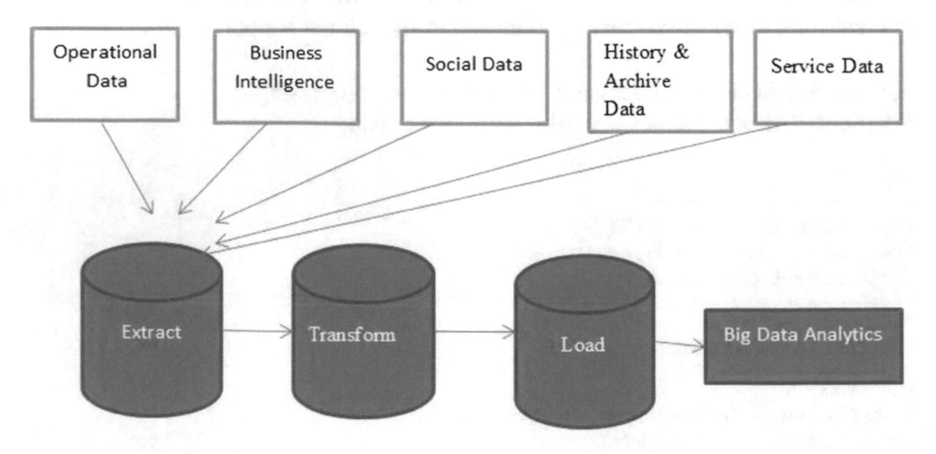

Fig. 1 Bigdata batch processing

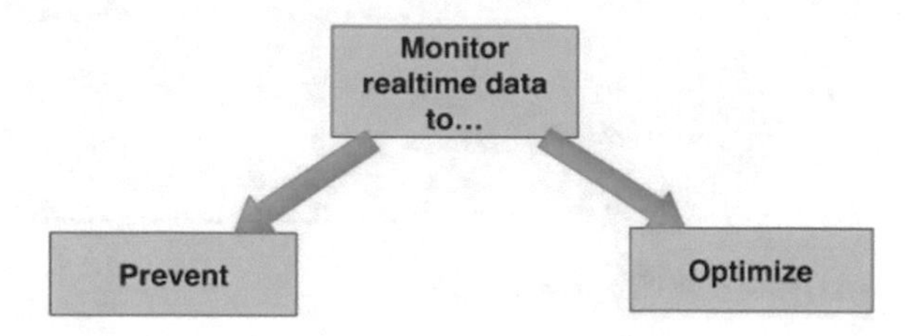

Fig. 2 Patterns driving most streaming use cases

Operational Intelligence (OI) uses CEP and RTDP in order to gain intuition to a process by successively analyzing the query against live nourishes and occasion information. Operational Intelligence is NRTP to the operational data and it also provides prominence for many of the data sources. The main aspect of this is to attain NRT intuition with continuous analytics to permit the system to take extant action [4, 5] (Fig. 2).

2 Problem Statement

The data is furnished in separate files. The incessant reason in connecting the data in these documents is the title of the screenplay, along with the making year to represent various distinctive forms, e.g., Fast and Furious part 1 and part 2. These files come in various formats, which have no delegations for instance Comma Separated Values (CSV). The data used here is not machine readable, instead it is set out to be human readable.

3 Shortcoming of MapReduce

- Forces your data processing into Map and Reduce
  - Other workflows missing include join, filter, flatMap, groupByKey, union, intersection,
- Based on Acyclic Data Flow from Disk to Disk (HDFS) [5, 6]
- Read and write to Disk before and after Map and Reduce (stateless machine)
  - Not efficient for iterative tasks, i.e., Machine Learning [3]
- Only Java natively supported
- Only for Batch processing.

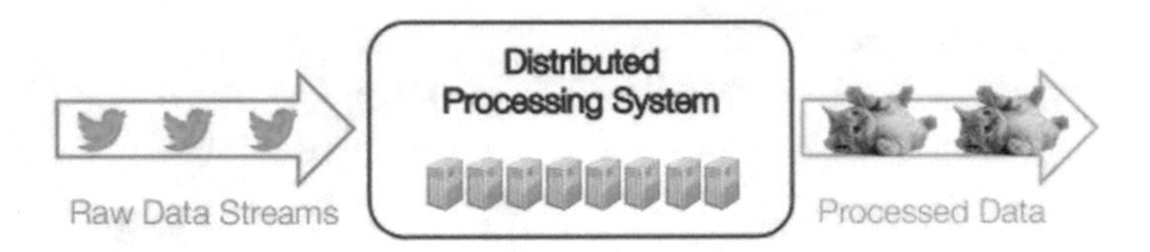

Fig. 3 Process big streaming data

4 Methodology

4.1 RDD'S

1. RDD's Data Containers
2. All the different processing components in Spark share the same abstraction called RDD [7]
3. As applications share the RDD abstraction, you can mix different kind of transformations to create new RDDs
4. Created by parallelizing a collection or reading a file
5. Fault tolerant [7] (Fig. 3).

5 Experimental Evaluation

(See Figs. 4 and 5).

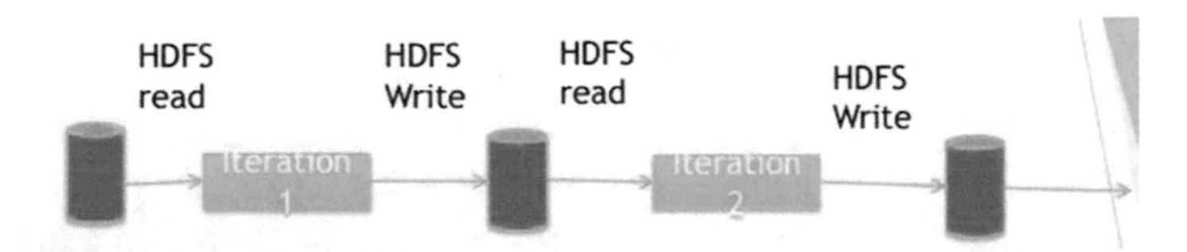

Fig. 4 Hadoop: use disk for data sharing

Fig. 5 Spark: in-memory data sharing

5.1 Solution is Apache-Spark

- This Apache-Spark is capable of ascendancy in Hadoop Ecosystem [7], e.g., HDFs, YARN, Hbase…
- In this workflows it consists functions like join, filter, flatMapdistinct, groupByKey, reduceByKey, sortByKey, collect, Count…
- Memory caching of data (for iterative, machine learning algorithms, etc.)
- It runs a program up to 100× faster than Hadoop Mapreduce in memory [8],
- DAG engine (Directed Acyclic Graph) optimize workflow.

5.2 Types RDD's Transformation Applied

There are generally different types of transformation

1. map()
2. flatmap()
3. filter()
4. distinct()
5. sample()
6. union, intersection, subtraction.

- We used two functions
- **map**: This applies a reduce function to entire RDD's Data, if we need to extract data from one field and perform action to multiply row and expecting the result as per required.
- **flatmap**: It is same as map but here this is mainly used not meant for onc-to-one rows relationship. Here we are looking for individual lines and applying for multiple lines for transformation.

5.3 Techniques Applied for Extracting Ratings

- SET UP OUR CONTEXT
  val = sc = new SparkContext("local[*]", "RatingsCounter")
- LOAD THE DATA

Cust-ID	Mov-ID	Raings	Time
196	242	3	881,567,899
186	206	2	886,978,945
22	377	3	88,567,893

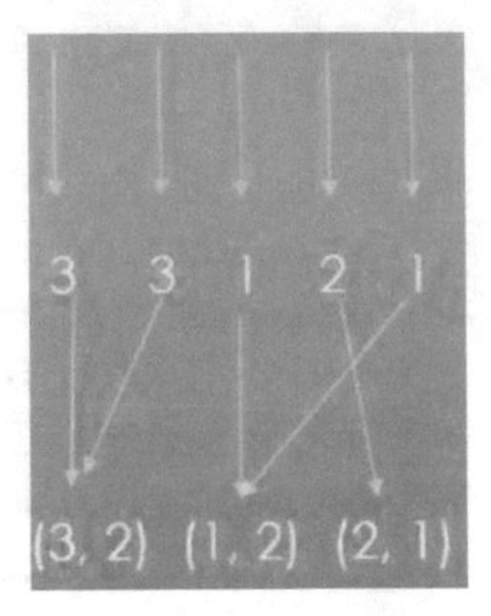

Fig. 6 Spark: execution flow of mapping in distributed manner

val lines = sc.textFile("../m1-100k/u.data")

- EXTRACT(map) THE DATA WE CARE ABOUT
  val ratings = lines.map(x => x.to.String().spilt("\t")(2)) 3 2 3
- PERFORMA AN ACTION: COUNT BY VALUE

  val results = ratings.countByValue()
  3 (3, 2)
  2 (2, 1)
  3

- SORT AND DISPLAY THE RESULT

  val sortedResults = results.toSeq.sortBy(_._1);
  sortedResults.foreach(println)
  1, 0
  2, 1
  3, 2

- AN EXECUTION PLAN IS CREATED FROM RDD's

  - **textFile**() This imports the raw data from the RDD
  - **map**() This maps the movie ratings from the data
  - **countByValue**() This count the total values from each rows (Fig. 6).

6 Optimization of the Data by Shuffling

- The job is broken into stages based on when data needs to be reorganized
- Stage 1 Combining the textFile() and map()
- Stage 2 countByValue()
- Each stage is broken into Tasks (Which may be distributed accross a Cluster) (Fig. 7).

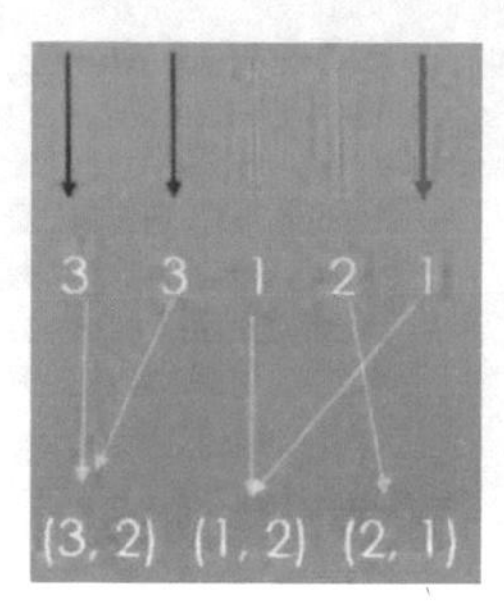

Fig. 7 Spark: distributed across a cluster

Table 1 Results tested on hadoop and spark framework

Sl. no	Hadoop	Spark
Data size (Tb)	102.5	100
Elapsed time (min)	72	23
Nodes	2100	206
Cores	50,400 physical	6592 virtualized
Cluster disk throughput (GB/s)	3150	618
Network (Gbps)	Dedicated data 10	Virtualized 10
Sort rate (Tb/min)	1.42	4.27
Sort rate/node (Gb/min)	0.67	20.7

7 Results Comparison

Table 1.

References

1. Han, J., Kamber, M.: Data Mining Concepts and Techniques. Morgan Kaufmann Publishers, San Francisco (2001)
2. Rahnama, A.H.A., Distributed real-timesentiment analysis for big data social streams. In: IEEE International Conference on Control, Decision and Information Technologies (CoDIT), pp. 789–794. https://doi.org/10.1109/codit.2014.6996998, Nov 2014
3. Neurosoft, S.A.: Neurosoft Envisioner. www.neurosoft.gr/products/envi.asp (1999)
4. Morales, G.D.F., Bifet, A.: SAMOA: scalable advanced massive online analysis. J. Mach. Learn. Res. **16**, pp. 149–153 (2015)
5. Bifet, A., Morales, G.D.F.: Big data stream learning with SAMOA. In: 2014 IEEE International Conference on Data Mining Workshop (ICDMW), pp. 1199–1202. 978-1-4799-4275-6
6. Singh, D., Reddy, C.K.: A survey on platforms for big data analytics. J. Big Data, pp. 2–8. https://doi.org/10.1186/s40537-014-0008-6 (2014)
7. Gradvohl, A.L.S., Senger, H., Arantes, L., Sens, P.: Comparing distributed online stream processing systems considering fault tolerance issues. J. Emerg. Technol. Web Intell. **6**(2), 174–179. https://doi.org/10.4304/jetwi.6.2.174-179 (2014)

8. De Francisci Morale, G.: SAMOA: a platform for mining big data streams. In: 22nd International Conference on WWW 2013 Companion, May 1317. ACM, Rio de Janeiro, Brazil. 978-1-4503-2038-2/13/05
9. Maske, M., Prasad, P.: A real time processing and streaming of wireless network data using storm. Int. J. Adv. Res. Comput. Sci. Softw. Eng. (IJARCSSE) **5**(1), 506–510. ISSN:2277 128X
10. Chen, M., Mao, S., Li, Y.: Big data: a survey. Mob. Netw. Appl. Springer Science+Business Media, New York, **19,** 171–209. https://doi.org/10.1007/s11036-013-0489-0 (2014)

Classification of Osteoarthritis-Affected Images Based on Edge Curvature Analysis

Ravindra S. Hegadi, Trupti D. Pawar and Dattatray I. Navale

Abstract Osteoarthritis is inflammation occurring in knee joint caused due to the loss of cartilage. This paper proposes the classification of knee joint X-ray images as normal or affected with osteoarthritis. The input images are converted to gray scale and noise is reduced by applying Wiener filter. Synovial cavity region is segmented using active contours and their edges are obtained. Curvature values are computed for longer edges, which correspond to the synovial cavity region. Abnormal images having osteoarthritis yield high variation in curvature as compared to normal images. The mean and standard deviation are computed for curvature values and used for further classification. k-nearest neighbor classifier is used to classify images as normal or abnormal. Ten normal and eight abnormal images are used for experimentation. The proposed method is successfully able to classify all the images to their respective class.

Keywords Osteoarthritis · Curvature analysis · Active contours · Image segmentation · Feature extraction · Classification

1 Introduction

Osteoarthritis commonly occurs in the weight-bearing joints of the hips, knees, and spine. It also affects the fingers, thumb, neck, and large toe, etc. It occurs due to loss of articular cartilage in joint. The common symptoms of osteoarthritis are joint aching and soreness, especially with movement pain, after long periods of inactivity stiffness, and joint swelling. It is generally categorized in five grades, namely grade

R. S. Hegadi (✉) · T. D. Pawar · D. I. Navale
School of Computational Sciences, Solapur University, Solapur 413255, India
e-mail: rshegadi@gmail.com

T. D. Pawar
e-mail: pawar1016@gmail.com

D. I. Navale
e-mail: navaledatta@gmail.com

P. Nagabhushan et al. (eds.), *Data Analytics and Learning*,
Lecture Notes in Networks and Systems 43,
https://doi.org/10.1007/978-981-13-2514-4_10

0 to grade 4, depending on its severity. The diagnosis of osteoarthritis in the stage of higher grade is very complex. X-ray images offer a better approach for detection of changes in inflammation at the joints affected by osteoarthritis leading to the identification of existence of osteoarthritis [1].

The possibility of occurrence of osteoarthritis is unpredictable but the chance of developing the disease increases with age. Most people over age 60 have osteoarthritis to some degree, but its severity varies. Even people in their 20s and 30s can get osteoarthritis, although there is often an underlying reason, such as joint injury or repetitive joint stress from overuse. In people over age 50, more women than men have osteoarthritis.

In the present scenario, the treatment of patients suffering from osteoarthritis is quite expensive. Development of advanced imaging technologies for early detection of the disease and diagnosis will significantly reduce the financial burden on the patient and will also help in early treatment.

In the past, many researchers worked toward solving the problem of grading of levels in knee osteoarthritis using X-ray, CT scan, and MRI images. Different preprocessing techniques, segmentation methods, and classifiers were used by these researchers to detect and segment the synovial cavity region and to identify whether the image is normal or abnormal and further grade these images.

Navale et al. [1] identified the knee osteoarthritis using block-based texture analysis approach and SVM classification technique. In their work, they divided images into nine equally sized blocks and extracted different texture features, namely skewness, kurtosis, standard deviatio,n and energy from each of these 9 blocks leading to 36 feature vectors for each image. SVM classifier was used to classify images as normal and affected images. The accuracy of their algorithm is 80% for normal images and 86.7% for the images affected by osteoarthritis. Duryea et al. [2] developed a semi-automated method for measuring loss of knee cartilage of patients suffering from osteoarthritis. Variation in the cartilage volume at five regions were recorded over time period. Measures like average change in volume, standard deviation, and standardized response time were used for their study.

Classification of osteoarthritis images using self-organizing map was proposed by Anifah et al. [3]. Authors used Contrast-Limited Adaptive Histogram Equalization and template matching to decide whether an image belongs to left knee or right knee. To identify the synovial cavity region, they segmented knee image using Gabor kernel, template matching, row sum graph, and gray level center of mass method. GLCM features are used for further classification of data in five grades of OA. They reported high rate of accuracy for grade 0 and grade 4 images but for images of grade 1 to grade 3, the accuracy of classification is very poor.

Our algorithm segments synovial cavity region from knee X-ray images and further, these images are classified as normal or abnormal. Section 2 describes the methodologies used, experimental results are presented in Sect. 3 and conclusions are given in Sect. 4.

2 Methodology

The input images are preprocessed to reduce the noise and segmentation is performed using active contours. The boundaries of synovial cavity region are extracted using morphological boundary extraction technique. The curvatures are computed for larger edges and their mean and standard deviation are calculated. The abnormal images generate edges with high curvature leading to high value of standard deviation. The k-nearest neighbor classifier is used to classify images as normal and abnormal. The detailed description of the methodology is presented in the following subsections.

2.1 Preprocessing

The input image will be in color format; it is converted to grayscale image as shown in Fig. 1a, b. The acquired X-ray images have noise, which is added during image

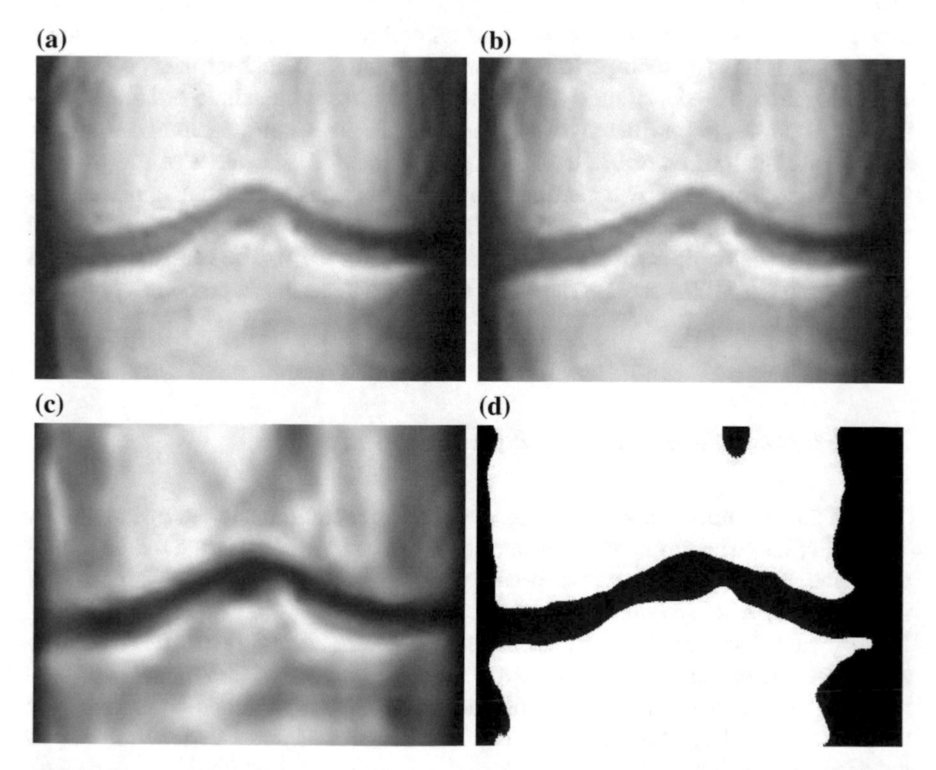

Fig. 1 Stages of image preprocessing **a** original image **b** converted to gray scale **c** image after noise removal and **d** segmentation using active contours

capturing process itself. Wiener filter is used to reduce the noise. It is a 2D adaptive noise removal filter which adapts itself for local variance in the image [4]. For higher local noise variance, filter performs lesser smoothing and for lower local noise variance, filter performs high smoothing. The filter performance is better as compared to the linear filters and it performs filtering selectively, that is, it preserves edges and high varying frequencies in the image. The filter is quite efficient in the time required for computation as compared to the linear filtering. It performs best for the white additive noise like Gaussian noise. The mean square error of this filter is optimal, which means the mean square error is minimized during filtering and noise reduction process. The Wiener filtering is optimal in terms of the mean square error. In other words, it minimizes the overall mean square error in the process of inverse filtering and noise smoothing. The Wiener filtering is a linear estimation of the original image. The approach is based on a stochastic framework. The orthogonality principle implies that the Wiener filter in Fourier domain can be expressed as follows:

$$W\left(f_1, f_2\right) = \frac{H^*\left(f_1, f_2\right) S_{xx}\left(f_1, f_2\right)}{\left|H\left(f_1, f_2\right)\right|^2 S_{xx}\left(f_1, f_2\right) + S_{\eta\eta}\left(f_1, f_2\right)}' \quad (1)$$

where $S_{xx}\left(f_1, f_2\right)$ and $S_{\eta\eta}\left(f_1, f_2\right)$ are the power spectrum of input image and additive noise, and $H\left(f_1, f_2\right)$ is filter used for smoothing the image. Basically, Wiener filter contains a part which performs inverse filtering and another part performing noise reduction. The size of the filter selected for the experiment is 15×15. After filtering the contrast of the filtered image is enhanced using Contrast-Limited Adaptive Histogram Equalization (CLAHE). Through this method, the values of image pixel intensities are enhanced by operating into smaller regions rather than entire image. The neighboring enhanced regions are joined using bilinear interpolation technique to avoid generation of boundaries for the processed blocks. The block size closed for applying CLAHE is 8×8. The result of these enhancement processes is shown in Fig. 1c.

2.2 Segmentation Using Active Contours

The image after noise removal is subjected to segmentation using active contours. The region-based energy model presented by Chan and Vese [5] is used for the implementation of active contours which is an approximation for multi-stage segmentation model. It uses level sets for the formulation of curve evolution as shown in Fig. 2. Basically, level set has an implicit contour in which the evolving curve is treated at zero level line of the level set function. The basic purpose of this model is to segment the input image into two regions known as object region and background by embedding boundary of object using zero level curve of 3D level set function. If ϕ is the level set function, then the Chan-Vese energy minimization function is given as

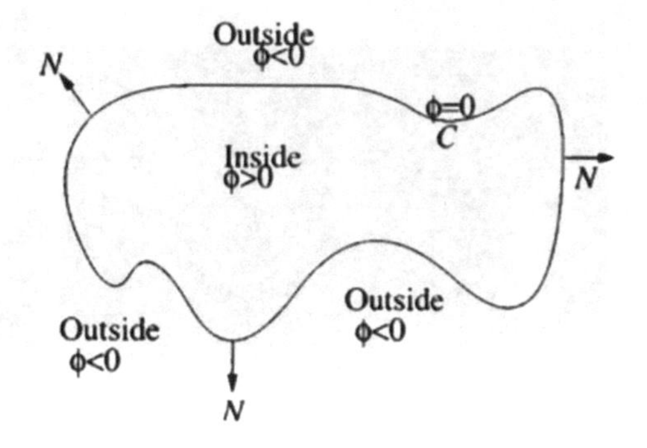

Fig. 2 Formulation of curve evolution using level set method

$$E(c_1, c_2, \phi) = \lambda_1 \int_{\Omega} (f - c_1)^2 H(\phi)\, dx + \lambda_2 \int_{\Omega} (f - c_2)^2 (1 - H(\phi))\, dx + \mu \int_{\Omega} |\nabla H(\phi)|\, dx \tag{2}$$

where $\lambda_1, \lambda_2 > 0$ and $\mu \geq 0$ are fixed parameters. The length parameter μ may be treated as a scale parameter because it plays vital role in measuring the relative importance of the length term. By decreasing the value of parameter μ, smaller objects and regions may be detected easily. The result of segmentation using active contours is shown in Fig. 1d.

2.3 Boundary Extraction

After segmentation of preprocessed image using active contours, the image will be in binary form. Simple morphological boundary extraction using 3×3 structuring element is applied to extract the inner boundary of segmented region, which separates the synovial cavity region from the bone parts. Suppose I is the segmented image, its boundary can be extracted using the following operation:

$$I_{Boundary} = I - (I \ominus B) \tag{3}$$

where B is 3×3 structuring element and $\ominus$ is morphological erosion operation. The result of extracted boundaries is shown in Fig. 3a and the boundaries in original image are shown in Fig. 3b.

2.4 Edge Curvature Computation

The basic purpose of extracting boundaries from the segmented images is to locate the boundary of synovial cavity region. But there are possibilities of generation of false object boundaries outside synovial cavity region. Generally, these edges will be

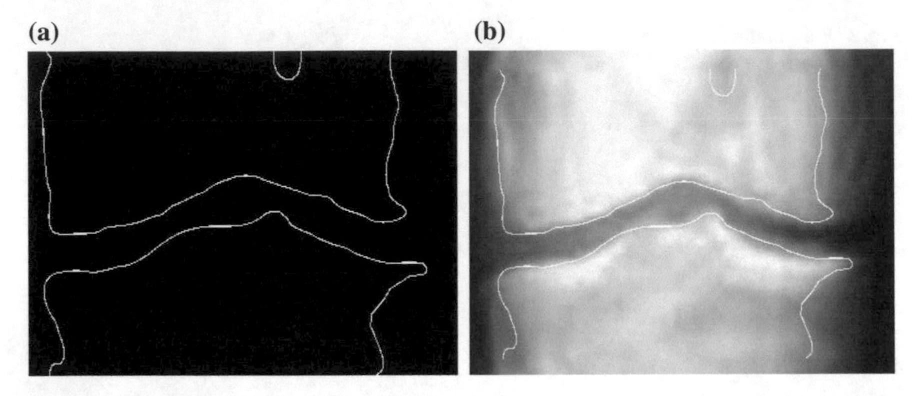

Fig. 3 Boundary extraction **a** extraction of boundary using morphological technique, **b** boundary over synovial cavity region

smaller in size. Such false contours are eliminated and larger edges are retained for further processing. The properties of synovial cavity region will help in classification of image as normal or affected with OA. For normal images, the edges would be smooth with less variation in curvature values, whereas for the abnormal images, the edges will have high local curvature. Derivatives are used for the computation of curvature [6]. For a continuous curve C, expressed as $\{x(s), y(s)\}$, where s is the length of edge points, its curvature can be expressed as:

$$k(s) = \frac{\dot{x}\ddot{y} - \ddot{x}\dot{y}}{\left(\dot{x}^2 + \dot{y}^2\right)^{3/2}} \tag{4}$$

where $\dot{x} = dx/ds$, $\ddot{x} = d^2x/ds^2$, $\dot{y} = dy/ds$ and $\ddot{y} = d^2y/ds^2$. The above expression of curvature is in continuous form and for digital implementation curvature, values may be treated as set of equally spaced grid samples in Cartesian coordinate system. The derivatives in Eq. 4 are represented using first-order and second-order derivatives as $\dot{x} = x_i - x_{i-1}$, $\dot{y} = y_i - y_{i-1}$, $\ddot{x} = x_{i-1} - 2x_i + x_{i+1}$ and $\ddot{y} = y_{i-1} - 2y_i + y_{i+1}$.

Figure 4a shows large edges remaining after elimination of false contours and Fig. 4b is the curvature profile of these edge segments.

2.5 Classification

The mean and standard deviation are computed from the array of curvature values as

$$\bar{x} = \frac{\sum_{i=1}^{n} x_i}{n}, \ \sigma = \sqrt{\frac{\sum_{i=1}^{n} (x_i - \bar{x})^2}{n}} \tag{5}$$

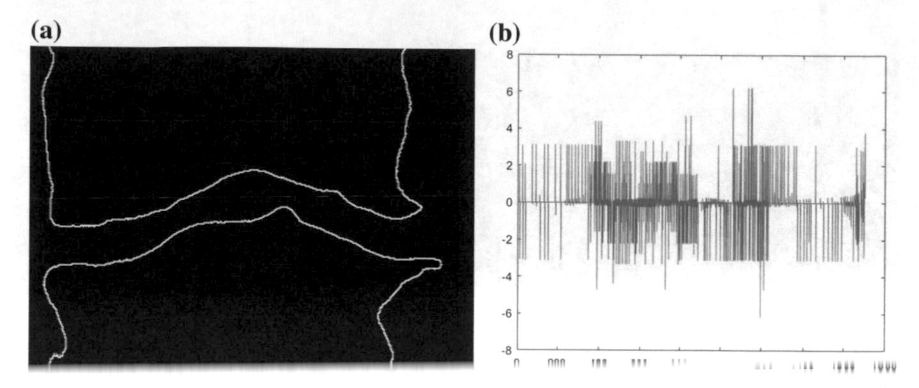

Fig. 4 **a** Large edge segments and **b** curvature profile of edges

The mean and standard deviation for curvature profile shown in Fig. 4 are 0 and 1.26, respectively. Mean and standard deviation values are computed for all the images and k-NN (k-Nearest Neighbors) classifier is used to classify the images into two groups: normal and affected with OA. k-NN classifier is a nonparametric method used for classification as well as regression. It has two phases, training and classification. The result of k-NN classifier is a class member. The classification of an object will depend on the majority of vote by its neighboring members and the object will be assigned to that class in which it is most nearest among the k nearest neighbors.

3 Experimental Results

For experimentation, 10 normal images and 8 images with different levels of osteoarthritis were collected from a local hospital. Guidance from radiologists was obtained for identifying whether the images belong to normal patients or for patients suffering from osteoarthritis. MATLAB R2016a software is used for the implementation of proposed work.

Figure 5 shows the result of the proposed method on an image of patient suffering from osteoarthritis, and its mean and standard deviation of edge curvature values are 0.05 and 1.47, respectively. The classification result using proposed method is 100% (Fig. 6).

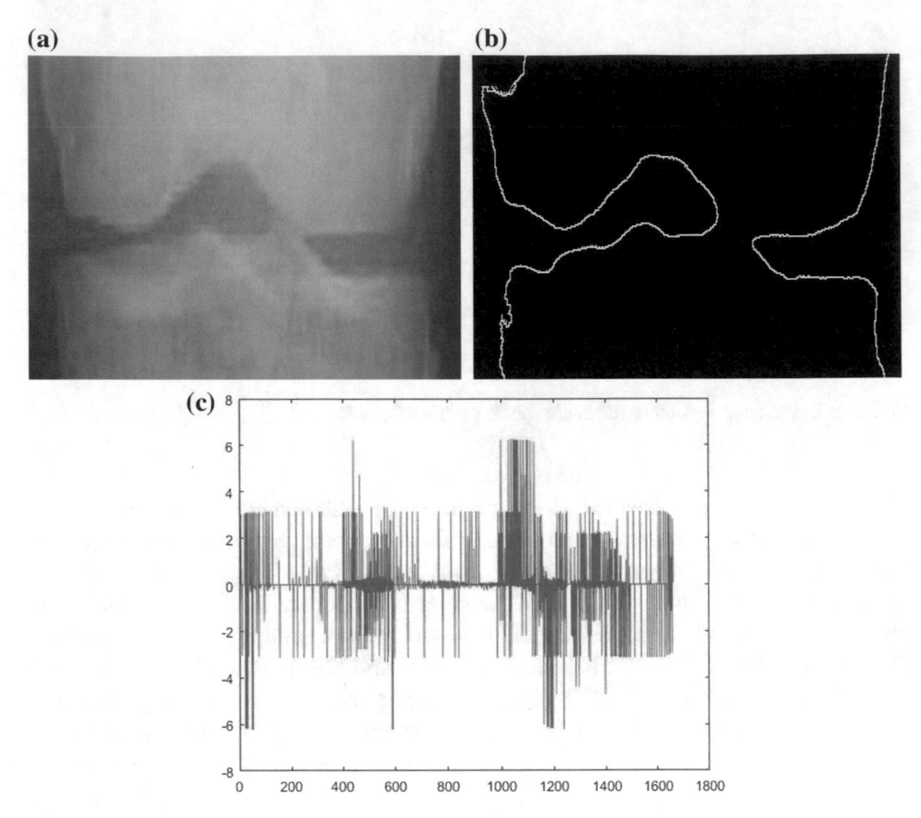

Fig. 5 Result of abnormal image **a** original image, **b** large edge segments and **c** curvature profile

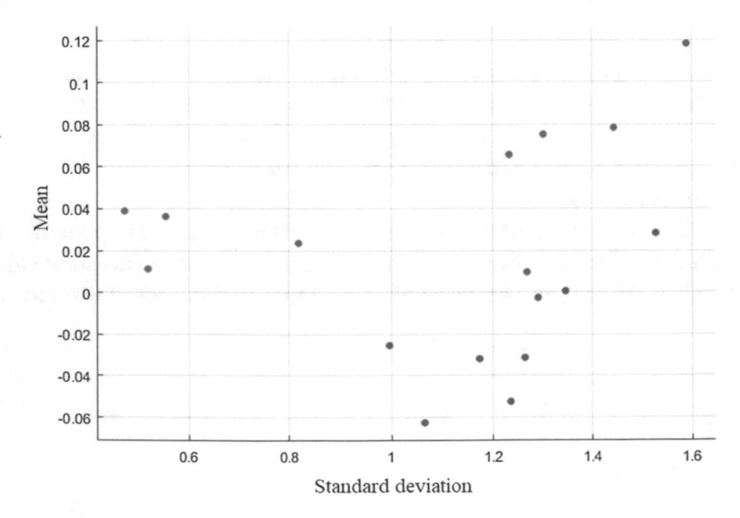

Fig. 6 Scatter plot showing classification results blue dots represents normal images and orange dots represents image with osteoarthritis

4 Conclusion

The development of automatic method for identification of osteoarthritis in knee images will lead to accurate and cost-effective treatment. Our proposed method could be able to successfully segment the synovial cavity region and classify the images into two classes as normal or affected with osteoarthritis. In the future, work can be carried for further classification of these images into different grades of OA.

Acknowledgements We would like to thank Chidgupkar Hospital Pvt. Ltd., a multispeciality hospital in Solapur, India, for providing us the X-ray images and giving permission to use these images for our experimentation.

References

1. Navale, D.I., Hegadi, R.S., Mendgudli, N.: Block based texture analysis approach for knee osteoarthritis identification using svm. In: IEEE International WIE Conference on Electrical and Computer Engineering, pp. 338–341. IEEE (2015)
2. Dureya, J., Iranpour-Boroujeni, T., Collins, J.E., Vanwynngaarden, C., Guermazi, A., Katz, J.N., Losina, E., Russell, R., Ratzlaff, C.: Local area cartilage segmentation: a semi-automated novel method of measuring cartilage loss in knee osteoarthritis. Arthritis Care Res. **66**(10), 1560–1565 (2014)
3. Anifah, L., Purnama, I.K.E., Hariadi, M., Purnomo, M.H.: Osteoarthritis classification using self organizing map based on gabor kernel and contrast-limited adaptive histogram equalization. Open Biomed. Eng. J. **7**, 18–28 (2013)
4. Brown, R.G., Hwang, P.Y.C.: Introduction to Random Signals and Applied Kalman Filtering, 3th edn (1996)
5. Chan, T.F., Vese, L.A.: Active contours without edges. IEEE Trans. Image Process. **10**(2), 266–277 (2001)
6. Hiremath, P.S., Dhandra, B.V., Hegadi, R.S., Rajput, G.G.: Abnormality detection in endoscopic images using color segmentation and curvature computation. In: International Conference on Neural Information Processing, pp. 834–841 (2004)

Identification of Knee Osteoarthritis Using Texture Analysis

Ravindra S. Hegadi, Umesh P. Chavan and Dattatray I. Navale

Abstract The major cause for the occurrence of osteoarthritis (OA) is due to wear and tear of protective tissue at the ends of cartilage. It occurs in the joints of hands, neck, lower back, knees, or hips. We propose a method to identify the existence of osteoarthritis in knee joint using X-ray images. The proposed methodology involves image enhancement using contrast-limited adaptive histogram equalization followed by identifying the location of center part of synovial cavity region, which exists between upper and lower knee bones. Four texture features namely, contrast, correlation, energy and homogeneity were extracted from synovial cavity of enhanced image and these features are used to classify the images into two classes: normal or affected with OA. The proposed method used cubic SVM and KNN as a classifier which gives robust results as compared to others.

Keywords Osteoarthritis · Texture analysis · Synovial cavity · Cubic SVM · KNN

1 Introduction

Arthritis is a commonly occurring disease but it is difficult to identify in the early stages. The pain in joint or some kind of disease in joint is referred as arthritis and it may be a combination of multiple diseases. More than 100 types of varieties are found among arthritis. Even though there is no discrimination in its occurrence as far as age, gender, and race are concerned, but the majority of sufferers are women as

R. S. Hegadi (✉) · U. P. Chavan · D. I. Navale
School of Computational Sciences, Solapur University, Solapur 413255, India
e-mail: rshegadi@gmail.com

U. P. Chavan
e-mail: umeshpchavan11@gmail.com

D. I. Navale
e-mail: navaledatta@gmail.com

P. Nagabhushan et al. (eds.), *Data Analytics and Learning*,
Lecture Notes in Networks and Systems 43,
https://doi.org/10.1007/978-981-13-2514-4_11

compared with men and probability of occurrence is higher in old age persons. The major cause of disability among Americans is arthritis. The people suffering from this disease are more than 50 million adults, whereas about 300,000 children are also suffering from this disease. The symptoms of arthritis occurring in joints are pain, stiffness, and swelling in joints, which may appear mild, moderate, or severe and may change status with change in time. Medication and physiotherapy may provide certain relief to the patient from suffering, but surgery may be needed to reduce pain and maintain joint movement.

Knee osteoarthritis (OA) is classified into five grades based on Kellgren and Lawrence system [1]. The authors proposed this classification in 1957 and later, the same was accepted by WHO in 1961. The classification provides grading of osteoarthritis from grade 0 to grade 4, with the absence of any OA being graded as 0 and knee with large osteophytes, severe sclerosis and definite bony deformity graded as 4.

A scheme to segment healthy knees from clinical MRI images was proposed by Fripp et al. [2]. Their work consisted of three stages, segmentation of bones region, extraction of interface between bone and cartilage, and finally cartilage extraction. For segmentation of bone, they used 3D active shape models. For extraction of bone and cartilage interface, prior knowledge based on points belonging to the region was used.

Antony et al. [3] proposed a different strategy to quantify the stages of osteoarthritis of the knee from the X-rays by means of deep convolution neural networks (CNNs). An automatic knee OA computer-aided diagnosis (KOACAD) algorithm [4] which digitizes knee X-rays as a DICOM file. In this procedure, for de-noising, the complete X-ray is subjected to filtering thrice by using a 3×3 square neighborhood median filter to which Robert's filter was used for extracting uneven edges.

Fully automatic knee OA computer-aided diagnosis system for quantification of major OA parameters on plain knee radiographs was developed by Oka et al. [5]. They validated the reproducibility and reliability of their system and further investigated the association between parameters and knee pain. Anteroposterior X-rays images of 1979 knees belonging to large-scale cohort population was analyzed by their proposed system as well as using conventional categorical grading systems. A novel method to segment synovial cavity region from knee X-ray images is presented by Hegadi and Navale [6]. The OA images were preprocessed using Wiener filter and thresholded. The synovial cavity region is located by identifying the distribution of intensity. The authors compared their results with the manual segmentation results.

In our work, we propose an automatic method to extract synovial cavity region and classification of these images as either normal or affected with osteoarthritis based on the analysis of texture properties. Section 2 presents the methodologies used for the proposed work, experimental results are presented in Sect. 3 and conclusions are given in Sect. 4.

2 Methodology

Image-based analysis and decision of osteoarthritis grading totally depends on the synovial cavity region. Hence, the extraction of exact synovial cavity region is very important. We proposed a methodology to automatically extract the center row of synovial cavity region and analyze the texture properties of pixels belonging to this region.

2.1 Preprocessing

The input X-ray image, as shown in Fig. 1a, is converted from RGB to gray scale which is shown in Fig. 1b. X-ray images are low in contrast, particularly there is no large gap in the intensities of bone and synovial cavity region. Along with low contrast X-ray, images are subjected to various noises. Since we attempt to obtain the texture features from the synovial cavity region, segmentation of synovial cavity region is essential. To separate synovial cavity region, enhancing the contrast of the entire image is a necessary step. A contrast-limited adaptive histogram equalization (CLAHE) [7] is applied to enhance contrast of the image. The CLAHE algorithm breaks the image into small regions called as tiles and performs enhancement. The size of each tile is 8×8 pixels. Contrast enhancement is performed on each tile depending on the local intensity variation. To bring the effect of continuation with neighboring tiles and remove the artifacts generated due to tiled enhancement, bilinear interpolation is performed [8]. The image after contrast enhancement is shown in the Fig. 1c. The enhanced image is thresholded to convert into gray scale image as shown in Fig. 1d. The area outside the bone region in the X-ray images are quite dark since these areas neither belong to bonc nor synovial cavity part. To avoid unnecessary complication in further segmentation due to presence of these darker portions, bounding box is drawn on the binary image as shown in Fig. 2a, which will draw a rectangle on all sides of the image such that every side of the box will touch minimum one white pixel in the binary image. Since synovial cavity region will also be darker, this step will help in locating the central part of synovial cavity region. Figure 2b is that portion of the enhanced image which is within the bounding box marked with blue color shown in Fig. 2a.

2.2 Locating Center of Synovial Cavity

Row histogram property is used to locate the center of synovial cavity region. In this step, the binary image is first inverted, due to which the pixels belonging to synovial cavity region appear as white and bone area pixels appear black. In the inverted image, maximum white pixel is located in the rows belonging to the synovial cavity region.

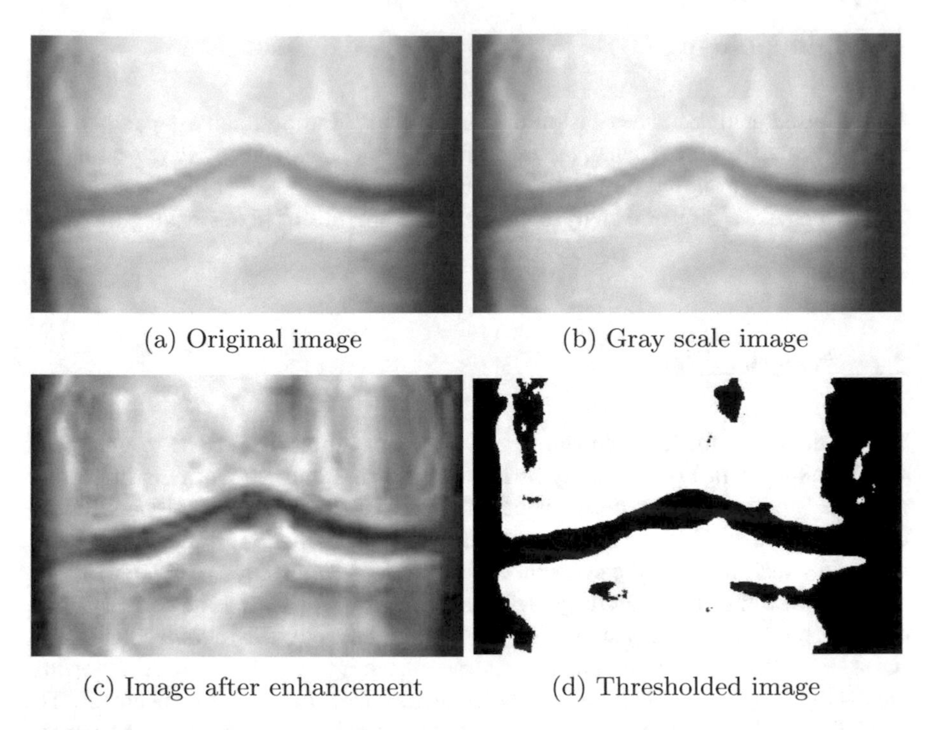

(a) Original image (b) Gray scale image

(c) Image after enhancement (d) Thresholded image

Fig. 1 Stages of preprocessing

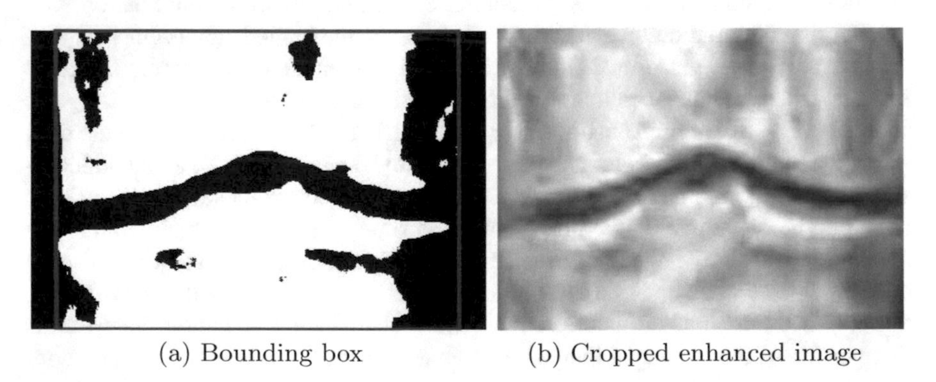

(a) Bounding box (b) Cropped enhanced image

Fig. 2 Cropping non bone and non synovial cavity regions

The row containing maximum white pixels is located from this image as shown in Fig. 3a with blue line. Figure 3b shows the same line in the enhanced image. The next task will be to remove unwanted region from the image by retaining region around synovial cavity. This is performed by empirically selecting 60 rows above and below the center line and extracting that region for the process of feature extraction, which is shown in Fig. 3c.

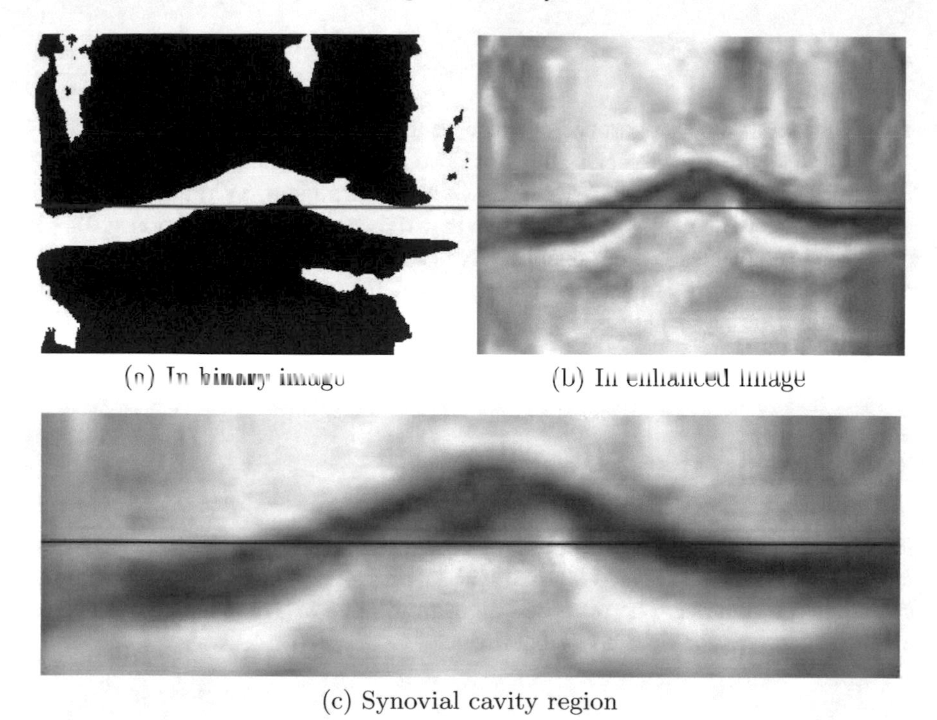

(a) In binary image (b) In enhanced image

(c) Synovial cavity region

Fig. 3 Location of center of synovial cavity region

2.3 *Feature Extraction*

Gray-level co-occurrence matrix (GLCM) [9] features from synovial cavity region is extracted for further classification. GLCM, also known as the gray-level spatial dependence matrix, is a statistical procedure to analyze texture of image, which considers the spatial relationship among image pixels. Based on the GLCM, texture properties of an image may be extracted, by calculating how often pairs of the pixel with specific values occur in an image and what type of spatial relationship exists among pixels may also be established. We extracted four image texture features based on GLCM contrast, correlation, energy, and homogeneity. The contrast, which is intensity contrast among entire image pixel and its neighbor is (0.0690, 0.1181) for image in Fig. 3c. The correlation is a statistical measure showing how a pixel is correlated with its neighbor in the entire image and it is in the range (0.9876, 0.9788) for the image in Fig. 3c. The energy, which is the sum of squared elements, is (0.1457, 0.1321) and homogeneity, which measures closeness of the elements distribution in GLCM to the GLCM diagonal, is (0.9655, 0.9410). The values obtained are the range values.

2.4 Classification

Two classifiers, cubic support vector machine (SVM) and k-nearest neighbors algorithm (KNN), are used for classification. The SVM algorithm generally used for pattern classification and regression [10]. During the training, SVM algorithm finds optimal linear hyperplane in such a way that the expected classification error for unknown test samples is minimized. KNN is used for classification and regression problem, moreover, it is a useful technique which is used to assign weight to the contributions of the neighbors. To cluster the data according to properties, KNN is the suitable algorithm [11].

3 Results and Discussion

The experimentation is carried out using MATLAB R2016a on PC having I7 processor with 8 GB memory. 14 normal and 17 abnormal images are used from the osteoarthritis initiative. The proposed algorithm failed to detect the center line of synovial cavity region for 3 abnormal images.

3.1 Results for Abnormal Image

Figure 4a shows image for patient with osteoarthritis and its enhanced image is shown in Fig. 4b. The result of cropping non bone region in binary form is shown in Fig. 4c and the tracing of center of synovial cavity region using proposed method is shown in Fig. 4d. The final segmented synovial cavity of this abnormal image is shown in Fig. 4e. GLCM features were extracted from the image shown in Fig. 4e and the results obtained are (0.0779, 0.0998) for contrast, (0.9725, 0.9652) for correlation, (0.2172, 0.2073) for energy and (0.9612, 0.9502) for homogeneity. On comparing these results with the results obtained for normal images it is noticed that there is a high variation in the energy feature among these two class of images. The normal image produced lower energy as compared to the abnormal image.

3.2 Comparison

Our proposed method could classify both normal and abnormal images correctly using KNN and cubic SVM classifier. The classification rate for normal images using KNN classifier is 100%, but for SVM, it is 79%. For abnormal images, KNN and SVM gives 100% accuracy. The overall classification accuracy of SVM classifier is 89%. These two classification methods validate with fourfold cross validation. The

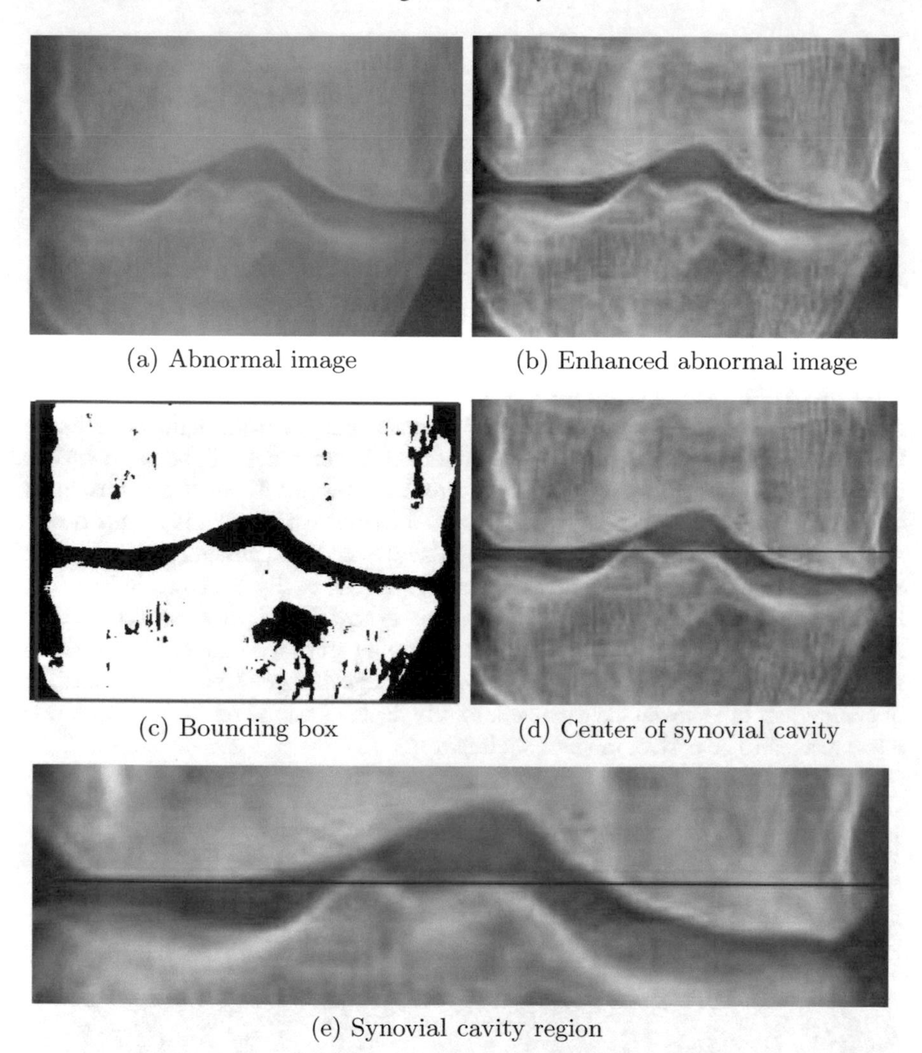

(a) Abnormal image
(b) Enhanced abnormal image
(c) Bounding box
(d) Center of synovial cavity
(e) Synovial cavity region

Fig. 4 Results of the proposed method on image having osteoarthritis

results are compared with the work proposed by GW Stachowiak et al. [12], which has six stages: (1) Selected Trabecular bone. (TB) texture ROIs. (2) Measurement of distances between TB texture images. (3) Generation of a classifier ensemble. (4) Selection of accurate classifiers. (5) Classification of the ROIs. (6) Classified ROIs. For the classification purpose, they have used dissimilarity- based multiple classifier (DMC) which gives 90.51% accuracy with twofold cross validation (Table 1).

Table 1 Classification results and comparison

Images	KNN (%)	Cubic SVM (%)	DMC based
Normal	100	79	–
Abnormal	100	100	–
Overall	100	89	90.51

3.3 Failure Analysis

As stated earlier, the proposed method successfully identified the central part of synovial cavity region but it fails to do so for three abnormal images. One of the reasons for its failure is the fact that in cases of images from patients suffering from osteoarthritis, the gap between the upper and lower bones will be narrow. With increase in the level of severity, this gap reduces further. In such circumstances the number of pixels belonging to the synovial cavity region will also come down. Because of this reason, the algorithm will falsely trace the synovial cavity region in some other part of bones other than the actual synovial cavity as shown in Fig. 5. In Fig. 5b it can be noticed that the center part of synovial cavity region, shown with blue colored line, is falsely identified at the bottom part of the image. The proposed algorithm also fails if the bone part in the X-ray image is skewed, since identification of central part of synovial cavity region totally depends on the position of bone and it is expected to be perfect in vertical direction.

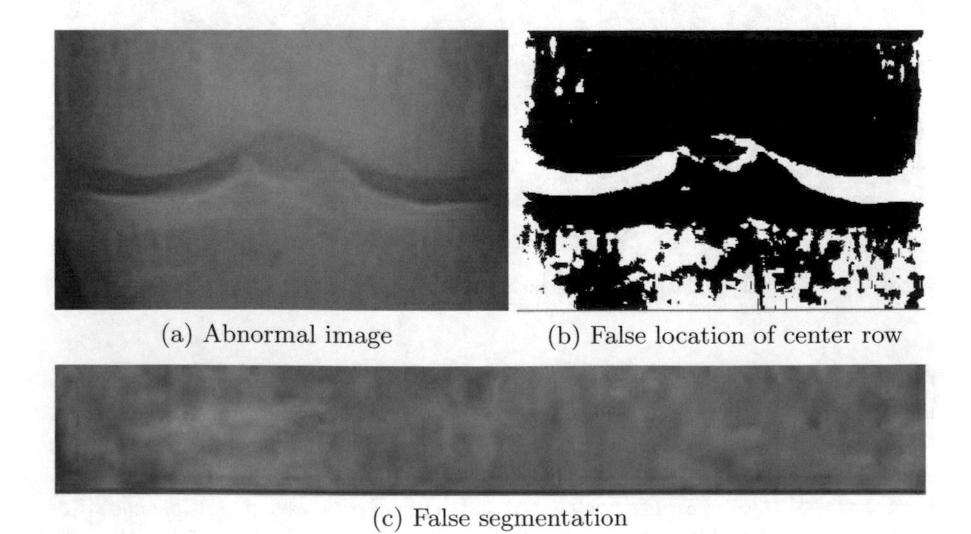

(a) Abnormal image (b) False location of center row

(c) False segmentation

Fig. 5 Failure to locate central part of synovial cavity region

4 Conclusion

A concept to locate the center part of synovial cavity region and computation of texture features were used in this paper to classify the knee X-ray images as normal or affected with OA. Two classifiers namely, cubic SVM and KNN are used for classification and results are compared with existing work. The performance of the proposed work is better than the existing algorithm. In certain cases, the algorithm fails to locate the central part of synovial cavity region, which may be considered for future enhancement.

Acknowledgements We would like to thank Chidgupkar Hospital Pvt. Ltd., a multispeciality hospital in Solapur, India, for providing us the X-ray images and giving permission to use these images for our experimentation.

References

1. Kellgren, J.H., Lawrence, J.S.: Radiological assessment of osteo-arthritis. Ann. Rheum. Dis. **16**(4), 494–502 (1957)
2. Fripp, J., Crozier, S., Warfield, S.K., Ourselin, S.: Automatic segmentation of articular cartilage in magnetic resonance images of the knee. In: International Conference on Medical Image Computing and Computer-Assisted Intervention, pp. 186–194. Springer, Berlin, Heidelberg (2007)
3. Antony, J., McGuinness, K., O'Connor, N.E., Moran, K.: Quantifying radiographic knee osteoarthritis severity using deep convolutional neural networks. In: 2016 23rd International Conference on Pattern Recognition (ICPR), pp. 1195–1200. IEEE (2016)
4. Canny, J.: A conferences approach to edge detection. IEEE Trans. Pattern Anal. Mach. Intell. **8**(6), 679–698 (1986)
5. Oka, H., Muraki, S., Akune, T., Mabuchi, A., Suzuki, T., Yoshida, H., Yamamoto, S., Nakamura, K., Yoshimura, N., Kawaguchi, H.: Fully automatic quantification of knee osteoarthritis severity on plain radiographs. Osteoarthr. Cartil. **16**(11), 1300–1306 (2008)
6. Hegadi, R.S., Navale, D.I.: Quantification of synovial cavity from knee X-ray images. In: International Conference on Energy, Communication, Data Analytics and Soft Computing, IEEE (2017)
7. Reza, A.M.: Realization of the contrast limited adaptive histogram equalization (clahe) for real-time image enhancement. J. VLSI Signal Process. Syst. Signal Image Video Technol. **38**(1), 35–44 (2004)
8. Zuiderveld, K.: Contrast limited adaptive histograph equalization. Graphic Gems **IV**, 474–485 (1994)
9. Jia, L., Zhou, Z., Li, B.: Study of sar image texture feature extraction based on glcm in guizhou karst mountainous region. In: 2012 2nd International Conference on Remote Sensing, Environment and Transportation Engineering (RSETE), pp. 1–4. IEEE (2012)
10. Navale, D.I., Hegadi, R.S., Namrata, M.: Block based texture analysis approach for knee osteoarthritis identification using SVM. In: 2015 IEEE International WIE Conference on Electrical and Computer Engineering (WIECON-ECE), pp. 338–341. IEEE (2015)
11. Kamble, P.M., Hegadi, R.S.: Geometrical features extraction and knn based classification of handwritten marathi characters. In: World Congress on Computing and Communication Technologies (WCCCT), pp. 219–222 (2017)
12. Stachowiak, G.W., Wolski, M., Woloszynski, T., Podsiadlo, P.: Detection and prediction of osteoarthritis in knee and hand joints based on the X-ray image analysis. Biosurface and Biotribology **2**(4), 162–172 (2016)

Logo Retrieval and Document Classification Based on LBP Features

C. Veershetty and Mallikarjun Hangarge

Abstract Logo is the strong entity for retrieval of Content-Based Information (CBI) from any complex document image. Logo is the primary and unique entity which is used to identify the ownership of the documents. Automatic logo detection and retrieval facilitates efficient identification of the source of the document and it is one of the interesting problems to the document retrieval community. Wes proposed a method based on Local Binary Pattern (LBP) for logo retrieval from document images. It is used to describe the logos both query and document logo. The candidate and query logos ares matched based on the cosine distance. Based on it, distance ranks are generated to estimate the relevance of the logo. Later, matched logos are retrieved at a selected threshold of 98%. The performance of the algorithm is experimentally validated and its efficiency is measured in terms of the mean precision at the rate 87.80%, and mean recall rate 88.20% as well as average F-measure 88.00%.

Keywords Logo retrieval · Cosine distance · Digital image processing (DIP) Document image retrieval (DIR)

1 Introduction

Logo is one of the important objects of the document image, which play a vital role in document categorization. And, signature and seal are the other entities used for the classification of the documents. Basically, the logo detection aims at efficient localization and extraction of the logo. Figure 1 displays sample logos of different types [1]. Research algorithms reported in [2–5] are mainly focused on logo recognition. In [6], the authors discussed many schemes of segmentation such as X-Y tree,

C. Veershetty (✉)
Department of Computer Science, Gulbarga University, Kalaburagi, India
e-mail: vshetty1180@gmail.com

M. Hangarge
Department of Computer Science, Karnatak Arts Science and Commerce College, Bidar, India

P. Nagabhushan et al. (eds.), *Data Analytics and Learning*, Lecture Notes in Networks and Systems 43,
https://doi.org/10.1007/978-981-13-2514-4_12

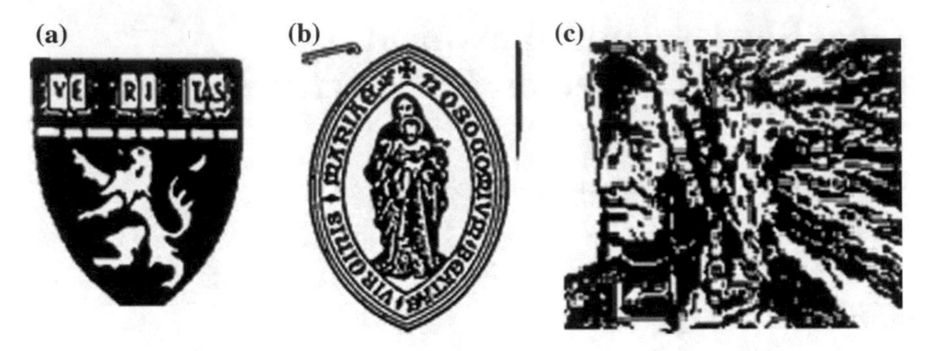

Fig. 1 Sample logos from Tobacco 800 database

hierarchical, and top-down approaches to analyze a binary document. Then, the features are computed on segmented components of the document for the classification of the components of the document as logos and non-logos. The X-Y cut method is not adoptive therefore, it cannot be generalized. In [7], a method is developed for gray scale document images, and in this method, it is assumed that the spatial density of the logo is high than non-logo regions and based on this assumption, logo detection is performed. In [8], the authors proposed a method to detect and extract logos from document images in a single framework. Coarse scale is employed to obtain gray scale blobs to reduce the resolution of the image so that the logo area becomes a single component. In this way, by merging all these connected components, the logo is obtained as a single component. Then, the merged connected component of the gray scale image size is computed, afterward successive cascade of classifiers are used to obtain a sub-logo candidate region. A novel logo detection and retrieval method based on LBP is presented here. The proposed algorithm has less time complexity and less false retrieval ratio are the main contributions of this paper.

The organization of this paper is as follows: Sect. 2 describes review of literature and in Sect. 3, the procedure of logo retrieval is presented. In Sect. 4, results and observations are discussed. Section 5 describes comparative analysis. Section 6 summarizes the work carried out.

2 Related Work

In one of the initial works on logo recognition, Doermann et al. [9] projected a technique based on global shape invariants for logo matching. Zhu and Doermann [10] used an OCR system to separate text and logos in the documents using a shape and profile features. After identifying the logo region in the document, the remaining text is discarded. On the opposite, in [11], areas having texts are first detected and observed as desired logo area; then, recognition is done based on color and shape properties. All the above approaches need logos to be on a clean white background.

In [12], SIFT descriptors are used to label the logos. The bag-of-words model is employed for the grouping of document. In [13], boosting classifier cascades are used for logo detection/localization across multiple image scales. Then, translation, scale, and rotation invariant shape descriptors are employed for logo matching. Identifying logos of different shapes and sizes are also discussed in [13]. Doermann et al. [9] have projected a multi-level phase method for logo recognition using global invariant features to crop the logo from the document image. Their method allows obtaining a logo which can be used for recognizing a variety of administrative documents. In [12] authors have examined algebraic and differential invariants features for detection and retrieval of logo from the document images. Projection profiles, normalized centroid distance, eccentricity, and numerous density features are proposed in [11] for logo recognition. A system of group membership is used for recognition of logos and is discussed in [14, 15]. In [16], a technique for detecting, segmenting and matching logo from document images is projected. First to identify logo and non-logo region, they used a fisher classifier. Then, using cascading of simple classifiers, they deleted false alarms which are present in the detected logo and then these spotted logos were kept in the database for future matching purpose. Then, the logos are compared with document image by using two technique such as representation of shape context method which is described in [17] and neighborhood graph matching method is also proposed.

3 Proposed Method

This section presents the proposed method for logo retrieval from document images. Figure 2 shows our methodology. Noise removal and logo extraction from document image is carried out by performing preprocessing steps. Then to extract the features from each logo, LBP is applied and feature vectors are generated. Then, the feature vector of document image and query logo are matched. Then, the existence of the target logo is detected using similarity matrix. Finally, relevant existences of logo and documents are retrieved.

3.1 Preprocessing and Segmentation of Logo

The logo extraction from the document image is achieved in three phases, viz., using Otsu's method, the document image is binarized [18]. Noise and symbols such as double quotation, commas present in the document are eliminated using a morphological operation, meanwhile, using various lines structuring elements horizontal and vertical dilation are performed to become a logo area as single connected component [19], based on the connected components rule each logo is extracted from the document. The procedure of logo extraction from documents can be seen in Fig. 3.

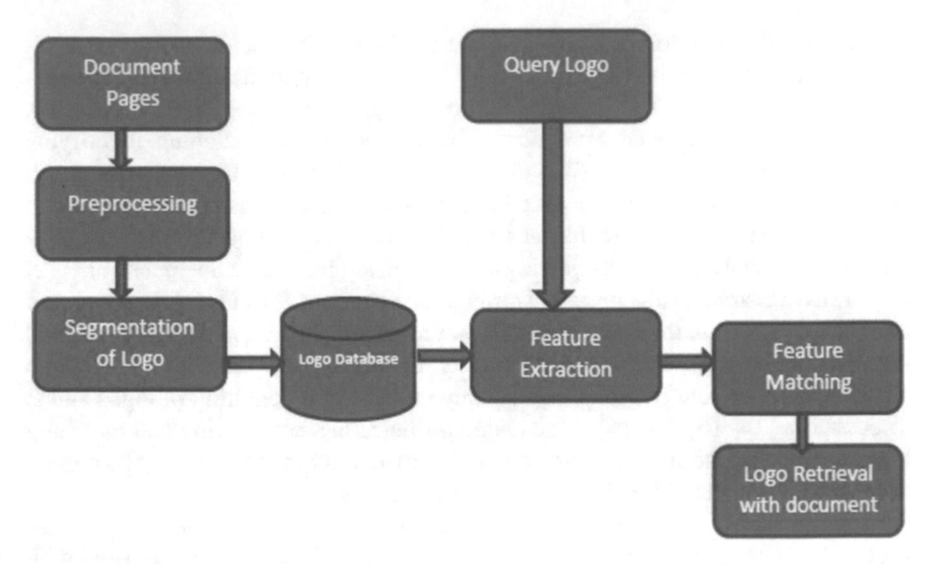

Fig. 2 Flow diagram of the proposed method

3.2 Feature Extraction

Local Binary Patterns (LBPs) have been presented by Ojala et al. [20] and demonstrated its significant role in pattern recognition area. LBP is used to describe image texture very efficiently. The procedure to capture the minute and discriminating texture properties of the image is explained as below (Fig. 4).

1. Divide the underlying image window into cells (e.g., 8×8 pixels for every cell).
2. For every pixel in a cell, equate the pixel to each of its eight neighbor's pixel. Follow the pixels along a circle, i.e., clockwise or counterclockwise.
3. Consider the center pixel value of the image window and compare with its neighbor's value. Write "0", if the given value is greater than the center pixel, otherwise, write "1". In this way, a binary number of 8-digit is obtained and it is converted to decimal for ease.
4. Compute the histogram of each cell of the window and calculate the occurrences or frequency of each "number". This histogram can be realized as a 256-dimensional feature vector.
5. Then normalize the given histogram.
6. Combine the (normalized) histograms of each cell. This gives a feature vector for the whole window.

we can obtain LBP_R^p with respect to center pixel (Ic) is given below

$$LBP_R^p(Ic) = \sum_{n=0}^{p=1} S(In - Ic)2^n \tag{1}$$

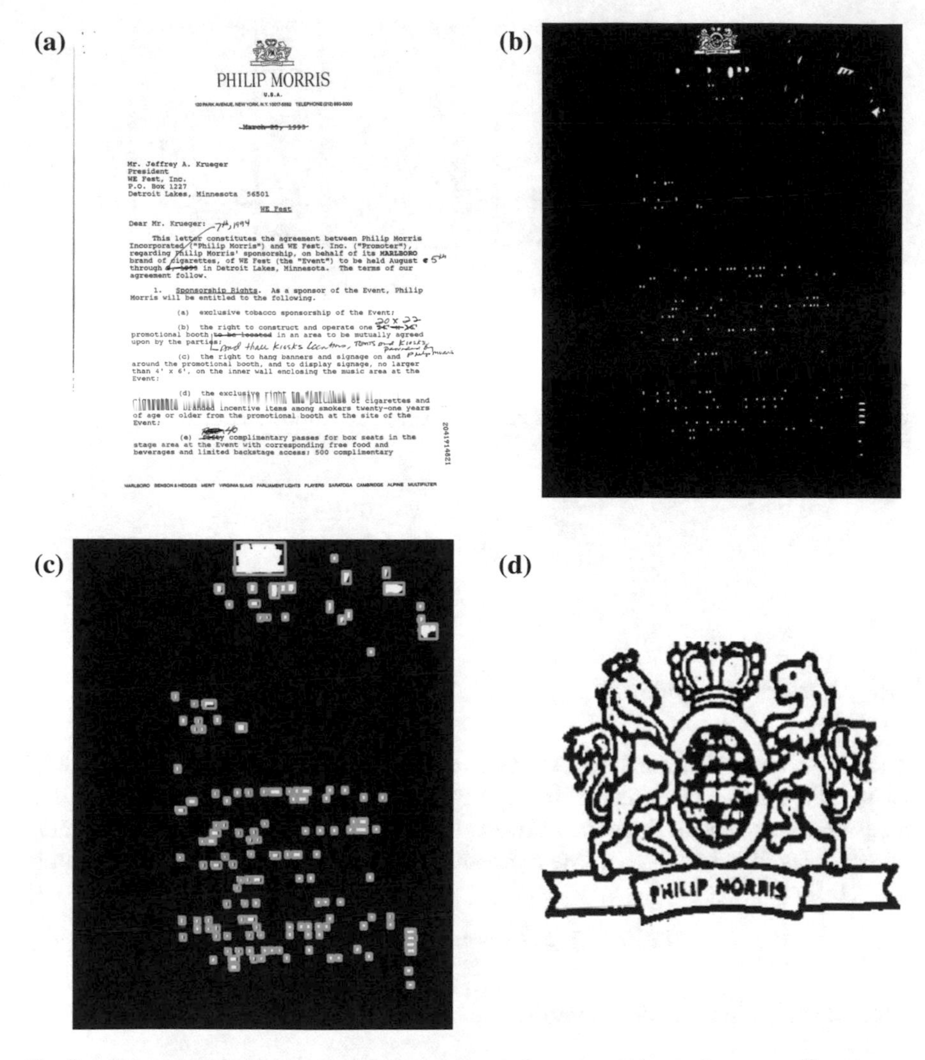

Fig. 3 **a** Document page **b** binarized image **c** connected component image **d** segmented logo image

30	20	15
77	50	59
11	66	8

Thresholding

0	0	0
1		1
0	1	0

Binary pattern
00010101

Fig. 4 Local binary operator

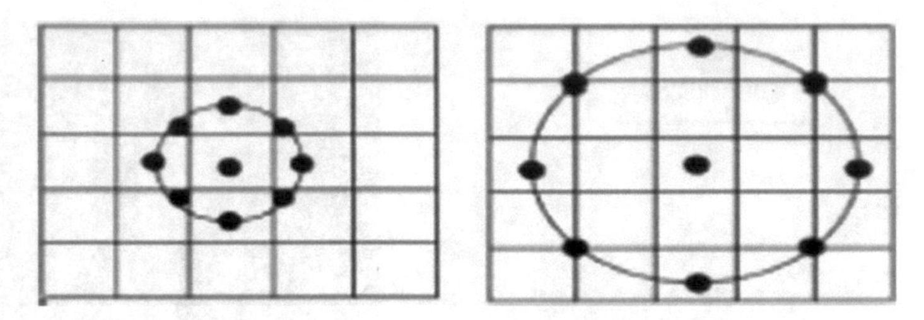

Fig. 5 Linear binary operator for the circular neighborhood PR

where (In and Ic) are equivalent to the values of center pixel and neighborhood pixels. Figure 5 shows the neighborhood of center pixel in a window. The value of s(x) can be acquired by Eq. 2 as shown below.

$$s(x) = \begin{cases} 1 \text{ for } x < 0 \\ 0 \text{ for } x < 0 \end{cases} \tag{2}$$

3.3 Similarity Measure

The cosine distance is employed to match query and candidate logos. Cosine similarity is the amount of closeness between two vectors of an inner product space that measures the cosine of the point between them. The cosine of 0° is 1, and for any other angles, it is less than 1. The mathematical equation of the cosine function is given below.

$$Similarity(a, b) = \cos\theta = \frac{a \cdot b}{||a|| * ||b||} \tag{3}$$

where x and y are the feature vectors.

4 Experiment

4.1 Dataset

For experimentation, we have employed the standard dataset, i.e., Tobacco 800. which is a subgroup of the IIT CDIP Test group [21]. Tobacco 800 dataset contains a total of 1290 document images composed with a special and different kinds of group, and out of these, 200 images are taken in the proposed work.

4.2 Evaluation Protocol

We have used three most commonly quoted measures, average precision, average recall, and average f-measure, to estimate the performance of the algorithm. Average Precision (AP) is the retrieval scheme that ranks relevant logo documents first. Average recall gives the exact ranking among the retrieved relevant documents. And finally, f-measure gives a complete performance of all queries. These measures are defined below:

Let N be the total number of logo instances for every key logo, M be the total number of identified logo instances, and CR be the correctly detected logo instance [2, 21]

$$Recall = \frac{CR}{N} * 100 \tag{4}$$

$$Precision = \frac{CR}{M} * 100 \tag{5}$$

$$F\text{-}measure = \frac{2 * (RC * PR)}{(PR + RC)} \tag{6}$$

4.3 Results and Discussion

The results achieved for detecting and retrieval of query logo instances in the document image are discussed here. The experiment is carried out on a Tobacco 800 dataset. As per Zhus and Doermann [22], if the detected region contains more than 75% pixels of a groundtruthed logo, it indicates the existence of logo. For the performance evaluation, we define the evaluation protocol as defined above; experimental results of the logo retrieval are presented in Table 1. And further, we have extended our experiment to classify the documents based on the query logos and its results are presented in Table 2. The following Figs. 6 and 7 show the examples of detected logos from Tobacco 800 database and retrieved document images with respect to query logo, respectively.

Local binary pattern technique is used to capture visual discriminating properties of the underlying image. The performance of these features are validated by classifying the documents based on logo detection as discussed above. LBP is preferred because it is efficient in characterizing linear visual patterns of the logos. Initially, we have carried out experiments based on traditional shape and structural features and noticed their weak performance. Meanwhile, LBPs invariant property to scale and rotation showed good performance, however, structural and shape features are sensitive to rotation and scale. But, LBP has more time complexity and miss some minute structural properties in computation. Therefore, LBP (see Table 2) shows poor performance in the case of degraded logos as it misses some minute properties of

Fig. 6 Examples of detected logos from the Tobacco 800 dataset

the broken logos, whereas outperforms in other cases. The comparative statistics of LPB and structural features are not reported in this paper.

Table 1 Results of the proposed method

Query logo	Precision (%)	Recall (%)	F-measure (%)
	70.00	70.00	70.00
	92.00	94.00	93.00
	77.00	77.00	77.00
	100.00	100.00	100.00
	100.00	100.00	100.00
Average	87.80	82.20	88.00

Table 2 Results of document retrieval with respect to query logo from Tobacco 800 database

Query logo	Precision (%)	Recall (%)	F-measure (%)
Query 1	70.00	70.00	70.00
Query 2	92.00	94.00	93.00
Query 3	77.00	77.00	77.00
Query 4	100.00	100.00	100.00
Query 5	100.00	100.00	100.00
Average	87.80	82.20	88.00

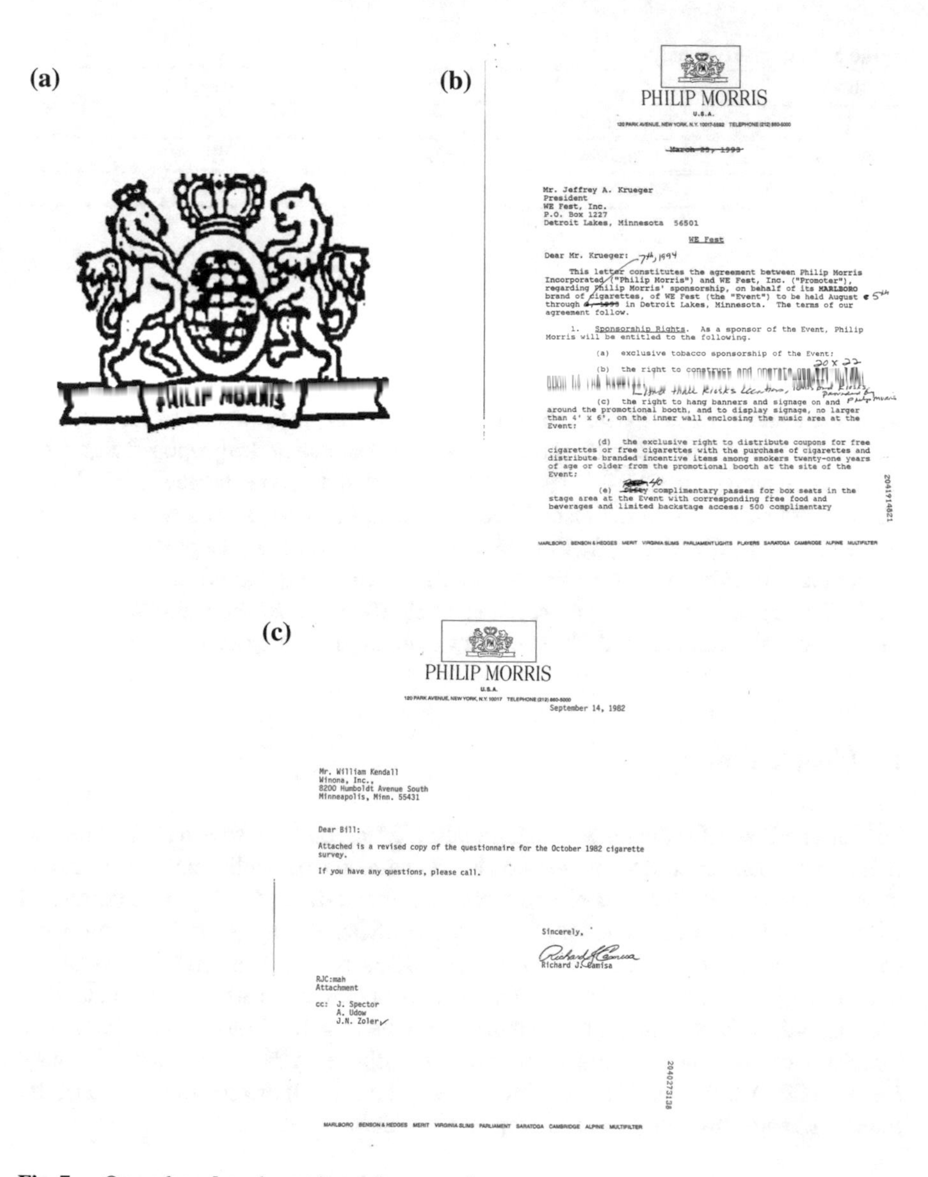

Fig. 7 **a** Query logo **b** and **c** retrieved document images

Table 3 Comparative study

Method	Dataset	Precision (%)	Recall (%)
Ref. [8]	Tobacco 800	73.5	NR
Ref. [13]	Tobacco 800	82.6	NR
Proposed method	Tobacco 800	87.80	88.20

5 Comparative Study

A good comparative analysis is that where the environment of experimentation will remain the same. In such cases, the comparison of the performance of the different algorithms is meaningful. In this context, the performance of the proposed algorithm is compared with [8] and [13]. These authors have used the same dataset, i.e., Tobacco 800 for automatic logo detection and document classification. They have reported the precision of logo detection (P) as 73.5% and 82.6%, respectively. Reported results are computed using shape context distance matrices. However, the proposed algorithm exhibited a remarkable average precision as 87.80% against the same dataset with the cosine distance metric. Table 3 contains comparative study analysis report.

6 Conclusion

LBP is employed for documents classification based on logo retrieval. The performance of the LBP features are validated in comparison with traditional structural and shape features. LBP features are efficient in characterizing the logos as compared to structural features. As a result, the logo classification results are high compared to the earlier reported results which are computed on the same publicly available dataset, i.e., Tobacco 800. The LBP features are invariant to rotation and scale. But, LBP algorithm is not efficient in capturing minute structural patterns of the image. Therefore to overcome this limitation, we are working on the use of Central Binary Pattern (CBP) technique. In the future, we make this algorithm more generic by attending major limitations of the proposed problem.

References

1. Kuo, S.-S., Agazzil, O.E.: Keyword spotting in poorly printed documents using pseudo 2-D hidden Markov models. In: IEEE, pp. 0162–8828 (1994)
2. The legacy tobacco document library (LTDL) at UCSF (2006). http://legacy.library.ucsf.edu/
3. Doermann, D., Rivlin, E., Weiss, I.: Logo recognition using geometric invariants. In: Proceedings of International Conference on Document Analysis and Recognition, pp. 897–7 (1993)
4. Doermann, D., Rivlin, E., Weiss, I.: Applying and differential invariants for logo recognition. Mach. Vis. Appl. **9**(2), 73–86 (1996)

5. Neumann, J., Samet, H., Soffer A.: Integration and global shape analysis for logo classification. Pattern Recognit. Lett. **23**(12), 1449–1457 (2002)
6. Suda, P., Bridoux, C., Kammerer, B., Maderlechner, G.: Logo and word matching using a registration. In: Proceedings of International Conference on Document Analysis and Recognition, pp. 61–65 (1997)
7. Pham, T.: Unconstrained logo detection in document images. Pattern Recognit. **36**(12), 3023–3025 (2003)
8. Zhu, G., Doermann, D.: Automatic document logo detection. In: Conference on Document Analysis and Recognition, pp. 864–868 (2007)
9. Doermann, D., Rivlin, E., Weiss, I.: Logo recognition using geometric invariants. In: Proceedings International Conference on Document Analysis and Recognition, pp. 897–903 (1993)
10. Zhu, G., Doermann, D.: Logo matching for document image retrieval. In: ICDAR, pp. 606–610 (2009)
11. Neumann, J., Samet, H., Soffer, A.: Integration and global shape analysis for logo classification. Pattern Recognit. Lett. **23**(12), 1449–1457 (2002)
12. Rusinol, M., Liados, J.: Logo spotting by a bag-of-words approach for document categorization. In: Proceedings of the International Conference on Document Analysis and Recognition (ICDAR), pp. 111–115 (2009)
13. Zhu, G., Doermann, D.: Logo matching for document image retrieval. In: Proceedings of the International Conference on Document Analysis and Recognition (ICDAR), pp. 606–610 (2009)
14. Pham, T.: Unconstrained logo detection in document images. Pattern Recognit. **36**(12), 2025–3023 (2003)
15. Chen, J., Leung, M.K., Gao, Y.: Noisy logo recognition using line segment Hausdorff distance. Pattern Recognit. **36**(4), 943–955 (2003)
16. Zhu, G., Doermann, D.: Logo matching for document image retrieval. In: Proceedings of the International Conference on Document Analysis and Recognition (ICDAR), pp. 606–610 (2009)
17. Dhandra, B.V., Hangarge, M..: On seperation of english numerals from multilingual document images. Int. J. Multimed. (JM) **2**(6), 26–33 (2007). ISSN 1796-2048. Academy Publisher, Oulu, Finland
18. Otsu, N.: A threshold selection method from gray-level histograms. In: IEEE PAMI-79, pp. 62–66 (1979)
19. Hangarge, M., Dhandra, B.V.: Script identification in indiandocument images based on directional morphological filters. Int. J. Recent Trends Eng. **2** (2009)
20. Ojala, T., Pietikainen, M.: Multiresolution gray-scale and rotation invariant texture classification with local binary patterns. IEEE Trans. Pattern Anal. Mach. Intell. **24**(7), 971–987 (2002)
21. The IIT complex document image processing (CDIP) test collection (2006). http://ir.iit.edu/projects/CDIP.html
22. Zhu, G., Doermann, D.: Automatic document logo detection. In: Conference on Document Analysis and Recognition. pp. 864–868 (2007)

Semantic Relatedness Measurement from Wikipedia and WordNet Using Modified Normalized Google Distance

Saket Karve, Vasisht Shende and Swaroop Hople

Abstract Many applications in natural language processing require semantic relatedness between words to be quantified. Existing WordNet-based approaches fail in the case of non-dictionary words, jargons, or some proper nouns. Meaning of terms evolves over the years which have not been reflected in WordNet. However, WordNet cannot be ignored as it considers the semantics of the language along with its contextual meaning. Hence, we propose a method which uses data from Wikipedia and WordNet's Brown corpus to calculate semantic relatedness using modified form of Normalized Google Distance (NGD). NGD incorporates word sense derived from WordNet and occurrence over the data from Wikipedia. Through experiments, we performed on a set of selected word pairs, and we found that the proposed method calculates relatedness that significantly correlates human intuition.

Keywords Semantic relatedness · WordNet · Brown corpus · Wikipedia · NGD Natural language processing

1 Introduction

Natural Language Processing (NLP) is the ability of computer program to understand human (natural) language as it is spoken. NLP has found a wide range of applications to redefine the ways human and computers interact. The challenges in NLP applications are owing to the fact that human language is not precise as opposed to highly structured and definite machine-readable languages. Human language contains slangs, multiple dialects, and it depends on the context and many such complex

S. Karve (✉) · V. Shende · S. Hople
Veermata Jijabai Technological Institute, Mumbai, India
e-mail: saketk1.sk@gmail.com

V. Shende
e-mail: vasishtshende77@gmail.com

S. Hople
e-mail: hopleswaroop@gmail.com

P. Nagabhushan et al. (eds.), *Data Analytics and Learning*,
Lecture Notes in Networks and Systems 43,
https://doi.org/10.1007/978-981-13-2514-4_13

variables. Complex NLP applications largely consider the meaning of words from their context and usage along with other words based on the English literature available as dataset. This makes some of the applications like text summarization [1, 2], keyword extraction [3, 4] difficult to analyze as the output varies with human understanding.

Over the years, more words are being added to the literature. NLP uses semantic relatedness [5–7] as a metric to understand the word sense. Semantic relatedness is a technique to measure the semantic distance between a pair of words according to the context, ontology, and their occurrences in a corpus. Semantic relatedness quantifies human intuition and calculates the ideological closeness in the meaning of the words and their combined usage in the literature. The relatedness of a pair of words which are unrelated in meaning and are rarely used together should be minimum. For example, {Launch, Towel}. A set of words which finds close relation as per human intuition should have high relatedness value. For example, {Bird, Fly}. A machine could "read" a human language better when it can understand slangs and vernaculars which are extensively used but are not available in traditional dictionaries. For example, perceptron, SpaceX, etc. An updated corpus provides diverse context to calculate relatedness in the words, for example, {Apple, Computer}.

In this paper, we propose a modified form of Normalized Google Distance [8z] which can be used to quantify the relatedness between two words irrespective of their source. The paper is organized as follows. With a brief introduction here in Sects. 1, 2 it covers the details of the existing knowledge-based approaches which is used to calculate semantic similarity. The proposed method is detailed in Sect. 3. The results obtained have been presented in Sect. 4 along with a detailed analysis of the observations. Finally, conclusions are drawn in Sect. 5.

2 Existing Similarity Measures

There have been many text similarity and/or relatedness measures used in various applications of Natural Language Processing. The earlier methods incorporated a simple lexical analysis of words and a score determined based on the number of matching lexical units [9]. These basic methods were improved by considering more sophisticated lexical properties like word stemming, part-of-speech tagging, removing stop words, etc. However, it is found that these lexical similarity measures fail to consider the contextual meaning and semantics of the words in determining the similarity score. Later, semantic networks like WordNet [10] were developed which presents a huge knowledge base of words in the form of a hierarchy. Words are interconnected on the basis of six types of relations as defined in WordNet. Semantic relatedness scores are calculated based on the path connecting the two words of concern in this knowledge base. Various methods to calculate relatedness using WordNet (or similar) corpus have been proposed. Path similarity, Leacock et al. (lch) [11] and

Wu and Palmer (wup) [12] quantify the relatedness between two words based on the length of the path and/or the depth of the nodes in the hierarchy. Resnik (res) [13], Lin [14], and Jiang and Conrath (jcn) [9] produces relatedness scores based on the information content of the words and its Least Common Subsumer (LCS) with reference to a standard corpus available in WordNet. These knowledge-based methods consider the relations among various concepts in the literature when calculating the relatedness scores. However, they fail to take into account the changing meanings of words over the years. Jargons and proper nouns like the name of an organization, names of great personalities, etc., are not updated in such knowledge base. Corpus-based methods like Pointwise Mutual Information (PMI) [15], Latent Semantic Analysis (LSA) [16], and Normalized Google Distance (NGD) [8] take into account the probability of occurrence of words over a large data store like Google, Wikipedia, and similar. This helps capture more information about words exclusively from a large corpus based on the usage of the words.

Pointwise Mutual Information: PMI [15] is an unsupervised relatedness measure. The relatedness score is calculated based on the mutual co-occurrence of words over a large corpus. PMI similarity for two words w_1 and w_2 are calculated using the following formula,

$$PMI\left(w_1, w_2\right) = log_2\left(\frac{p(w_1\,AND\,w_2)}{p(w_1) \times p(w_2)}\right) \tag{1}$$

where

$$\begin{aligned} p(w_1\,AND\,w_2) &= PNEAR(w_1\,AND\,w_2) \\ &= \frac{hits(w_1\,NEAR\,w_2)}{WebSize} \end{aligned}$$

The probabilities are calculated based on the total hits of the given words as obtained from the corpora. The numerator represents the probability of the two words occurring together and the denominator is the product of their independent probabilities of occurrence.

Normalized Google Distance: NGD [8] also uses the number of hits of a word from a large corpus but calculates the semantic distance between the two words. The relatedness score can be estimated from the distance as its reciprocal. NGD between two words x and y are calculated as

$$NGD(x, y) = \frac{max\{log(f(x)), log(f(y))\} - log(f(x, y))}{log\,M - min\{log(f(x)), log(f(y))\}} \tag{2}$$

where M is the total number of web pages searched (generally a constant), $f(x)$ is the number of hits of the word x in the corpora, and $f(x, y)$ is the number of times the two words occurred together in one web page. The above formula gives a normalized score which ranges from 0 to ∞. A distance of zero indicates the two words are identical and a distance of ∞ indicates absolutely no relation between the two words.

NGD has been found to be better than PMI on most of the datasets. PMI tends to be more biased toward word pairs, which co-occur more often than independently. NGD tends to produce an unbiased outcome.

The number of hits can be calculated from various APIs like Google Custom Search API [17], Wikipedia API [18], and Yahoo API [19].

Google API (Google Custom Search API). We use Application Program Interface (API) provided by various search engines to calculate hits of a particular word, which means the total number of times the particular word appears in the search engine. Google provides an API through which we can calculate the total hits of a particular word. Google Search API enables us to collect data from various websites, blogs, etc. However, this API is highly unreliable. The number of hits of a given word changes every time the API is called. This results in a different outcome on every execution.

Yahoo API (Yahoo Partner ads). This API is also referred to as BOSS search API. This API uses the Yahoo search engine to calculate the number of hits of a given word. But this API was discontinued from March 31, 2016 and an alternative Yahoo Partner ad was. However, the Yahoo search API is for business analysis purpose and its integration is paid.

Wikipedia Search API (MediaWiki Search API). Wikipedia provides a REST API which allows accessing all the Wikipedia articles. It is highly reliable because the context of the word always remains the same. Hence, the total count of the word remains the same unless a new article is added containing the word of our concern. Moreover, it is available for free and provides easy access to resources. Therefore, we have used Wikipedia's API.

The modified form of NGD derives a similarity metric based on the Information Content (IC) of the words obtained from WordNet's Brown Corpus [20] and the probability of the words from Wikipedia. IC considers natural language sense of the words and Wikipedia provides the occurrence statistics of the words. Integration of these two factors gives the semantic distance between the words that incorporates traditional usage and definition from WordNet along with the newly formed word sense provided by Wikipedia.

3 Modified Normalized Google Distance

Knowledge base like WordNet has corpora like the brown corpus which manifest semantic information. This information cannot be ignored while quantifying the relatedness between words. Corpus-based methods like NGD consider only the way the words have been used over a large corpus and tend to ignore the semantics of the language. In order to make sure the semantics of words are preserved while quantifying relatedness, we propose a modified form of the Normalized Google

Distance. The information content of the words present in the WordNet corpus is considered instead of the probabilities calculated based on the number of hits from a more general source like the internet. Data from the internet can be skewed at times and can lead to an incorrect estimation of the relatedness score. Including information from WordNet tries to balance the negative impact of this skewness and gives a more unbiased estimate.

The original formulation of NGD is based on the natural logarithm of the number of hits of a given word. However, the information content as provided by WordNet is the natural logarithm of the probability of occurrence of the given word. Hence, in order to incorporate this data, we modify the formulation to consider the logarithm of probabilities instead of absolute hits of the given word.

Equations 6, 8, and 10 represent the modified form of NGD. This novel form integrates information from two sources and calculates the distance between words. Relatedness of a word pair is the reciprocal of its distance. Information Content (IC) retains word sense of word pair from WordNet's hierarchy and the context of words from Brown Corpus. Brown Corpus is a set of text documents containing news articles. Hence, it has words with common usage and current literary use.

A word can be dictionary defined or a slang, jargon, vernacular, etc. The latter is not defined in WordNet. Therefore, calculating relatedness of such words is not possible. Modified NGD is formed considering such words and is implemented for three cases.

Both words in WordNet. IC of both words is independently obtained from WordNet. Since WordNet does not provide IC for a word pair, i.e., for words occurring together in the corpus, it is estimated from the common hits obtained from Wikipedia. The estimated IC hits as obtained from Wikipedia, IC of individual words as obtained from WordNet and hits of individual words as obtained from Wikipedia is assumed to be proportional to each other. The proportionality constant used for the estimation is the average of the ratio of hits from Wikipedia to hits obtained from WordNet for the two words as shown in Eqs. (3) and (4).

$$\alpha_x = \frac{p(x)}{e^{-ic_x}}; \alpha_y = \frac{p(y)}{e^{-ic_y}} \tag{3}$$

$$\alpha_{avg} = \frac{\alpha_x + \alpha_y}{2} \tag{4}$$

This estimated proportionality constant is used to get an estimate for the IC of both the words together as shown in Eq. (5).

$$ic_{xy} = log\left(\frac{p(x\,AND\,y)}{\alpha_{avg}}\right) \tag{5}$$

Equation 6 gives the NGD for such word pair

$$NGD_{WN}(x, y) = \frac{max\{(-ic_x), (-ic_y)\} - (ic_{xy})}{-min\{(-ic_x), (-ic_y)\}} \tag{6}$$

Both words in Wikipedia. When both words are not present in the WordNet corpus, IC cannot be used to calculate NGD of the words. Therefore, IC is replaced by its counterpart hit probability. Wikipedia provides hits of a word and hits of word pair. Hit probability is calculated as the ratio of hits to articles searched. Wikipedia had 5,554,235 articles in the English language [21] at the time of conducting this research. Total documents searched being a constant hit probability is given as,

$$p(x) = \frac{hit(x)}{5,553,245} \tag{7}$$

where

x word(s) to be searched
$hit(x)$ hits of word x as per Wikipedia
$p(x)$ Hit Probability of word x

Modified form of NGD for this case of word pair is given as

$$NGD_{Wiki}(x, y) = \frac{max\{log(p(x)), log(p(y))\} - log(p(x\, AND\, y))}{-min\{log(p(x)), log(p(y))\}} \tag{8}$$

Only one word in wordnet. A word pair where only one word has its meaning defined in WordNet requires the relatedness calculation to be based on two different corpora. For the word which is available in WordNet, IC is retrieved from WordNet's Brown corpus. For the other word, hit probability is calculated from the data obtained from Wikipedia. Relatedness calculation of such words is done by estimating a comparable parameter that is available for both words. IC is calculated as the negative logarithm of word probability, i.e., negative likelihood from the documents in the Brown Corpus. Therefore, the counterpart of hit probability of words from Wikipedia is the concept probability viz., IC from Brown Corpus is used as a substitute to word probability for such words.

$$IC = -log(concept\, probability) \tag{9}$$

Modified NGD for such hybrid combination is given as

$$NGD_{hybrid}(x, y) = \frac{max\left\{log(p(x)), -ic_y\right\} - log(p(x\, AND\, y))}{-min\left\{log(p(x)), -ic_y\right\}} \tag{10}$$

4 Results and Analysis

Experiments were conducted using a few word pairs from the standard Rubinstein dataset [22] and the remaining have been picked such that a proper comparison of the performance can be made. All words from this dataset can be found in WordNet.

Hence, in order to test the validity of other cases, we have considered word pairs which are best suitable for this task. We test the performance of the algorithm against human intuition for the relatedness among the two words and similar existing algorithms for the respective cases.

When both words are available in WordNet, we compare the results with Resnik (res), Jiang and Conrath (jcn), and Wu and Palmer (wup) similarity measures along with human intuition. The output of the proposed algorithm represents the distance between the two words and not the relatedness between them directly. Words which are closer to each other have less distance between them and are more related to each other. The relatedness is calculated by taking the reciprocal of the output. Modified NGD model returns relatedness by taking the reciprocal of the NGD distance. All other measures used, directly calculate the relatedness and/or similarity between the two words. Table 1 shows the results obtained for this case.

As seen from the results in Table 1, the output of Wu and Palmer (wup) is normalized between a range from 0 to 1. Resnik (res) and Jiang and Conrath (jcn) produces output in the range from 0 to ∞. Resnik (res) produces output comparable to the human intuition in their respective ranges. However, in the case of a few word pairs, the results deviate a lot from expectations. Jiang and Conrath (jcn) tends to show random behavior than the expectation. Wu and Palmer (wup) does a comparatively better job and is quite close to the human intuition. The output of the proposed NGD model is normalized while generating the graphs for ease in comparison. The proposed NGD model gives results much closer to the human intuition in almost every case as can be seen from the Fig. 1.

When both words are not found in WordNet, we need to calculate the relatedness using data from Wikipedia. In this case, we compare the results of our proposed algorithm with the results obtained from the PMI similarity measure and human intuition. The results obtained can be found in Table 2.

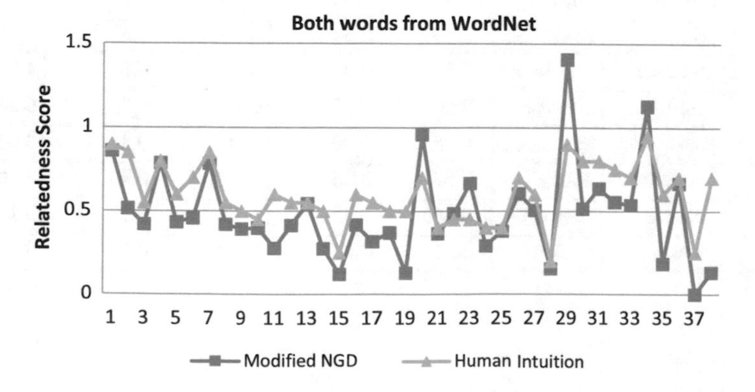

Fig. 1 Comparison of modified NGD and human intuition—both words from WordNet

Table 1 Comparison of word relatedness measures—both words from WordNet

Sr. no.	Word 1	Word 2	res	jcn	wup	Human intuition	Modified NGD model
1	car	automobile	7.59	e^{300}	1.0	0.9	4.424779
2	gem	jewel	2.31	0.05	0.58	0.85	3.012048
3	boy	lad	5.33	0.14	0.67	0.55	2.617801
4	coast	shore	9.42	1.39	0.91	0.8	4.132231
5	asylum	madhouse	3.98	0.07	0.63	0.6	2.673797
6	magician	wizard	2.33	0.06	0.63	0.7	2.777778
7	midday	noon	11.06	e^{300}	1.0	0.85	4.115226
8	furnace	stove	2.31	0.06	0.45	0.55	2.617801
9	food	fruit	0.80	0.09	0.29	0.5	2.5
10	bird	cock	0.80	0.06	0.21	0.45	2.531646
11	bird	crane	2.22	0.06	0.6	0.6	2.03252
12	tool	implement	5.87	1.52	0.93	0.55	2.597403
13	brother	monk	2.33	0.07	0.57	0.55	3.134796
14	lad	brother	2.33	0.07	0.60	0.5	2.020202
15	crane	implement	1.53	0.06	0.47	0.25	1.3947
16	journey	car	0.0	0.07	0.09	0.6	2.610966
17	monk	oracle	2.33	0.06	0.57	0.55	2.202643
18	cemetery	woodland	1.29	0.05	0.43	0.5	2.415459
19	food	rooster	0.81	0.06	0.21	0.5	1.420455
20	coast	hill	5.88	0.13	0.67	0.7	4.830918
21	shore	graveyard	1.29	0.05	0.43	0.4	2.398082
22	monk	slave	2.33	0.07	0.67	0.45	2.881844
23	coast	forest	0.0	0.05	0.17	0.45	3.636364
24	lad	wizard	2.33	0.07	0.67	0.4	2.09205
25	forest	graveyard	0.0	0.05	0.13	0.4	2.457002
26	shore	voyage	0.0	0.05	0.13	0.7	3.378378
27	bird	woodland	1.29	0.07	0.40	0.6	2.967359
28	furnace	implement	2.31	0.08	0.59	0.2	1.529052
29	acid	chemistry	0.60	0.05	0.21	0.9	6.666667
30	cat	pet	4.95	0.09	0.64	0.8	3.012048
31	contract	policy	0.596	0.06	0.24	0.8	3.508772
32	course	subject	0.596	0.07	0.31	0.75	3.174603
33	wasp	cricket	2.22	0.05	0.57	0.7	3.095975

(continued)

Table 1 (continued)

Sr. no.	Word 1	Word 2	res	jcn	wup	Human intuition	Modified NGD model
34	cricket	bat	4.94	0.07	0.58	0.95	5.524862
35	locust	centipede	7.78	e^{-300}	0.75	0.6	1.650165
36	mercury	fever	0.60	0.04	0.25	0.7	3.636364
37	factorial	smile	0.60	e^{-300}	0.24	0.25	0.896861
38	extrema	maximum	0.0	0.05	0.15	0.7	1.43472

Table 2 Comparison of word relatedness measures—both words from Wikipedia

Sr. no.	Word 1	Word 2	PMI	Human intuition	Modified NGD model
1	perceptron	hyperloop	−15.75	0.2	1.841621
2	renault	mahindra	−16.49	0.75	2.105263
3	baleno	kwid	−14.2	0.7	2.083333
4	cout	printf	−12.52	0.9	3.344482
5	cout	cin	−16.35	0.8	1.782531
6	ariane	perceptron	∞	0.15	0
7	ariane	spaceX	−19.02	0.6	1.457726
8	perl	scanf	−13.57	0.55	2.03252

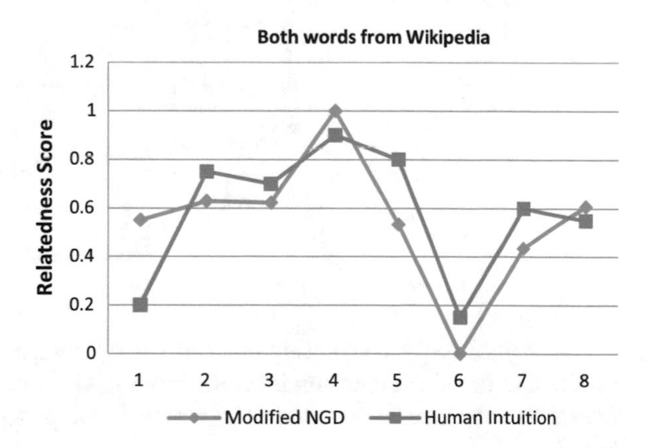

Fig. 2 Comparison of modified NGD and human intuition—both words from Wikipedia

PMI tends to produce output similar to our proposed modified NGD model. However, it has a wider range on the negative side of the number line. This makes it difficult to interpret and use as a relatedness measure. Figure 2 shows the comparison of the proposed model with human intuition in a normalized range.

Table 3 Comparison of word relatedness measures—only one word from WordNet

Sr. no.	Word 1	Word 2	Human intuition	Modified NGD model
1	passenger	hyperloop	0.75	5.91716
2	company	mahindra	0.9	10.86957
3	car	scanf	0.15	1.519757
4	output	printf	0.75	3.436426
5	input	cin	0.75	3.484321
6	neural	perceptron	0.55	2.544529
7	rocket	spacex	0.75	3.921569
8	computer	scanf	0.6	2.994012
9	rocket	renault	0.25	2.55102
10	pearl	baleno	0.3	4.651163
11	pearl	perl	0.25	3.389831
12	airline	ariane	0.4	3.236246
13	standard	cout	0.5	2.03666

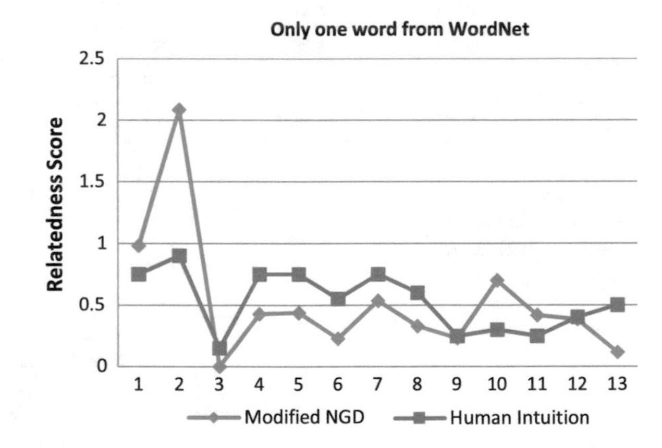

Fig. 3 Comparison of modified NGD and human intuition—only one word from WordNet

When just one word is present in WordNet, we compare the result with just human intuition as no other algorithm is found to work with such word pairs. Table 3 shows the results obtained in this case. The results are depicted in a graphical manner in Fig. 3.

5 Conclusion

In this paper, we have developed a novel approach, modified NGD that incorporates non-dictionary terms to calculate relatedness of a set of words considering their word sense and their occurrences in Wikipedia. After testing the modified form of NGD against other semantic relatedness measures, the following conclusions are observed.

1. Incorporating Wikipedia helps consider jargons, proper nouns, and words with modified definitions which are not reflected in the traditional WordNet Corpus. For example, Apple Inc. Wikipedia reflects the new meaning between two words which is not available in the English literature.
2. Modified NGD works at par with human intuition when tested against a set of word pairs which contain one word from the WordNet corpus and the other is not present in WordNet.
3. Modified NGD outperforms PMI-IR for quantifying relatedness between two words. Modified NGD gives a normalized [illegible]
4. However, it has been found that the modified NGD works better when both the words are present in Wikipedia or WordNet. More work needs to be done in the case when both the words are not from the same source.

References

1. Gupta, V., Lehal, G.S.: A survey of text summarization extractive techniques. J. Emerg. Technol. Web Intell. **2**(3), 258–268 (2010)
2. Das, D., Martins, A.F.: A survey on automatic text summarization. Lit. Surv. Lang. Stat. II Course CMU **4**, 192–195 (2007)
3. Hasan, K.S., Ng, V.: Automatic keyphrase extraction: a survey of the state of the art. In: Proceedings of the 52nd Annual Meeting of the Association for Computational Linguistics (Volume: Long Papers), vol. 1, pp. 1262–1273 (2014)
4. Lott, B.: Survey of Keyword Extraction Techniques. UNM Education, 50 (2012)
5. Gabrilovich, E., Markovitch, S.: Computing semantic relatedness using wikipedia-based explicit semantic analysis. In: *IJcAI*, vol. 7, pp. 1606–1611, Jan 2007
6. Budanitsky, A., Hirst, G.: Evaluating wordnet-based measures of lexical semantic relatedness. Comput. Linguist. **32**(1), 13–47 (2006)
7. Gomaa, W.H., Fahmy, A.A.: A survey of text similarity approaches. Int. J. Comput. Appl. **68**(13)
8. Cilibrasi, R.L., Vitanyi, P.M.B.: The google similarity distance. IEEE Trans. Knowl. Data Eng. **19**(3), 370–383 (2007)
9. Jiang, J.J., Conrath, D.W.: Semantic similarity based on corpus statistics and lexical taxonomy (1997). arXiv:cmp-lg/9709008
10. Miller, G.A.: WordNet: a lexical database for English. Commun. ACM **38**(11), 39–41 (1995)
11. Leacock, C., Miller, G.A., Chodorow, M.: Using corpus statistics and WordNet relations for sense identification. Comput. Linguist. **24**(1), 147–165 (1998)
12. Wu, Z., Palmer, M.: Verbs semantics and lexical selection. In Proceedings of the 32nd Annual Meeting on Association for Computational Linguistics, pp. 133–138. Association for Computational Linguistics (1994)
13. Resnik, P.: Using information content to evaluate semantic similarity in a taxonomy (1995). arXiv:cmp-lg/9511007
14. Lin, D.: An information-theoretic definition of similarity. In: Icml, vol. 98, no. 1998, pp. 296–304, July 1998
15. Turney, P.D.: Mining the web for synonyms: PMI-IR versus LSA on TOEFL. In: European Conference on Machine Learning, pp. 491–502. Springer, Berlin, Heidelberg Sept 2001
16. Landauer, T.K., Foltz, P.W., Laham, D.: An introduction to latent semantic analysis. Discourse Process. **25**(2–3), 259–284 (1998)

17. https://developers.google.com/custom-search/json-api/v1/overview
18. https://www.mediawiki.org/wiki/API:Search
19. https://developer.yahoo.com/boss/search/
20. Marcus, M.P., Marcinkiewicz, M.A., Santorini, B.: Building a large annotated corpus of English: the penn treebank. Comput. Linguist. **19**(2), 313–330 (1993)
21. https://en.wikipedia.org/wiki/Special:Statistics
22. Rubenstein, W.B., Kubicar, M.S., Cattell, R.G.G.: Benchmarking simple database operations. ACM SIGMOD Rec. **16**(3) 387–394. ACM

Design and Simulation of Neuro-Fuzzy Controller for Indirect Vector-Controlled Induction Motor Drive

B. T. Venu Gopal and E. G. Shivakumar

Abstract This paper displays a unique adaptable Neuro-Fuzzy Controller (NFC)-based speed control for three-phase induction motor drive. The suggested NFC integrates fuzzy logic idea with a four-layer Artificial Neural Network (ANN). Speed and change in speed are sent as input to Neuro-Fuzzy Controller and it winds up noticeably fit for real-time electromechanical drives. The complete simulation model for indirect vector control of induction motor including the suggested NFC is developed. Induction motor assumes an imperative part in the field of electric drives. Without genuine controlling of the speed, it is difficult to accomplish required errand for specific application. AC motors are solid, less cost, reliable, and maintenance free. In light of absence of capacity of regular control strategies like PID and PI controllers to work in a broad range of operations, AI-based controllers are extensively utilized as a part of the industry like Neural Networks, Fuzzy Logic, and Neuro-Fuzzy controller. The principle issue with the typical fuzzy-based controllers is that the parameters related with the membership functions and the rules depend predominantly on instinct of the specialists, fuzzy logic cannot naturally get the rules utilized for settling on the decision, however, great at clarifying the decision. To overcome from this issue, Neuro-Fuzzy Controller [ability to learn without anyone else alongside decision-making] is recommended. Keeping in mind the end goal to prove the predominance of the proposed Neuro-Fuzzy Controller, the results of the suggested NFC technique is compared with the results of PI controller. NFC-based control of induction motor will end up being more trustworthy than other conventional control techniques.

Keywords Neuro fuzzy · Fuzzy logic control · Indirect vector control
PI or PID control

B. T. Venu Gopal (✉) · E. G. Shivakumar
Department of Electrical Engineering, UVCE, Bangalore University,
Bengaluru, India
e-mail: btvgopal@gmail.com

E. G. Shivakumar
e-mail: shivaettigi@gmail.com

P. Nagabhushan et al. (eds.), *Data Analytics and Learning*,
Lecture Notes in Networks and Systems 43,
https://doi.org/10.1007/978-981-13-2514-4_14

1 Introduction

The AC motors are the generally utilized motors in modern drives, because of its effortlessness, tough development, unwavering quality, and simple construction. The improvement of fast semiconductor control switches and DSPs have empowered AC machines to accomplish greater in modern drive applications. The vector control strategy is used with the goal that the induction motor can accomplish the high unique execution capacity of the independently energized DC machine, however, holding upsides of the ac machines over DC machines [1]. Superior motor drives utilized as a part of electromechanical technology, moving plants and machine apparatuses require dynamic speed reaction and brisk recuperation of the speed from any unsettling influence and vulnerabilities.

Over quite a while, the regular PI and PID controllers have been used by the industry for flexible speed drives. Be that as it may, the outlines of these PI/PID controllers rely upon correct machine model with exact parameters. Acquiring the correct parameters of the induction motor prompts unmanageable plan approach. The ordinary PI and PID are exceptionally touchy to disturbances [1]. The outlines of fuzzy and neuro-fuzzy controllers need not bother with the correct numerical model of the system. In this way, an artificial intelligent controller requests outstanding consideration for the speed control of high-performance variable speed drives.

Adaptive neuro-fuzzy interface systems (ANFIS) is a blend of two delicate registering methods of neural systems and fuzzy logic. Neuro-fuzzy strategy depends on a fuzzy system which is prepared by a taking in calculation acquired from neural systems. It does not require past information of relationships of data. It makes them tune, self-learning, and self-sorting out abilities. NN are great at perceiving examples and they are bad at clarifying how they achieve their decisions. Neuro-Fuzzy crossbreed systems can be dependably (sometime recently, after and during training) deciphered on an system of fuzzy rules.

Points of interest of Neural Networks:

- Require less statistical training
- It can be utilized as a part of nonlinear issues
- Execute numerous preparation calculations
- Variable change is automatic during computation.

Benefits of Fuzzy Logic based Controller:

- It can corporate human insight in control calculation.
- No culminate scientific model of the procedure plant is essential.
- It can work viably both for direct and non-linear systems
- Speed of reaction is high and overshooting is less.
- Linguistic factors are utilized as a part of place of numerical variable.
- Degree of exactness is high.
- Provides preferable outcome over PI controller.

2 Methodology

In this paper, NFC-based novel speed control technique for an induction motor drive is exhibited. NFC is created on motor dynamics, neural system, and fuzzy methods [2]. Backpropagation scheming is utilized to limit the error amongst wanted and genuine strategy. A four-layer neural system (NN) is used to prepare the parameters of the Fuzzy Logic Control which disposes of a major deal of undesirable trial-and-error technique, which is the situation for fuzzy control (Fig. 1).

2.1 Field-Oriented Control or Vector Control

It comprises of adjusting the stator currents represented by a vector. This control depends on estimations that change a three-phase speed and time-dependent system into a two-phase (dq coordinate) time-invariant system. These transformations and projections prompt a comparative control arrangement to that of a DC motor. Vector control needs two quantities as input references: torque component (lined up with q axis) and the flux component (lined up with d axis). The three-phase currents, fluxes, and voltages of induction motors can be analyzed as space vectors. This transformation is separated into two stages: (a, b, c) → (α, β), (Clarke transformation), which gives two coordinate time variation systems. (α, β) → (d, q), (Park transformation) which gives two coordinate time-invariant system. In Clarke transformation process, three-phase quantities of any currents or voltages fluctuating in time, along the axis a, b, and c can be scientifically changed into twophase currents/voltages shifting in time with respect to axes α and β [3].

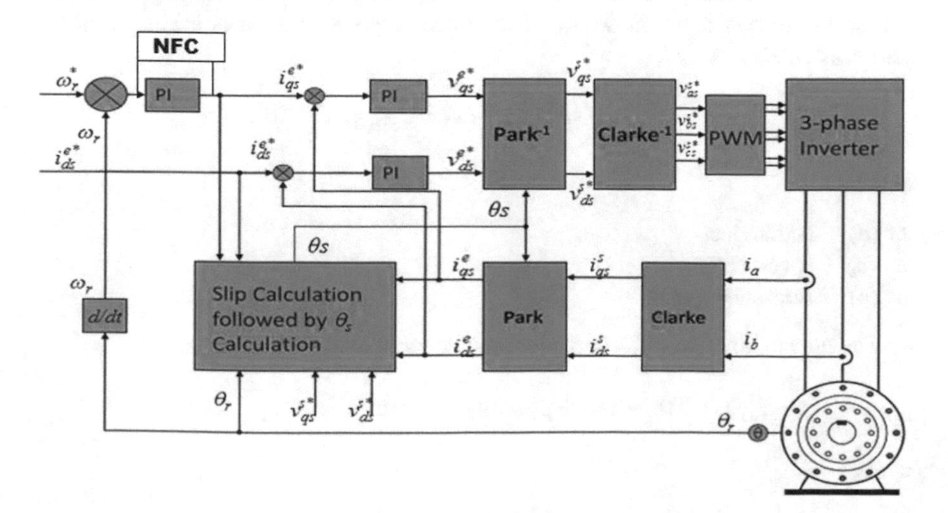

Fig. 1 Block diagram of indirect vector-controlled induction motor drive

The significant error of direct vector control technique is the need of such a large number of sensors. Settling such huge numbers of sensors in a machine is a dreary work and also costlier. A few different issues like drift on account of temperature, poor transition detecting at low speeds likewise continues. Because of these harms, indirect field-oriented control is utilized. In case of indirect vector control, the rotor position is computed from the speed input signal of the motor. This procedure dispenses with the vast majority of the issues, which are related with flux sensors as they are truant.

Focal points of Vector Control

- Improves torque reaction
- Torque control at low frequencies and low speed.
- Dynamic speed control.
- Motor size is diminished, minimal effort, and low power utilization.

Ventures in Vector Control

- Measure the currents which are as of now flowing in the motor.
- Then contrast the present and wanted current and produce the error signal.
- Amplify the error signal to create the revision voltage.
- Set the redress voltage on to the motor terminals.

3 Control Design

3.1 PI Controller

PI controller is broadly utilized as a part of the industry because of its simple structure and straightforwardness in design. The successive speed error is evaluated at nth sampling instant as

$$\omega e(n) = \omega r(n) * -\omega(n)$$

where

ωr (n) rotor speed
ωe (n) speed error
ωr (n)* reference speed

The new estimation of torque reference is calculated by

$$T(n) = T(n-1) + Kp\{\omega e(n) - \omega e(n-1)\} + K1\{\omega e(n)\}$$

where

ωe (n − 1) speed error of the past interval
ωe (n) speed error of the working interval.

Kp and K1 PI controller constants

Kp and K1 constants are calculated by Ziegler–Nichols method [4].

3.2 Neuro-Fuzzy Controller Design

The proposed Neuro-Fuzzy Controller comprises of four-layer ANN structure and fuzzy logic with the learning design for the neural system is appeared in Fig. 2. In order to closely match the desired execution of the drive, the learning algorithm simplifies the Neuro-Fuzzy Controller. The [illegible] discussions on various layers of the NFC is shown below

Input layer: The standard speed error is given as input for the NFC. Speed error is acquired by the contrasting the deliberate speed and the set speed.

Fuzzification (second) Layer: It quantifies the estimation of inputs. To decide the fuzzy number for the input layer, membership functions like triangular, straight, and trapezoidal depends on fuzzy set utilized.

Rule (third) Layer: Each governs in this layer plays out the precondition coordinating of the fuzzy set tenets. Since there is just single input to the layer, no AND logic is never required in this layer. Each node in this layer decides the weights of fuzzy layer which were institutionalized. It takes in itself from the training illustrations.

De-fuzzification (fourth) layer: Here, the centroid point of gravity strategy was utilized to figure or specify the output of NFC. This layer was likewise called as

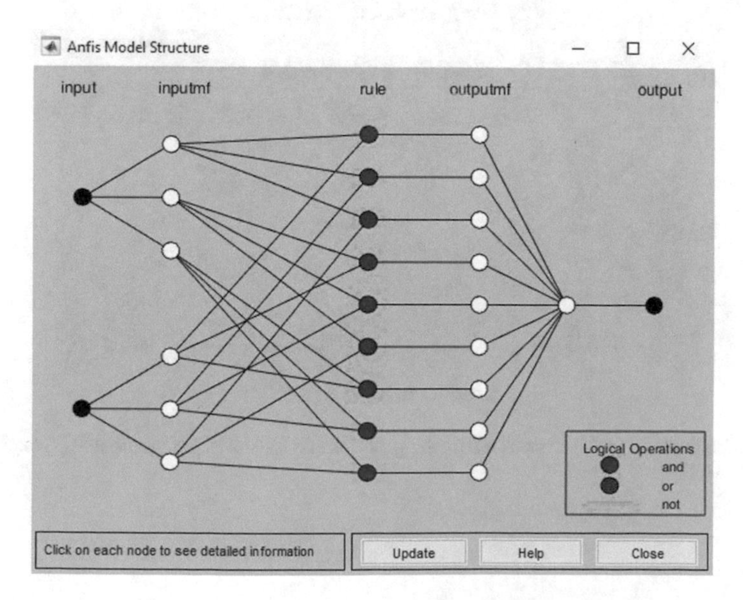

Fig. 2 Basic structure of self-tuned proposed NFC with four-layer structure

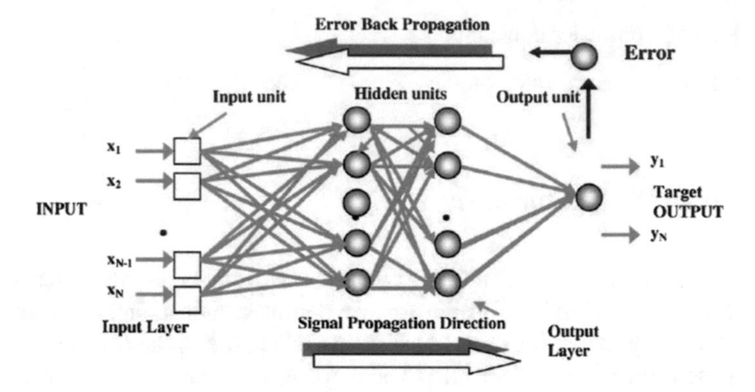

Fig. 3 Back propagation algorithm of ANFIS

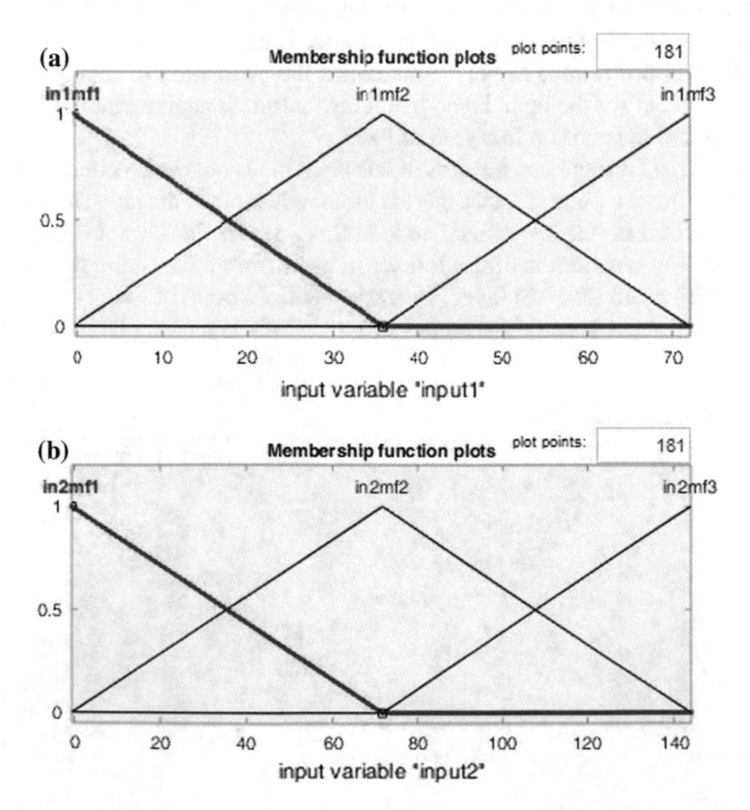

Fig. 4 **a** Input membership function 1 of the proposed ANFIS. **b** Input membership function 2 of proposed ANFIS

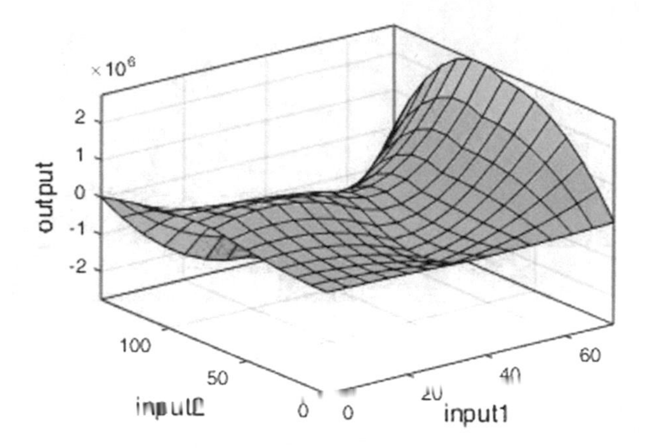

Fig. 5 Output surface window of the proposed ANFIS

output layer where it aggregates up the data sources which were originating from the layer 3 and changes the fuzzy classification results comes into crisp values. A scale mapping, which changes over the scope of ranges of values of output into relating universe of discourse [4].

The aggregate error between the coveted and genuine state factors may then be back engendered as appeared above in Fig. 3, to alter the weights of the neural model, with the goal that the output of this model tracks the real output. At time of training and this training is finished, the weights of the neural system should relate to the parameters in the actual motor.

The proposed ANFIS had the following features: Type- Sugeno; AndMethod-product; OrMethod- probor (probabilistic or); DefuzzMethod- wtaver (weighted average); No. of input- 2; No. of input membership function- 3; input membership function type-Triangular; No. of output- 1; No. of output membership function-9; No. of rules- 9.

Figure 4 and Fig. 5 shows input membership functions and output surface of proposed ANFIS, respectively.

4 Development of Simulink Model

The suggested NFC-based hybrid vector control technique is implemented in MATLAB. Then, the speed performance of NFC technique was tested with induction motor 50 HP/400 V. The speed performance of the NFC technique was compared with PI controller. From the developed model, the speed, torque, current, and voltages were analyzed. The Simulink model of Neuro-Fuzzy controller and PI control system were demonstrated in Fig. 4 and Fig. 5, respectively, (Figs. 6 and 7) [5–13, 3, 14].

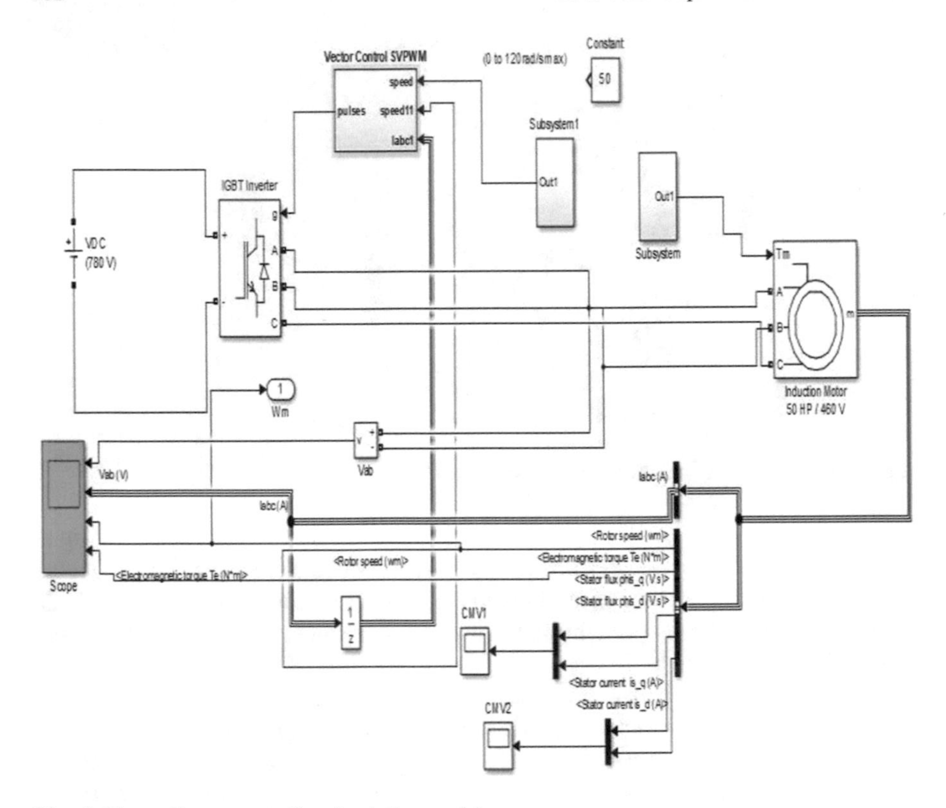

Fig. 6 Neuro-Fuzzy controller simulation model

5 Results

In Fig. 8, the comparison performance of proposed ANFIS speed controller, the pi speed controller are described. PI controller speed point is also set to 72 rad/sec. Neuro-fuzzy controller is trained to a particular speed of 72 rad/sec for a specific application, i.e., for blowers. From comparative analysis, the proposed Neuro-Fuzzy controller delivers better speed control performance when compared to pi controller. Overshoot is very less in case of NFC. PI controller does not reach the steady-state set speed. Therefore, the suggested NFC speed controller is better than PI controllers. The performance of current, voltage, torque, speed, torque ripples, and comparison of speed are depicted from Figs. 9, 10, 11, and 12.

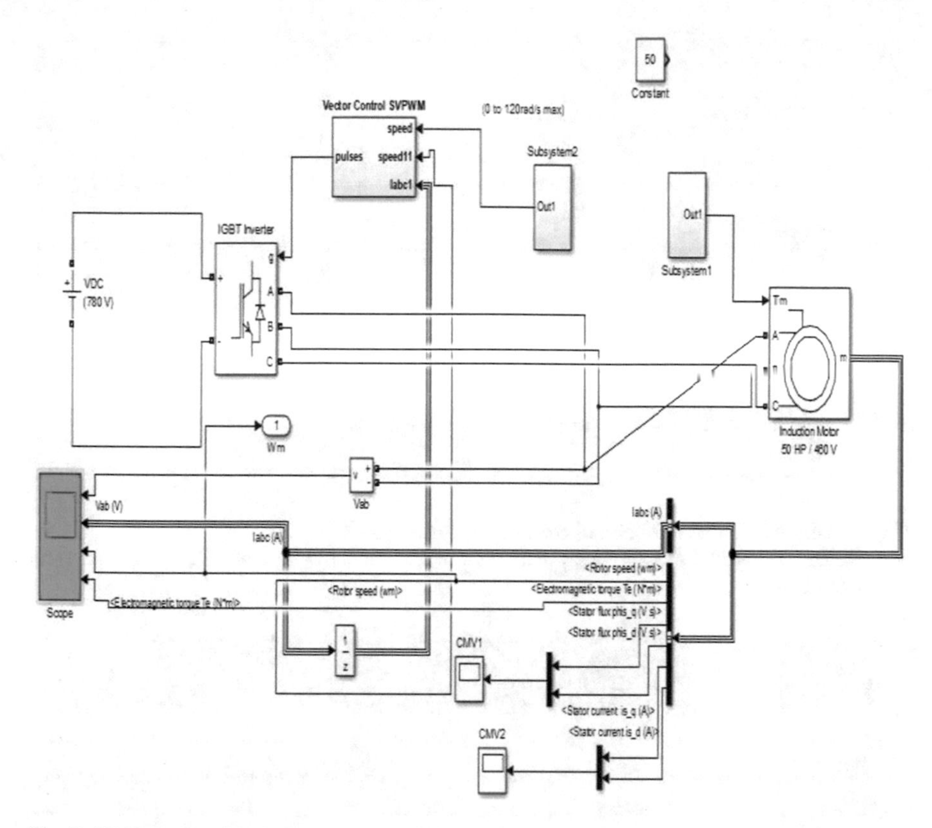

Fig. 7 PI controller simulation model

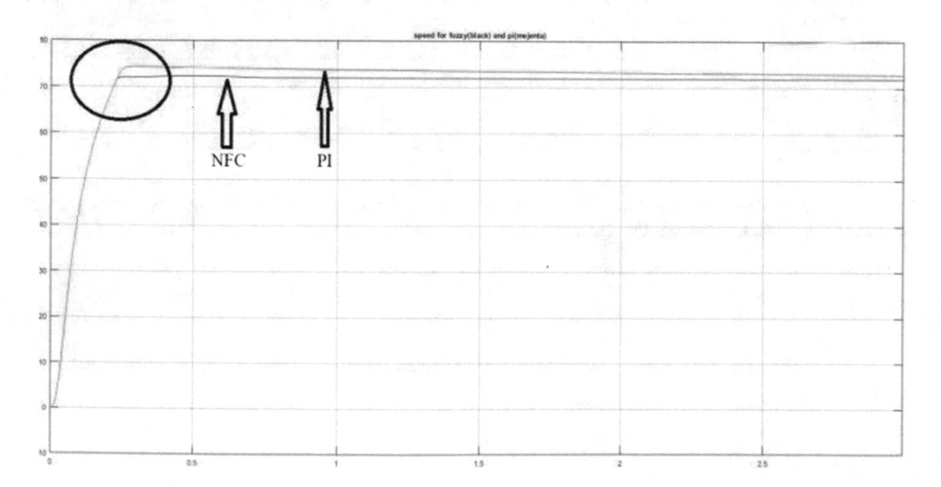

Fig. 8 Speed response of NFC and PI controller

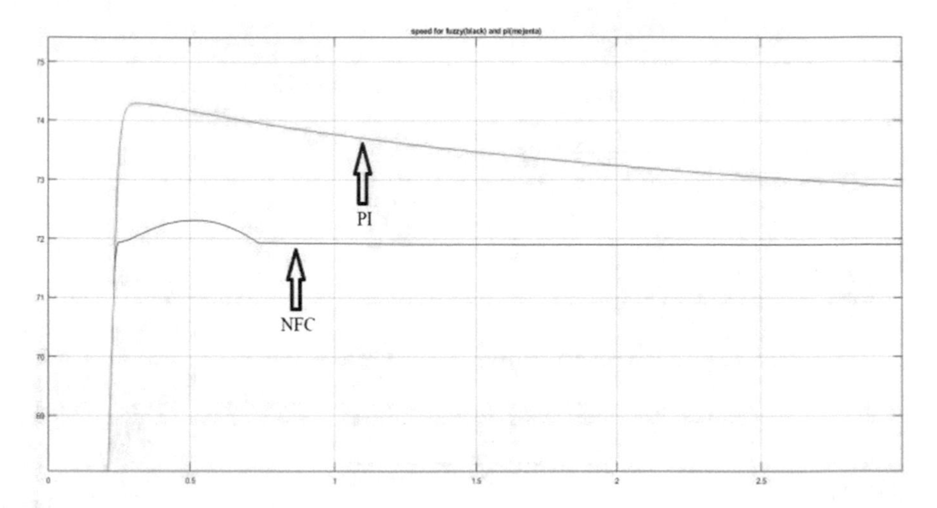

Fig. 9 Enlarged view of NFC and PI controller speed response.
Note Reference Speed: 72 rad/s steady-state error: +2.06 rad/s (for PI controller) −0.3 rad/s (for NFC controller)

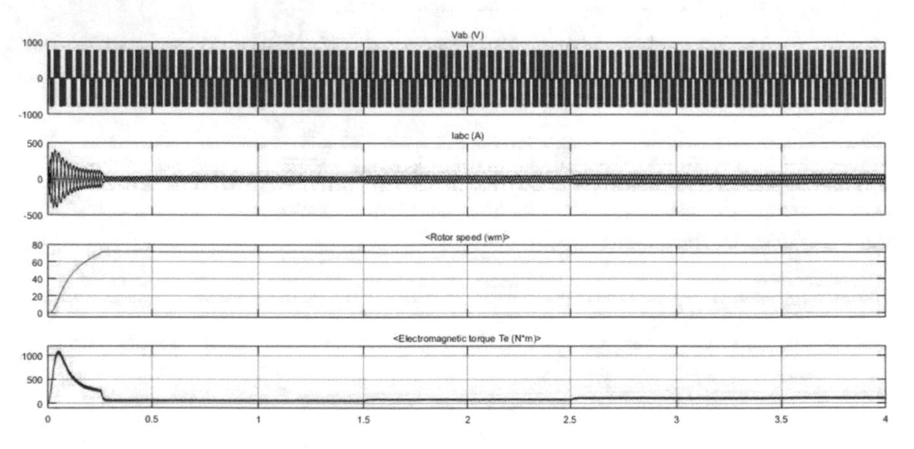

Fig. 10 Line-to-line voltage Vab, Phase Currents Iabc, Rotor speed (wm) and Electromagnetic torque (Nm) of Neuro-Fuzzy Controller

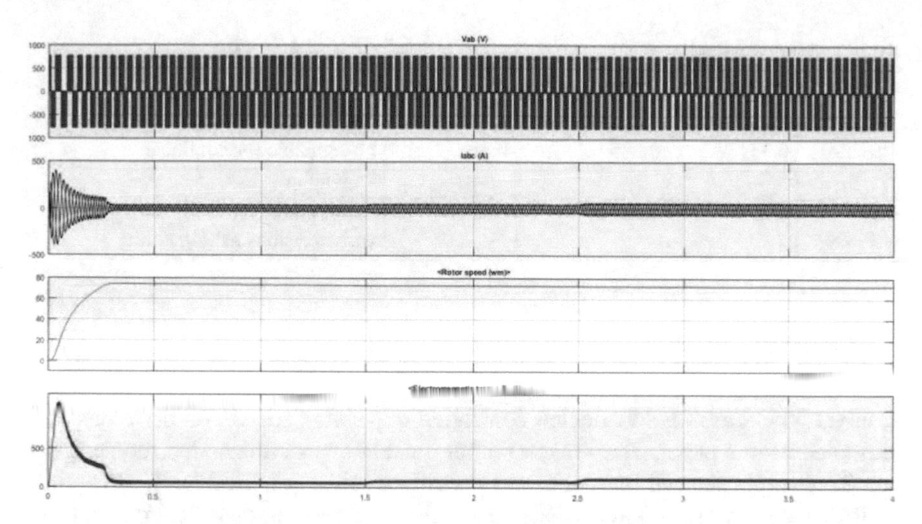

Fig. 11 Line-to-Line voltage Vab, Phase Currents Iabc, Rotor speed (wm) and Electromagnetic torque (Nm) of PI controller

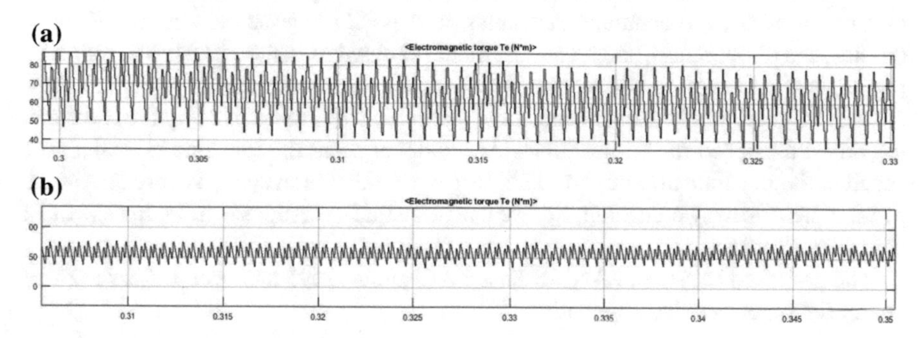

Fig. 12 **a** Performance characteristics of drive with PI speed controller: torque ripples **b** Performance characteristics of drive with NFC speed controller: torque ripples

Table 1 Result analysis

Controller	Peak time (in s)	Settling time (in s)	Rise time (in s)	Point of view
PI	0.331	0.335	0.22	Not suitable for nonlinear systems
NFC	0.272	0.2881	0.21	Self-learning and self- tuning with decision-making ability

6 Conclusion

A novel NFC-based field-oriented controlled adjustable speed AC drive has been presented in this paper. The system output closely approximates the required output. The results showed that NFC was capable of closely reproducing the optimal performance. This technique is fast in execution, efficient in identification, and easy in implementation for the structure and the parameters of induction motor. The suggested model is less complicated approach in determining the identification of induction motor with good accuracy. Advantages of NFC are lessened number of rules, quick speed of operation, and no requirement for changes in membership function by regular trial-and-error technique for ideal reaction. In view of the set point and feedback, NFC produces legitimate control signal which will be utilized by the inverters, to control the speed of AC machines. At whatever point the machine is loaded, the speed of the machine dropped to some degree yet this fall in speed is very less in the event of neuro-fuzzy controller. So, we can state that overshoot is more in case of PI controller (Table 1).

The proposed self-tuned ANFIS-based AC motor drive has been observed to be hearty for superior drive application.

Appendix

Parameter of Induction Motor: Nominal power 50HP, Line-to-line voltage 400 volts, Frequency 50 Hz, Stator resistance 0.087 O, Stator Inductance 0.8 mH, Rotor resistance 0.228 O, Rotor inductance 0.8 mH, Mutual Inductance 34.7 mH, Inertia 1.662 kg m^2, Friction 0.1 Nms, and Number of poles 2. Parameters of PI speed controller: Proportional constant (Kp) 150, and Integral constant (Ki) 50.

References

1. Wen, H., Nasir Uddi, M.: Development of a Neuro Fuzzy Controller For Induction Motor. CCECE 2004 CCGEI 1004, pp. 1225–1228
2. Rushi Kumar, K., Sridhar, S.: A genetic algorithm based neuro Fuzzy controller for the speed control of induction motor. Int. J. Advan. Res. Electr. Electron. Instrument. Eng. **4**(9), 7837–7846 (2015)
3. Bose, B.K.: Modern Power Electronics and Motor Drives, Advances and Trends (2009)
4. Varatharaju, V.M., Mathur, B.L.: Adaptive neuro Fuzzy speed controller for hysterisis current controlled PMBLDC motor drive. IJAET 212–223
5. Menghal P.M., Jaya laxmi, A.: Adaptive Neuro Fuzzy based dynamic simulation of induction motor drives. In: 2013 IEEE international Conference on Fuzzy Systems (2013)
6. Li, H., Qiuyun, M., Zhilin, Z.: Research on Direct Torque control of induction motor based on genetic algorithm and Fuzzy PI controller. In: 2010 International Conference on Measuring Technology and Mechotronics Automation, pp. 46–49 (2010)
7. Ahmed, A.M., Eisa Bashier, M., Tayeb, A., Alim, T., Habiballh, A.H.: Adaptive neuro fuzzy interface system identification of an induction motor. Eur. J. Sci. Eng. **1**(1), 26–33 (2013)
8. Nasir Uddin, M., Wen, H.: Development of Self Tuned Neuro Fuzzy Controller for Induction Motor Drives. IEEE, pp. 2630–2636 (2011)
9. Meghal, P.M., Jaya Laxmi, A.: Adaptive Neuro Fuzzy Based Dyanamic Simulation of Induction Motor Drives
10. Rajaji, L., Kumar, C.: Adaptive neuro fuzzy interface system into squirrel cage induction motor drive: modeling, control and estimation. In: 5th International Conference of Electrical and Computer Engineering ICECE, Dhaka, pp.162–169 (2008)
11. Sujatha, K.N., Vaisakh, K.: Implementation of adaptive neuro fuzzy interface systems in speed control of induction motor drives. J. Intell. Learn. Syst. Appl. **2** 110–118 (2010)
12. Kumar, R., Gupta, R.A., Surjuse, R.S.: Adaptive Neuro Fuzzy Speed Controller for Vector Controlled Induction Motor Drive, pp. 8–12
13. Bonde, S.D., Dhok, G.P.: ANFIS control scheme for the speed control of the induction motor. Int. J. Eng. Res. Appl. **4**(3) (Version 1) ISSN 2248-9622, 35–39 (2014)
14. Reddy, K.H., Ramasamy, S., Ramanathan, P.: Hybrid Adaptive Neuro Fuzzy based speed Contol for Brushless DC Motor, pp. 93–110

Entropy-Based Approach for Enabling Text Line Segmentation in Handwritten Documents

G. S. Sindhushree, R. Amarnath and P. Nagabhushan

Abstract Determining text and non-text regions in an unconstrained handwritten document image is a challenging task. In this article, we propose a novel approach based on entropy for enabling the text line segmentation. A document image is divided into multiple blocks and entropy is calculated for each block. Entropy would be higher in the text region when compared to that of non-text region. Separator points are introduced accordingly to separate text from non-text part. Further correspondence between these separators would enable text line segmentation. The proposed algorithm works with an order of $O\ (m \times n)$ in worst case and requires less buffer space, since it is based on unsupervised learning. Benchmark ICDAR-13 dataset is used for experimentation and accuracy is reported.

Keywords Separators · Entropy · Correspondence · Text line segmentation

1 Introduction

The unconstrained handwritten document images have varied text orientation, different alignment, and unequal spacing between consecutive text lines [1]. Spotting the separator points for text line and non-text lines in the document images thus facilitates the text line segmentation of the document image [2].

Though the text line segmentation of printed text documents is seen as a solved problem, the text line segmentation of unconstrained handwritten documents endures hurdles, among which few are listed below.

G. S. Sindhushree (✉) · R. Amarnath
Department of Studies in Computer Science, University of Mysore, Mysore, Karnataka, India
e-mail: sindhugs91@gmail.com

R. Amarnath
e-mail: amarnathresearch@gmail.com

P. Nagabhushan
Indian Institute of Information Technology, Allahabad, India
e-mail: pnagabhushan@hotmail.com

P. Nagabhushan et al. (eds.), *Data Analytics and Learning*,
Lecture Notes in Networks and Systems 43,
https://doi.org/10.1007/978-981-13-2514-4_15

(a) Text in handwritten documents is curvilinear and is shown in Fig. 1.
(b) Lines will be tilted, touching, and overlapping.
(c) Start and end points of a sentence will be nonuniform in the entire document and the distance between two consecutive text lines will not be alike.

While there exist several methods for text line segmentation of handwritten documents, motivation behind this work is to compute the energy change (entropy) of every divided block of the document image and find the correspondence between the text and non-text regions.

In the proposed method, the document image is divided into equal number of blocks (horizontal slice) and the horizontal entropy is computed for each block. The entropy will be high in the text regions when compared to that of entropy in the non-text regions. Once the entropy is computed, the midpoint of every block in the document for both text and non-text regions is determined. Subsequently, the mean of midpoints for both text and non-text regions is calculated. The calculated midpoints help in finding the correspondence between text and non-text regions of the document which enables the text line segmentation as shown in Fig. 2.

The paper is organized as follows: Sect. 2 describes the related work in the field of text line segmentation and document image analysis. Section 3 explains the terminologies used. Section 4 describes the proposed idea in detail. Section 5 consists of the results of experiments conducted on benchmark dataset. Results and discussions are included in Sect. 6. Conclusion and future work are presented in Sect. 7.

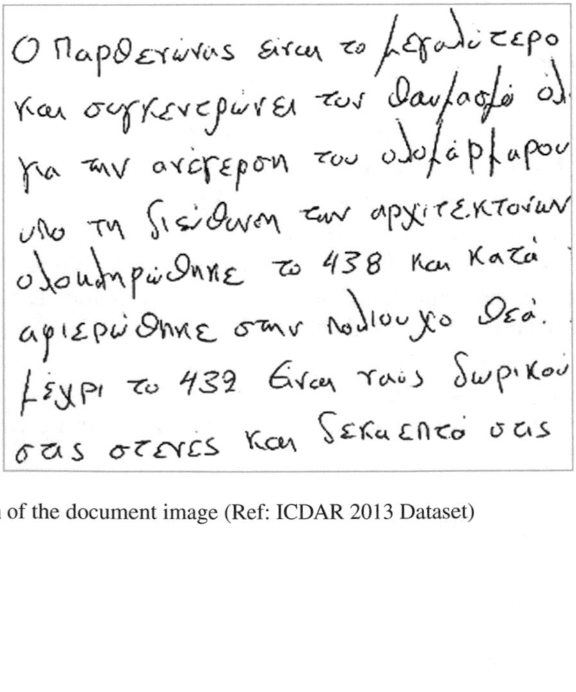

Fig. 1 A portion of the document image (Ref: ICDAR 2013 Dataset)

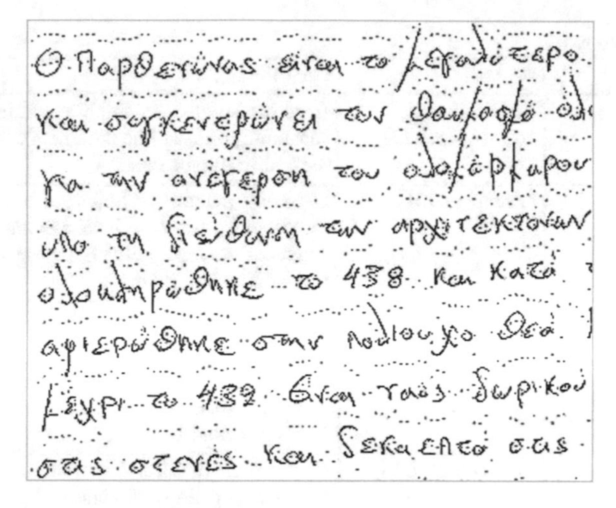

Fig. 2 Enabling text line segmentation based on entropy

2 Related Work

This section lists the related work with respect to text line segmentation. Most of the contributions are in the field of text line segmentation of historical documents, text, and word segmentation of handwritten documents and offline text line segmentation. Motivation behind this work is the need of effort in finding the correspondence between the text and non-text regions, which further facilitates the text line segmentation of the document (Table 1).

3 Terminologies

3.1 Entropy

Entropy is a measure of degree of irregularities in an image (0–1, 1–0). In other words, it is defined as energy change. The entropy for the document image (shown in Fig. 1) is illustrated in Fig. 3.

3.2 Separator Points

These are the points that facilitate the separation of text line from one another. Separator points are depicted in Fig. 4.

Table 1 Related work

Research area	Authors	Contribution
Text line segmentation	Amarnath and Nagabhushan [2]	Used run length encoding for spotting the separator points at line terminals which enables the text line segmentation
Text line segmentation	Alaei et al. [3]	Used new painting technique to smear the foreground portion of the document image, which enhances the separability between the foreground and background portions
Pattern recognition	Du et al. [4]	Used Mumford–Shah model for script independent text line segmentation
Pattern recognition	Louloudis et al. [5]	Applied Hough transform on a subset of the document image connected components for text line segmentation
Document analysis and recognition	Nicolaou et al. [6]	Used tracers to follow the white-most and black-most paths (bidirectional tracing) in order to shred the image into text line areas
Pattern analysis and machine intelligence	Li and Doermann [1]	Used density estimation and state-of-the-art image segmentation
Text line segmentation	Razak et al. [7]	It is comprehensive review of the methods for offline handwriting text line segmentation previously proposed by researchers

	1	0	0	0	0	1
	1	0	0	0	0	0
	1	1	0	0	0	0
	1	1	0	0	0	0
	1	1	1	0	0	0
	1	1	1	0	0	0
	1	1	1	0	0	0
	1	1	1	0	0	0
	1	1	1	1	0	0
	1	1	1	1	0	0

Fig. 3 Binary data

3.3 Correspondence

Correspondence refers to the association between the text-to-text and the non-text-to-non-text regions (Figs. 5 and 6).

where s refers to whitespace in between the text lines
t refers to the text line in the document image.

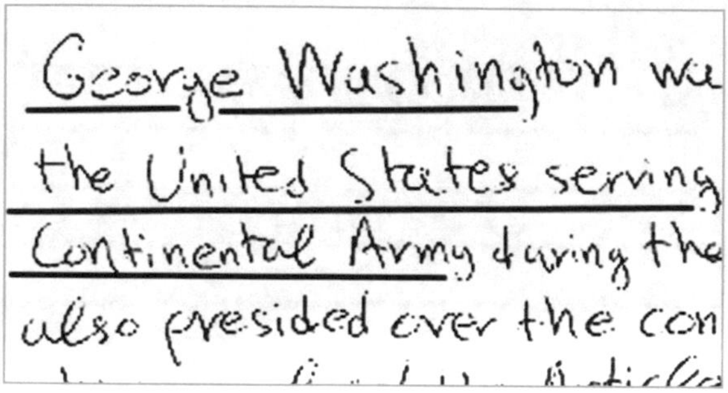

Fig. 4 Separator points for the text

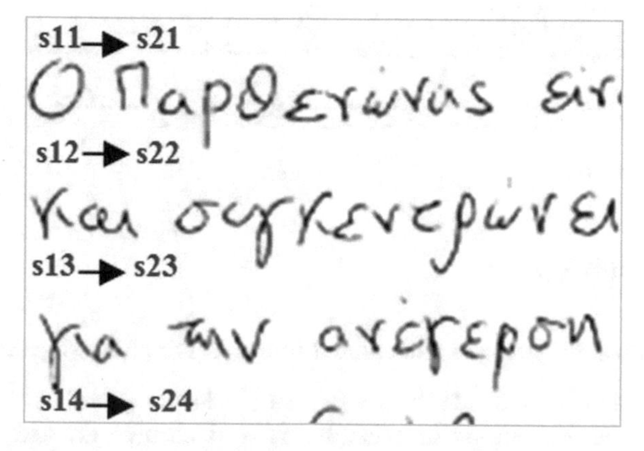

Fig. 5 Correspondence amid non-text region

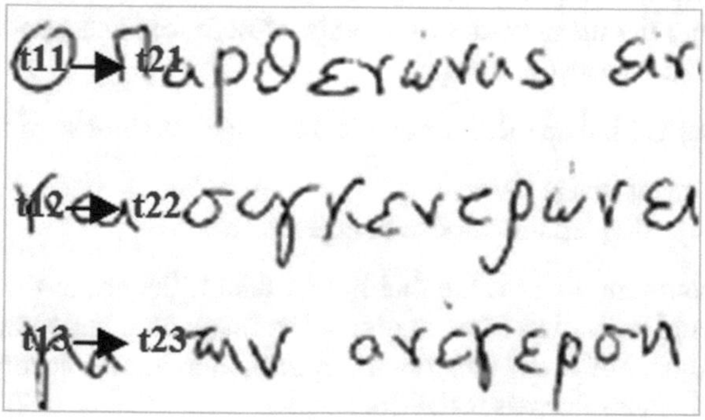

Fig. 6 Correspondence amid text region

Finding the correspondence between the separator points is not an easy task. Because the handwritten document images are unconstrained, they have different orientation, varied fonts and spacing. Correspondence for both the text and non-text regions of the image is considered, since they are relative in any Pattern Recognition systems.

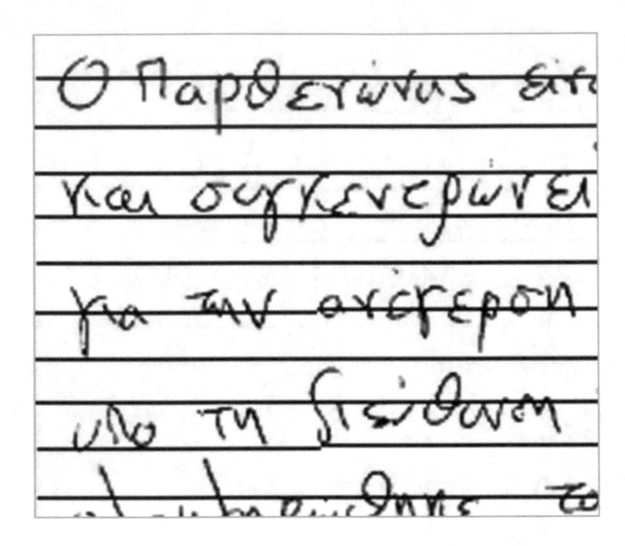

Fig. 7 Image division

4 Proposed Method

Finding the separator points in a document image has the following stages:

(i) To find an offset to divide the image into blocks of equal size.
(ii) To estimate the entropy of every block to determine the text and non-text regions.
(iii) To compute the midpoint of every block.
(iv) To compute the mean of the midpoints of both text and non-text regions, in order to find the correspondence.

(i) **Algorithm 1**: Find an offset to divide an image into blocks of equal size.

Input: Document image [h, w]
Output: Image divided into blocks of equal size

Choose a minimum offset value that is less than h (height of the image). Minimum value should be chosen so as to divide the image into large number of blocks having equal number of pixels which will, in turn, increase the accuracy of text line segmentation, though it increases the time.

Once the offset value is selected, choose a range of numbers that divides the offset with minimum remainder. Minimum remainder is selected, since it gives the quotient that specifies the number of pixels in each block and thus divide the document image into equal number of blocks. Dividing that quotient by width/height gives us the number of blocks.

```
Step 1:
begin
      ht = height_of_img
    oset = offset_value
Step 2: for i ←— 10 to 120
Step 3: if (mod(h,i)<ht)
            ht = mod(w,i);
            oset = i;
            endif
endfor
Step 4: no_of_blocks = |wd/oset|;
end
```

The time complexity to find an offset and divide the image into blocks of equal size is O(n), where n is the height of the input image.

(ii) **Algorithm**: Find the entropy of the divided blocks of the image

Input: Image divided into blocks of equal size [h, w]
Output: Entropy of each block of the image.

Entropy is calculated by initializing a counter. Whenever there is an entropy change with respect to the block, a counter is incremented.

```
Step 1: [h, w] = size_of_the_image
Step 2: for h ←— 1 to height_of_image
            counter ←— 0

Step 3: for w←— 1 to width
Step 4: if img[h, w] != img[h, w+1]
            counter ←— counter + 1;
          endif
        endfor

endfor
return ent
end
```

The time complexity to find the entropy of each block of image is O(m * n), where m is height of the image, n is width of the input image (Figs. 8 and 9).

(iii) a. **Algorithm**: Compute the midpoint of text lines

```
Input: Entropy computed for each block of the image for text
Output: Mid-point of every text region in the image.
for each divided block of the image
    find the non-zero value in each block;
        mark the indices as start and end.
              Compute the mid-point:
                    mid = |(start+end)/2|
```

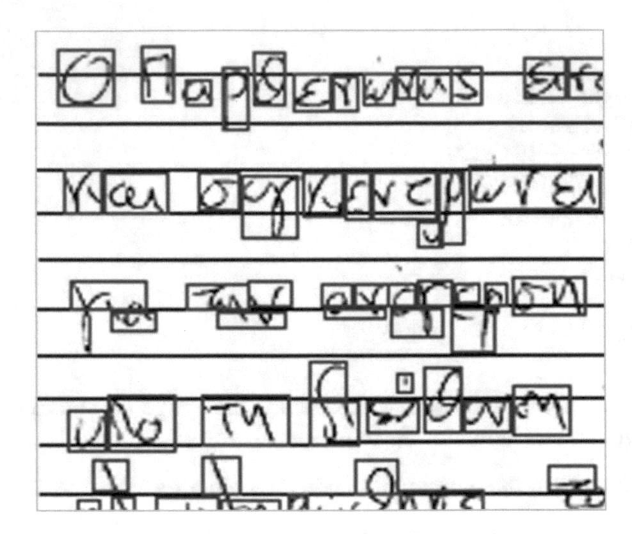

Fig. 8 Entropy in text region

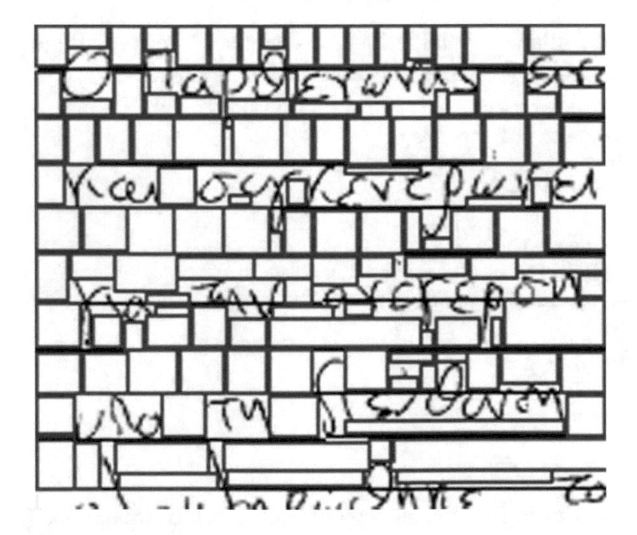

Fig. 9 Entropy in non-text region

(iii) b. Algorithm: Compute the midpoint of non-text regions

Input: Entropy computed for each block of the image
Output: Mid-point of every non-text region in the image.

```
for each divided block of the image
    find the null values in each block;
        mark the indices as start and end.
            Compute the mid-point:
                mid = |(start+end)/2|
```

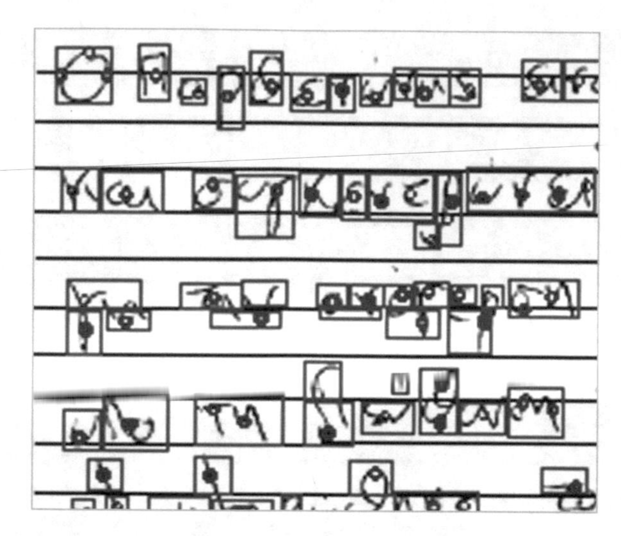

Fig. 10 Midpoint of the text regions in each block

The time complexity to compute the midpoint of text and non-text regions in each block is O(n), where n is the number of blocks in the document image (Figs. 10 and 11).

(iv) **Algorithm**: To compute the mean of the midpoints of text and non-text regions.

Input: Mid-point of the text and non-text regions of the image
Output: Mean of the mid-points of both text and non-text regions.

```
for i ← 1:no_of_blocks
        Compute mean of mid-points for text
        Compute mean of mid-points for non- text
endfor
```

The time complexity calculates the mean of mid-points of both text and non-text regions is O(n), where n is the number of blocks in the document image.

5 Experimental Analysis

Experimentation with benchmark ICDAR-13 dataset is performed and the results are recorded and tabulated below.

I. Table below shows the results of horizontal division of the image in which the horizontal entropy is calculated based on different offsets and angle of rotation.
(a) Different values for offset (Table 2).
(b) Different angles of rotation (Table 3).

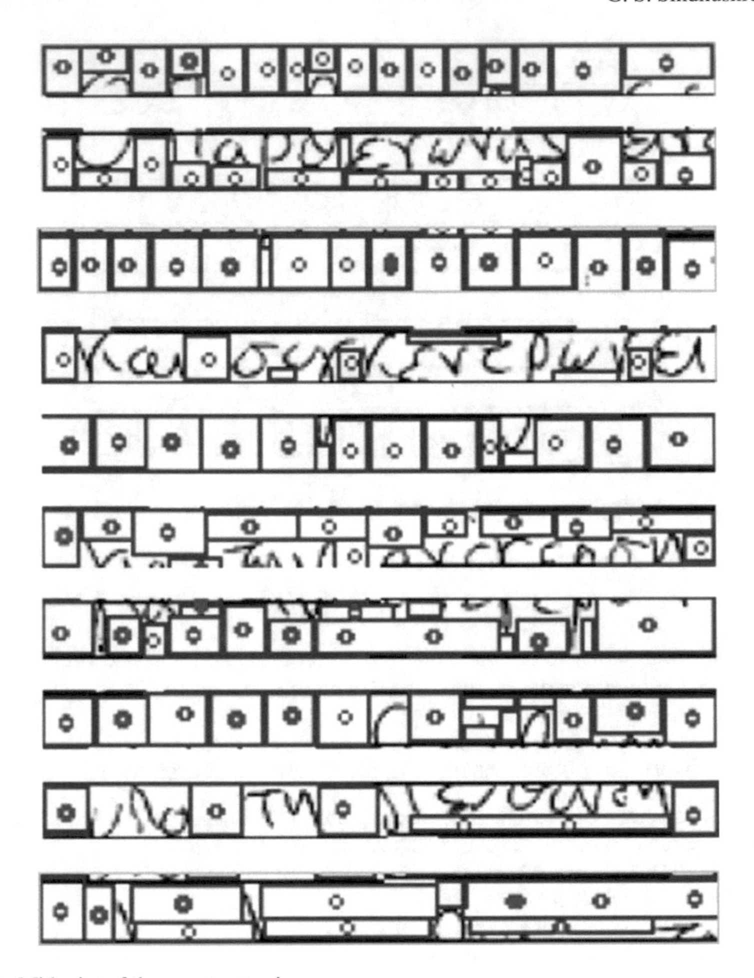

Fig. 11 Midpoint of the non-text regions

Table 2 Results of different values for offset

Block size/Number of blocks	Offset	Range	Angle of rotation
12/208	10	10–120	10
23/108	20	20–140	10
85/29	30	30–130	10
85/29	40	40–160	10
163/15	60	60–180	10
163/15	90	90–190	10
217/11	130	180–250	10

Table 3 Results for different angles of rotation of the document image

Block size/Number of blocks	Offset	Range	Angle of rotation
12/208	10	10–120	10
23/108	20	20–140	10
85/29	30	30–130	15
85/29	40	40–160	20
163/15	60	60–180	20
163/15	90	90–190	25
217/11	130	180–250	30

Table 4 Results of horizontal division of the image with computed vertical entropy

Block size/No. of blocks	Offset	Range	Angle of rotation
18/139	10	10–120	10
139/18	20	20–140	10
41/61	30	30–130	15
139/18	40	40–160	20
139/18	60	60–180	20
139/18	90	90–190	25
250/10	130	180–250	30

II. Table below shows the results of horizontal division of the image in which the vertical entropy is calculated based on different offsets and angle of rotation (Table 4).

From the above experimental analysis, it is observed that choosing minimum value for offset provides more number of blocks and also has large number of separator points, which results in higher accuracy. Choosing higher values for the offset will have lesser number of blocks and thus the entire text and non-text regions of the document may not be taken into consideration. Since the proposed method does not require any training, i.e., unsupervised learning, it requires less buffer space and thus the time required for computation will also be less.

Experimental analysis to find the correspondence between text and non-text regions in Fig. 7 is shown. The image is divided into 15 blocks and for each block, the midpoint is calculated, for both text and non-text regions. Merge operation is performed for each of the block (text and non-text regions) and is illustrated below.

$$\begin{array}{ll} \textbf{Block1}: & \\ \text{Text}: 17 & \text{Mean}: 17 \\ \text{Non}-\text{text}: [6; 36; 105] & \text{Mean}: 49 \end{array}$$

Merge a midpoint of text with the midpoint of the non-text value in a sorted order. Hence,

Hence,

Sequence after merge is: 6,17,36,105

Mid-point of Non-text region

Here, it is observed that two consecutive non-text regions are merged. In order to eliminate this situation, compute the difference of 105 and 36. $105 - 36 = 69$, which is greater than $49/2 = 24.5$. (The difference between consecutive non-text regions). Since 69 is greater than 24.5, remove 105. And hence, the new sequence will be:

New Sequence: 6, 17, 36

$$\begin{array}{ll} & \textbf{Block2}: \\ \text{Text}: [53; 93; 132; 156] & \text{Mean}: 108.5 \\ \text{Non}-\text{Text}: [4; 35; 71; 115; 150] & \text{Mean}: 51 \end{array}$$

Sequence: 4, 35, 53, 71, 93, 115, 132, 150, 156

Mid-point of Non-text region

Compute the difference: $35 - 4 = 31$

$51/2 = 25.5$

Since 31 is greater than 25.5, replace 4 and 35 with the average of mean computed. That is,

New Sequence: 26, 53, 71, 93, 115, 132, 150, 156

$$\begin{array}{ll} & \textbf{Block3}: \\ \text{Text}: [13; 54; 92; 127; 151] & \text{Mean}: 88.6 \\ \text{Non}-\text{Text}: [5; 33; 73; 108; 140] & \text{Mean}: 71.8 \end{array}$$

Sequence: 5, 13, 33, 54, 73, 92, 108, 127, 140, 151

$$\begin{array}{ll} & \textbf{Block4}: \\ \text{Text}: [15; 56; 130; 157] & \text{Mean}: 89.5 \\ \text{Non}-\text{text}: [4; 37; 76; 110; 145] & \text{Mean}: 74.4 \end{array}$$

Sequence: 4,15, 37, 56, 76, 110, 130, 145, 157

Mid-point of Non-text region

$76 - 110 = 34$ which is lesser than $74.4/2 = 37.2$. Hence, replace it with the mean.

New sequence: 4, 15, 37, 56, 74, 130, 145, 157

Block5 :

Text : [20; 54; 87; 143; 155] Mean : 91.8

Non − Text : [8; 36; 71; 107; 131] Mean : 70.6

Sequence: 8, 20, 36, 54, 71, 87, [107, 131], [143, 155] → Mid-point of Text region

↓

Mid-point of Non-text region

131 − 107 = 24, this is lesser than 70.6/2 = 35.3. Hence, remove 131 from the sequence.

155 − 143 = 12, this is lesser than 91.8/2 = 45.9.

Hence, remove 155 from the sequence. New sequence after merge will be,

New Sequence: 8, 20, 36, 54, 71, 87, 107, 143,

Block6 :

Text : [23; 53; 90; 126; 152] Mean : 88.8

Non − Text : [7; 41; 71; 108; 139] Mean : 73.2

Sequence: 7, 23, 41, 53, 71, 90, 108, 126, 139, 152

Block7 :

Text : [16; 58; 90; 128; 154] Mean : 89.2

Non − Text : [5; 37; 76; 109; 142] Mean : 73.8

Sequence: 5, 16, 37, 58, 76, 90, 109, 128, 142, 154

Block8 :

Text : [21; 54; 89; 117; 154] Mean : 87

Non − Text : [8; 36; 74; 101; 142] Mean : 72.2

Sequence: 8, 21, 36, 54, 74, 89, 101, 117, 142, 154

Block9 :

Text : [20; 54; 129; 155] Mean : 89.5

Non − Text : [8; 36; 72; 101; 115; 139] Mean : 78.5

Sequence: 8, 20, 36, 54, [72, 101, 115,] 129, 139, 155

↓

Mid-point of Non-text region

Replace with mean of non-text region.

New Sequence: 8, 20, 36, 54, 79, 129, 139, 155

$$\begin{array}{ll}\textbf{Block10}: & \\ \text{Text}: [20; 54; 88; 124; 150] & \text{Mean}: 87.2 \\ \text{Non}-\text{Text}: [8; 37; 67; 106; 135] & \text{Mean}: 70.6\end{array}$$

Sequence: 8, 20, 37, 54, 67, 88, 106, 124, 135, 150

$$\begin{array}{ll}\textbf{Block11}: & \\ \text{Text}: [18; 53; 91; 122; 153] & \text{Mean}: 87.4 \\ \text{Non}-\text{Text}: [6; 36; 61; 76; 106; 138] & \text{Mean}: 70.5\end{array}$$

Sequence: 6, 18, 36, 53, [61, 76,] 91, 106, 122, 138, 153

Mid-point of Non-text region

Replace 61 and 76 with the mean of non-text region. Hence, new sequence will be,

New Sequence: 6, 18, 36, 53, 71, 91, 106, 122, 138, 153

$$\begin{array}{ll}\textbf{Block12}: & \\ \text{Text}: [18; 56; 88; 119; 153] & \text{Mean}: 86.8 \\ \text{Non}-\text{Text}: [7; 35; 74; 102; 138] & \text{Mean}: 71.2\end{array}$$

Sequence: 7, 18, 35, 56, 74, 88, 102, 119, 138, 153

$$\begin{array}{ll}\textbf{Block13}: & \\ \text{Text}: [53; 94; 124] & \text{Mean}: 90.33 \\ \text{Non}-\text{Text}: [25; 70; 111; 143] & \text{Mean}: 87.25\end{array}$$

Sequence: 25, 53, 70, 94, 111, 124, 143

$$\begin{array}{ll}\textbf{Block14}: & \\ \text{Text}: [17; 52; 87; 122] & \text{Mean}: 69.5 \\ \text{Non}-\text{Text}: [6; 34; 70; 104; 142] & \text{Mean}: 71.2\end{array}$$

Sequence: 6, 17, 34, 52, 70, 87, 104, 122, 142

Block15 :

Text : [16; 52; 87; 124; 151] Mean : 86
Non – Text : [6; 33; 70; 106; 138; 157] Mean : 85

6, 16, 33, 52, 70, 87, 106, 124, 138, 151, 157

6 Results and Discussion

Below shown are the results of experimentation:

1. Offset: 10, range: 10-–120, angle of rotation: 10° (counter-clockwise).
2. Offset: 20, range: 20–140, angle of rotation: 20° (counter-clockwise)
3. Offset: 30, range: 30–130, angle of rotation: 0°
4. Offset: 40, range: 40–160, angle of rotation: −10° (Clockwise).

Below table shows the accuracy rate computed for ICDAR-13 dataset for different offsets (Table 5 and Fig. 12).

Advantages of the proposed system:

i. It requires less buffer space, since no prior training is required, i.e., unsupervised learning.
ii. Time complexity is in the order of O(m * n).
iii. In this approach, information about both text and non-text regions are established, which is not employed in other methodologies. This helps in text line segmentation of the document and this could enhance the text line segmentation.

Limitations of the proposed system:

i. The proposed system works only for the plain binary images without any figures, tables, and graphics.
ii. It works well up to an angle of rotation of 30°.

Table 5 Accuracy values for different offsets

Offsets	Accuracy with ICDAR-13 dataset
10	89.47
20	88.24
30	87.20
40	87.22
60	84.43
70	84.22
90	84.08
130	80.81

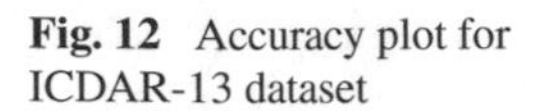

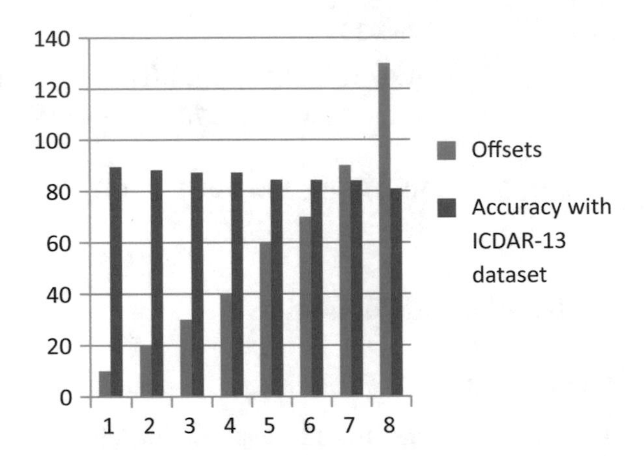

Fig. 12 Accuracy plot for ICDAR-13 dataset

7 Conclusion and Scope for Future Work

In this paper, a novel approach for finding the correspondence between the separator points of a document is proposed. Choosing the minimum value for offset helps in having more number of blocks in the image, this will, in turn, enable the text line segmentation with higher accuracy rate. Entropy is used to determine the text and non-text regions in the document. Experimentation was conducted with benchmark ICDAR-13 dataset and the results are recorded. It requires less buffer space since, no training is required in prior. In this paper, we have showcased that text line segmentation can be performed with just entropy calculations. The limitations, as well as other techniques along with entropy, can be considered for future work.

References

1. Li, Y., Doermann, D.: Script-independent text line segmentation in freestyle handwritten documents. IEEE Trans. Pattern Anal. Mach. Intell. **30**(8) (2008)
2. Amarnath, R., Nagabhushan, P.: Spotting separator points at line terminals in compressed document images for text-line segmentation. Int. J. Comput. Appl. **172**(4), 0975–8887 (2017)
3. Alaei, A., Pal, U., Nagabhushan, P.: A new scheme for unconstrained handwritten text-line segmentation. Pattern Recogn. **44**, 917–928 (2011)
4. Du, X., Pan, W., Bui, T.D.: Text line segmentation in handwritten documents using Mumford–Shah model. Pattern Recogn. **42**(12) (2009)
5. Louloudis, G., Gatos, B., Pratikakis, I., Halatsis, C.: Text line and word segmentation of handwritten documents. Pattern Recogn. **42**(12) (2009)
6. Nicolaou, A., Gatos, B.: Handwritten text line segmentation by shredding text into its lines. In: 10th International Conference on Document Analysis and Recognition (2009)
7. Razak, Z., Zulkiflee, K., Idris, M.Y.I., Tamil, E.M., Noor, M.N.M., Salleh, R., Yaakob, M., Yusof, Z.M., Yaacob, M.: Off-line handwriting text line segmentation: a review. IJCSNS Int. J. Comput. Sci. Netw. Secur. **8**(7) (2008)

Automated Parameter-Less Optical Mark Recognition

N. C. Dayananda Kumar, K. V. Suresh and R. Dinesh

Abstract Innovation and development of computer vision algorithms finds real-time application in automated evaluation of optical mark sheets. Most of the existing methods are template specific and requires parameter tuning based on the given template. Developing robust and low-cost solutions for optical mark recognition (OMR) is still a challenging problem. The need for parameter less and computationally efficient method that works in real time adds on to the existing challenges. To address these problems, robust and computationally efficient parameter-less OMR method is proposed. The main contributions of this paper are dynamic localization of optical mark region that makes the algorithm generic and independent of templates. Accuracy is improved by implementing some of the heuristics approaches which identifies marking regions that are missed during preprocessing. Marked regions are accurately detected by computing local threshold instead of global classification margin. The algorithm is evaluated on various OMR templates obtained from different scanners and smartphones. Experimental results depict the efficiency of the proposed method with the error rate less than 2% by processing 150 sheets per minute.

Keywords Optical mark recognition · Document analysis
Multiple choice questions

N. C. Dayananda Kumar (✉) · K. V. Suresh
Department of Electronics and Communication Engineering,
Siddaganga Institute of Technology, Tumkur, India
e-mail: dayanandkumar.nc@gmail.com

K. V. Suresh
e-mail: sureshkvsit@yahoo.com

R. Dinesh
Department of Information Science and Engineering, Jain University,
Bengaluru, India
e-mail: dr.dineshr@gmail.com

P. Nagabhushan et al. (eds.), *Data Analytics and Learning*,
Lecture Notes in Networks and Systems 43,
https://doi.org/10.1007/978-981-13-2514-4_16

1 Introduction

Optical mark recognition system evaluates the optical mark sheets which typically contains a grid of circles for marking the answers. These marked regions are detected and compared against the key answers in master data. The automated OMR system finds a wide range of applications in (a) Evaluating the exams based on multiple choice question (MCQ) patterns, (b) Banking and financial institutions, (c) Medical institutes for monitoring and preserving patient track records, and (d) Government and consumer surveys so on. Developing accurate and low-cost solution for these applications is still challenging and of great interest. Most of the existing systems assume that the regions of the marking circles are known prior. Hence, the reference template of mark sheet is required as input parameter which is a limitation.

In order to develop an efficient OMR system, a generic parameter less algorithm is proposed which is independent of answer sheet templates. The positions of the marking circles are dynamically computed by localizing the marking region. Segmented region of interest is obtained by imposing the constraint that, relative distance between marking circles is less than the marking regions. The obtained region of interest is processed individually by mark detection algorithm. The accuracy of mark detection is further improved by computing the local threshold along the row of marking circles. The proposed algorithm is evaluated on mark sheets with varying templates and obtained from different scanners. The performance analysis ensures that the system works in real time with reliable results.

In further sections of this paper, some of the major works related to OMR is discussed in Part 2. The proposed algorithm is briefly discussed in Part 3. Experimental results of the algorithm are shown in Part 4.

2 Related Work

Hussmann et al. [1] developed a prototype of high-speed OMR system used for marking multiple choice questions. To achieve the high processing speed, the complete system is implemented using single Field Programmable Gate Array (FPGA). This mark detection algorithm has been developed to achieve real-time performance. Hui Deng et al. [2] developed a system to process thin paper answer sheets with low-printing precision. The system key techniques include tilt correction, scanning error correction and mark recognition. Garima et al. [3] developed OMR system on java platform which provides a tool for the user to design the response sheet layout of required format. This template is filled and evaluated further to detect the marked regions. Houbakht et al. [4] developed a system for multiple choices question exam based on morphology and error rejection algorithm. This system includes image scanning, preprocessing and mark identification steps.

Pegasus Imaging Corporation presents a Software Development Kit (SDK) [5] for optical mark recognition of document images. The rectangle region with a specified number of rows and columns is considered as marking field to be evaluated. The

SDK scans the region horizontally and then vertically to locate the marking circles. Sanguansat et al. [6] proposed a system for the automated data entry for survey by using different pattern of questionnaires for each survey. This system claims the accuracy rate as 93.36% by using three different patterns of the templates. Nguyen et al. [7] proposed the method in which captured answer sheet image is processed using Hough transform and then skew-corrected into the proper orientation. The mark corresponding to the answer for each question is recognized by allocation of the mask which wraps the answer area. CheckIt [8]—a mobile phone-based OMR system is developed using open-source technology but requires prior information about the optical mark sheet layout. Four classifier-based approach is proposed by Haskins [9] which is based on the number of pixels, edges, patterns, and edit distance. Chui [10] Proposed a method which enables the users to customize and print their own answer sheets. It is able to process the images from ordinary document scanner without the need of expensive and specialized optical mark recognition scanners.

3 Proposed Method

The current state-of-the-art algorithms in optical mark recognition is proved to be efficient and works in real-time based on some constraints. However, in evaluating the key answers, an error rate of 2% still exists in most of the systems. Also, the increasing users and data emphasize the need for systems with higher processing capability. In order to address these issues of template specific processing and use of hardware components, an efficient novel algorithm is proposed which works in real time (Fig. 1).

Input optical sheet image to be processed is obtained from ordinary scanner or a smartphone. These images are preprocessed to align properly by performing skew correction. Marking circle regions are localized dynamically to locate the position of circles that need to be evaluated. The barcode reader module is used to decode the register number present in the OMR image. Local threshold is computed to compare the intensity level of marked circle. The detected circles are evaluated against key answers and the processed output string with barcode information is saved in a file. Further subsections briefly discuss about the algorithms involved in processing.

3.1 Preprocessing

The scanned optical mark sheets are converted to grayscale image and enhanced using un-sharp masking filter. These sheets may encounter rotation and tilting during scanning and hence they are not aligned. These sheets should be automatically adjusted and aligned properly by tilt correction. To perform skew correction (Fig. 2), the square markers are provided at the corner of the sheets as reference. These squares are detected from the scanned sheet and the angle of orientation is computed as explained below.

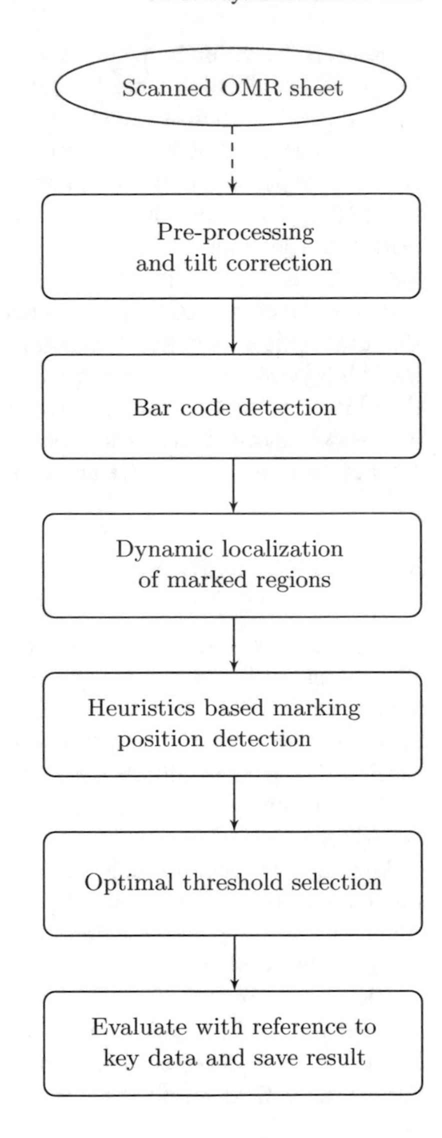

Fig. 1 Block diagram of the proposed OMR System

Let $p_1(x_1, y_1)$ and $p_2(x_2, y_2)$ be the center coordinates of the reference square markers on the top left and right corners of the mark sheets. The angle of orientation is computed as given in Eq. (1).

$$\theta = tan^{-1}\left(\frac{y_2 - y_1}{x_2 - x_1}\right) \quad (1)$$

The image is warped around the center with the computed angle for correcting the scan tilt. The tilt corrected and enhanced image are binarized by adaptive thresholding (Fig. 3).

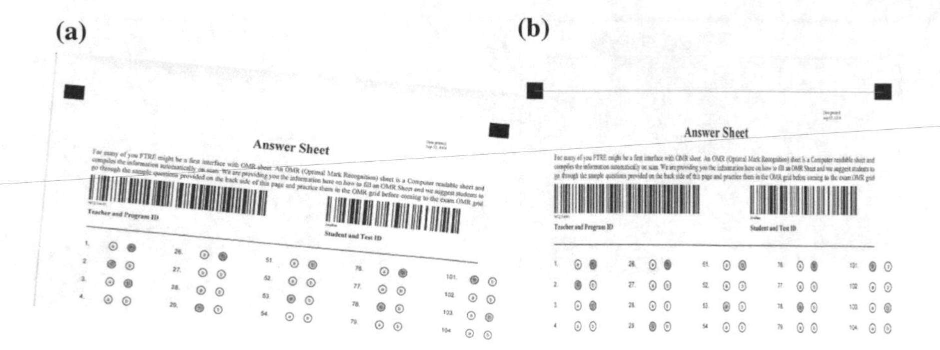

Fig. 2 **a** Scanned OMR sheet **b** Tilt Corrected OMR Sheet

Fig. 3 Binarized image

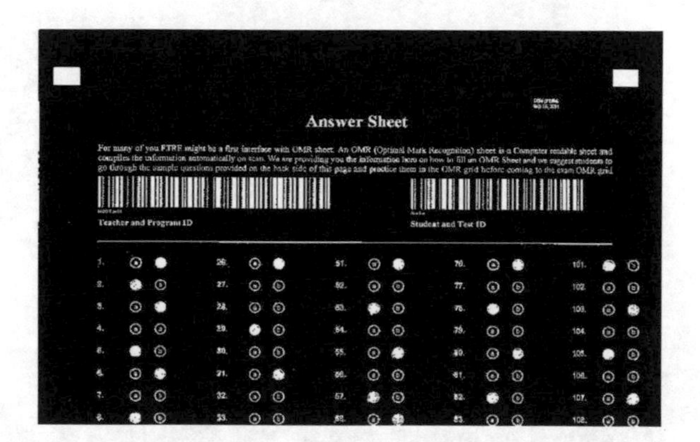

3.2 Dynamic Localization of Marked Regions

Most of the existing OMR systems are template specific which detects the marked region with reference to given template. In order to overcome the template specific constraint, the marked regions are localized dynamically. In the proposed method, the relative distance between the circles is assumed to be less than the distance between marking regions and dynamic Region of Interest (ROI) is detected as explained: (1) Perform morphological close operation on obtained binarized image. (2) Find contours through connected components and fit the bounding rectangle to detected bigger regions which constitutes the region of interest (Figs. 4 and 5).

3.3 Detecting Mark Position

The position of marking circles is detected by finding contours from obtained segmented region of interest. Due to thresholding and some artifacts, few of the marking circles may be missed. In order to address this issue, position of each marking circle is verified by forward and backward search based on initial position detection.

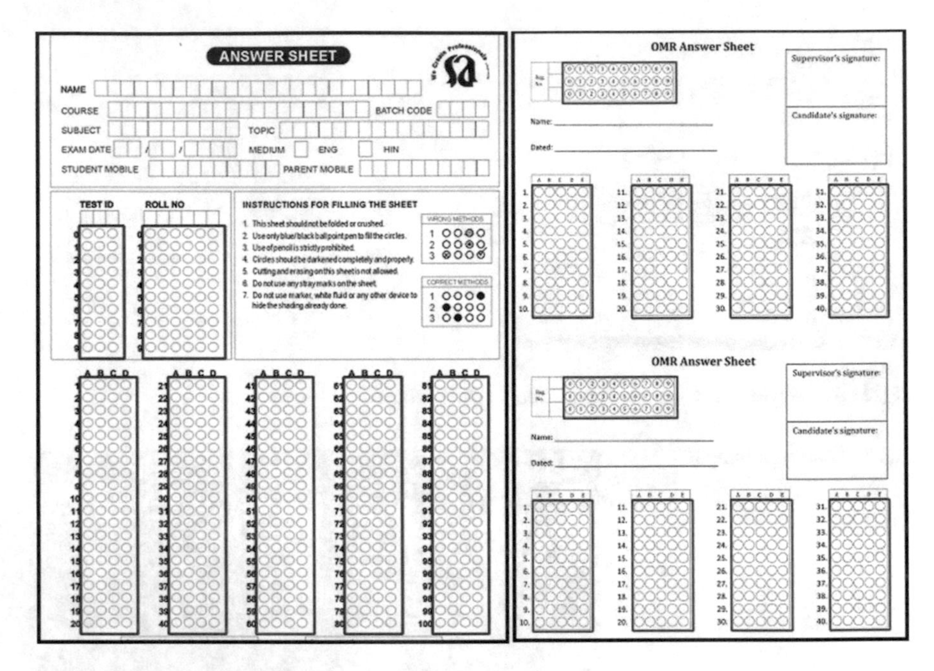

Fig. 4 Dynamic ROI obtained for templates 1 and 2

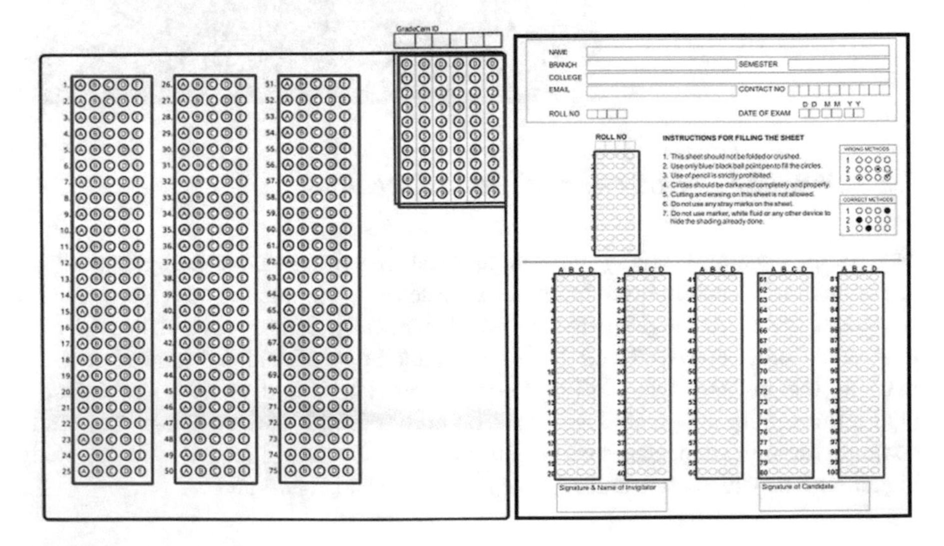

Fig. 5 Dynamic ROI obtained for templates 3 and 4

The detected regions as marked in Figs. 4 and 5 are added to a list and processed individually as explained below (Fig. 6).

a. For each dynamic region of interest, the contours are extracted and their bounding coordinates are added to the array called contourList.

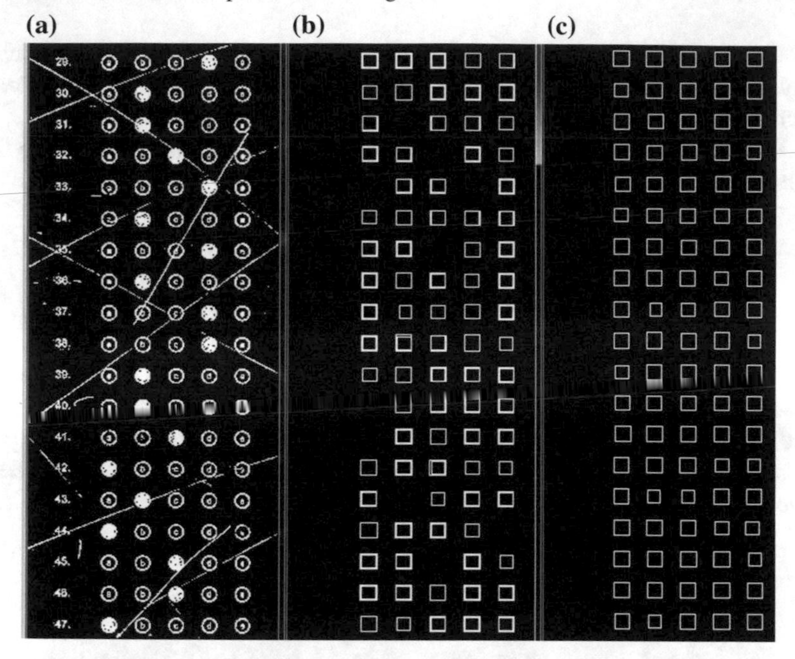

Fig. 6 **a** Indicates the binarized image, **b** Depicts the missed out contours, **c** Identified marking regions through the proposed method

b. Since the contours in this array are un-ordered, Sort the contourList along y coordinates.

c. Compute the difference in y coordinates of successive contours and compare with the suitable threshold to identify start and end of the contour rowList as below.

```
For index = 1 to sizeof(contourList)
If (contourList [index] - contourList [index - 1]
< 0.5 * contourList [index].height)
Add the contour to vector rowList[k]
Else
Add the contour to vector rowList[++k]
End
End
```

Hence, the contours are grouped into separate rowList based on their positions.

d. Since the template is unknown, identify the number of marking regions in each row. Initially set contourRowCount to zero and let M be the total number of rowList

```
M = Total number of rowList
For k = 1 to M
contourRowCount + = sizeOf(rowList[k])
End
contourRowCount = ceil (contourRowCount / M )
```

e. Compare the size of each rowList with obtained contourRowCount and check if any marking region is missed out in the rowList. Perform forward and backward search to identify the nearest rowList in which all regions are present.

f. Obtain missed out region by substituting the corresponding coordinates from identified rowList.

3.4 *Optimal Threshold Selection*

After detecting the accurate marking regions, the next step is to identify whether it is marked or not. Hence, the average intensity of the region within the marking coordinates is computed and compared with the suitable threshold value. To improve the accuracy, optimal threshold for each rowList is computed separately instead of single global threshold value (Fig. 7).

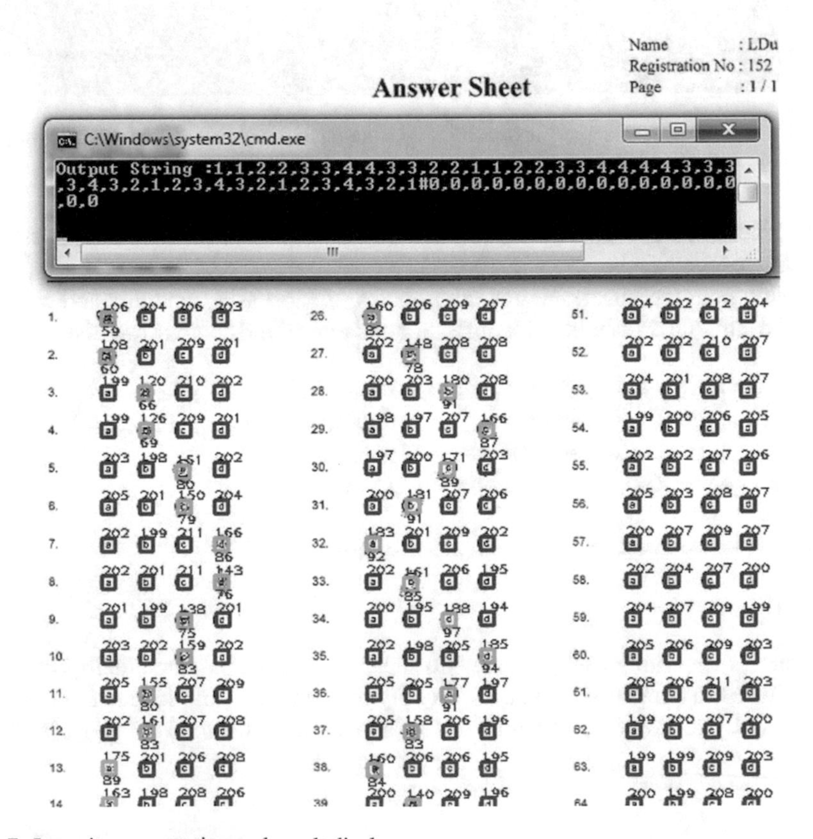

Fig. 7 Intensity computation and result display

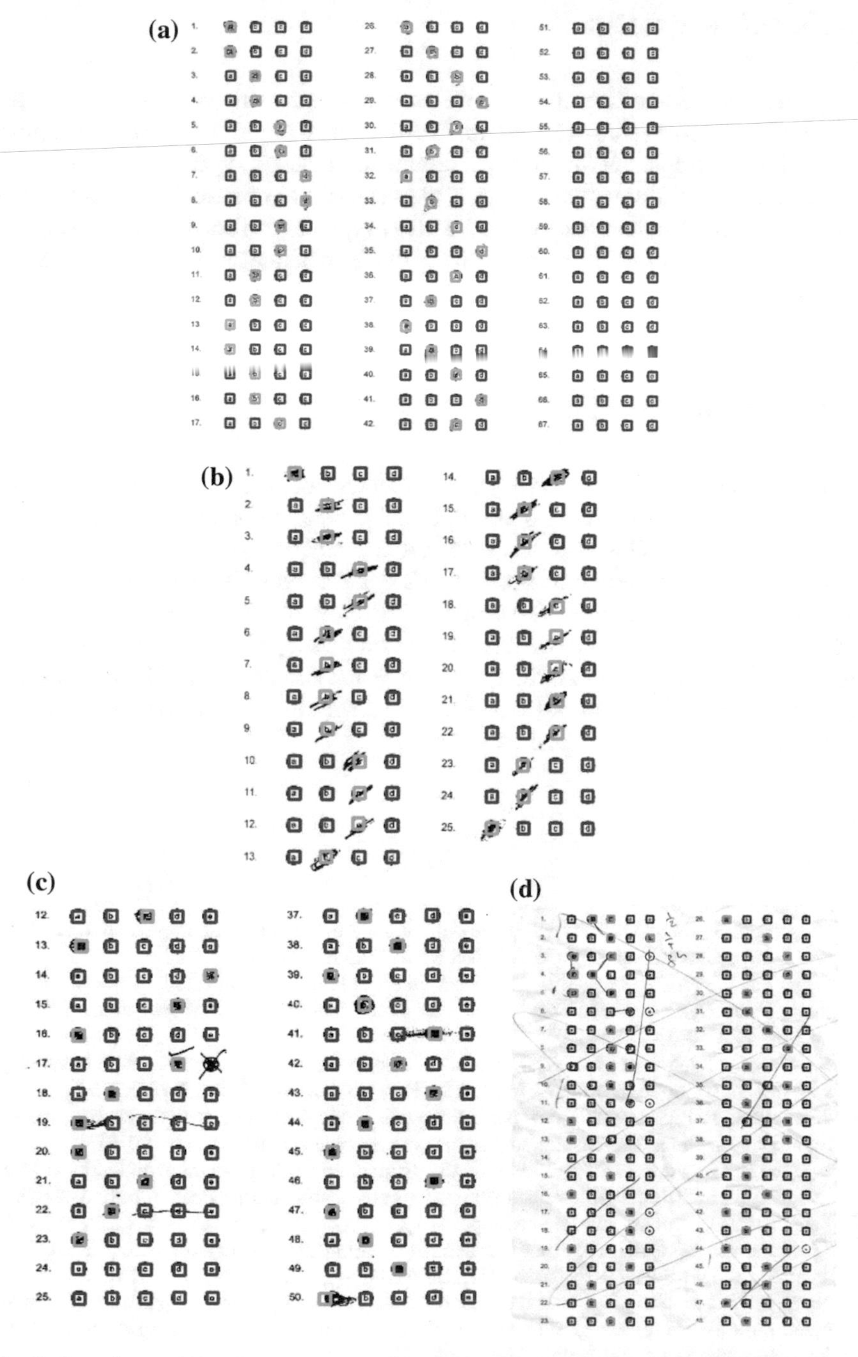

Fig. 8 Experimental Results: **a** Detection of marking circles. **b** Case 1: Sheets with improper markings. **c** Case 2: The noisy and distorted marking circles and, **d** Case 3: Scribbled and corrupted markings

4 Experimental Results

The OMR sheets scanned through two different scanners with varying resolution are obtained. In order to evaluate the system, five different templates with varying number of questions, multiple answer choices, dimension of marking circles, and patterns are processed. In addition, the system is also capable to process unclear markings, scribble in marked regions and the sheet with low quality and poor marking visibility. The marked circles are detected efficiently with the error rate less than 2%. The results of the proposed method are provided in Fig. 8 by considering different scenarios.

5 Conclusion

The proposed method focuses on template independent processing, heuristics-based identification of missing regions and optimal threshold computation for marking region classification. This system is tested on different types of templates for performance evaluation. The error rate is less than 2% and the algorithm is computationally efficient to work in real time. The scope of current work is to process the scanned OMR images in a PC-based environment. The future scope is to develop an Android-based smart phone application which is portable and less expensive. Hence, the algorithm should be ported to SDK-based Android environment with GPU optimization for mobile-based acquisition and evaluation.

References

1. Hussmann, S., Deng, P.W.: A high speed optical mark reader hardware implementation at low cost using programmable logic. Sci. Direct Real-Time Imaging **11**(1) (2005)
2. Deng, H., Wang, F., Liang, B.: A low-cost OMR solution for educational applications. In: International Symposium on Parallel and Distributed Processing with Applications, ISPA'08, pp. 967–970. 10–12 Dec 2008
3. Krishna, G., Rana, H.R., Madan, I., Kashif., Sahu, N.: Implementation of OMR technology with the help of ordinary scanner. Int. J. Adv. Res. Comput. Sci. Softw. Eng. (2013)
4. Attaran, H., Rad, F.S., Ashkezari, S.N., Rezazadeh, M.: Checking multiple choice question exams. International Symposium on Advances in Science and Technology (2013)
5. Pegasus imaging corporation (25 Mar 2013). Structured forms processing and Optical Mark Recognition SDK [Online]. Available https://www.accusoft.com/products/formfix/overview/
6. Sanguansat, P.: Robust and low-cost optical mark recognition for automated data entry. In: 12th International Conference on Electrical Engineering/Electronics, Computer, Telecommunications and Information Technology (ECTI-CON), pp. 1–5. 24–27 June 2015
7. Nguyen, T.D., Manh, Q.H., Minh, P.B., Thanh, L.N., Hoang, T.H.: Efficient and reliable camera based multiple-choice test grading system. In: Proceedings of 2011 Advanced Technologies for Communications (ATC), pp. 268–271 (2011)
8. Patel, R., Sanghavi, S., Gupta, D., Raval, M.S.: CheckIt–A low cost mobile OMR system. IEEE Region 10 Conference on TENCON 2015–2015, pp. 1–5. Macao (2015)

9. Haskins, B.: Contrasting classifiers for software-based OMR responses. In: Pattern Recognition Association of South Africa and Robotics and Mechatronics International Conference (PRASA-RobMech), vol. 2015, pp. 233–238 (2015)
10. Chai, D.: Automated marking of printed multiple choice answer sheets. In: 2016 IEEE International Conference on Teaching, Assessment, and Learning for Engineering (TALE), pp. 145–149. Bangkok (2016)

A Bimodal Biometric System Using Palmprint and Face Modality

N. Harivinod and B. H. Shekar

Abstract In this paper, we present a bimodal biometric identity assurance system by combining palmprint and face modalities. The Gabor features are extracted from both modalities and combined at the feature level. Kernel Fisher discriminant analysis is performed to obtain the features and thereby to create knowledge base. The experiments are carried for biometric identification and verification on various datasets. The combination of modalities improves the accuracy of results.

Keywords Multibiometrics · Gabor features · Feature fusion

1 Introduction

Nowadays, there is a increasing growth in maintaining the privacy in personalization and access control. In some cases, we need to identify the right person over a large population and in some other cases, we need to permit only the right person to access secured data. To achieve this, biometric systems are highly preferred as they offer high reliability and accuracy. Biometrics plays an important role on identity assurance; both in recognizing an individual in the group (identification) and verifying the truthfulness of individuals identity (verification or authentication). Various biometric traits used for this purpose are broadly classified into two categories; physical and behavioral. Some of the example for physical traits include face, palmprint, and iris and that of behavioral traits include signature, gait, keystroke dynamics, etc.

Single biometric-based identity assurance systems sometimes fail due to the failure of getting proper information [1]. The failure is due to the noise occurred in

N. Harivinod (✉)
Department of Computer Science and Engineering, Vivekananda College of Engineering and Technology, Puttur, Karnataka, India
e-mail: harivinodn@gmail.com

B. H. Shekar
Department of Computer Science, Mangalore University, Mangalore, Karnataka, India
e-mail: bhshekar@gmail.com

P. Nagabhushan et al. (eds.), *Data Analytics and Learning*,
Lecture Notes in Networks and Systems 43,
https://doi.org/10.1007/978-981-13-2514-4_17

the image acquisition, considerable variation in the image samples from the same individual (intra-class variation), similarity in the sensed data between two or more individuals (inter-class similarity), the biometric may be easily adaptable for spoof attacks or any other reasons. The drawbacks of single biometric trait can be reduced by using multiple traits together which are called as multi-biometric systems. Among the different available traits, face and palmprint are simple and easy to access as both are user-friendly and incorporate contactless image capturing system. Hence in our work, we have chosen face and palmprint modalities for the design and implementation of the bimodal biometric system.

In the literature, we can find many ways to combine the information from multiple traits [2]. These include sensor level, feature level, matching score level, and decision level fusion. Among these, feature level fusion is preferred as the information is fused at the early stage and it does not require special hardware sensors as that of sensor level fusion. Due to this superiority, we have used feature level fusion in our work.

The rest of the paper is organized as follows. A survey of bimodal biometric system using face and palmprint are presented in Sect. 2. The proposed methodology of biometric system is given in Sect. 3. Database description and experimental results are provided in Sect. 4. Conclusion is presented in Sect. 5.

2 Literature Survey

Here, we present some of the prominent recent works in palmprint and face biometric modalities. Initially, we present few works on score level fusion followed by feature and sensor level fusion are presented. Matching score fusion of face and palmprint modality is reported in [3, 4]. Poinsot et al. [3] proposed palmprint and face biometric fusion using binary features vectors. The information are fused using score level fusion. It particularly concentrates on small number of samples for the recognition which well suites for the modern embedded systems. In [4], an authentication system based on face and palmprint using fusion of matching score was proposed. Here, principal lines extracted from palmprint and eigenfaces computed from face are used as features. The matching scores of these features are fused and decision level thresholding is performed to get the final results.

Some of the works on palmprint and face that use sensor or feature level information fusion are presented in [1, 5–8]. A pixel level fusion of face and palmprint biometric is proposed in [7]. The images of face and palmprint are decomposed into bit-plane and pixel level fusion is applied by simple averaging. As the resulting features are high in dimensionality, principle component analysis is performed to reduce the dimension. Feedforward backpropagation neural network is used for the classification. In [8], a method to combine the feature using weighted feature level fusion was proposed. The binary variant of particle swarm optimization technique is used for the weightage computation. This method will select the most discriminant features from the face and palmprint images. In [5], DCT-based features of face and palmprint are extracted and concatenated to form feature vector. The low-frequency components of DCT are extracted from the top-left corner in zigzag fashion. Recognition experiments are carried out by varying number of DCT coefficients. A non-stationary feature fusion is proposed in [6] for identification purpose. Here, feature fusion is achieved by matrix interleaved concatenation approach, that uses the statistical properties and Gaussian Mixture Models.

A comparison of various levels of fusion of face and palmprint modalities was presented in [1]. The features are extracted by linear discriminant analysis and local phase quantization methods. The information fusion is done at sensor level (by using wavelet method), feature level (by applying min-max rule, Z-score, and Tanh), matching score level (by using max, min, and sum rule), and decision level (by using OR and AND rules).

3 Methodology

The proposed method consists of the following steps:

a. Palmprint and Face image acquisition followed by preprocessing
b. Feature extraction using Gabor filters and feature fusion
c. Kernel Fisher analysis and preparing knowledge base
d. Feature matching.

The block diagram of the proposed bimodal biometric system is given in Fig. 1.

3.1 Image Acquisition

We have used existing images from PolyU palmprint dataset [9], IIT Delhi Palmprint dataset [10], CASIA 2D face dataset [11] and ORL face dataset [12] for the experimentation and testing. More details on this dataset are presented in Sect. 4. The central region of palm (region of interest or ROI) from the palmprint image is extracted using the technique proposed in [13]. The ROI is resized to 128×128 for further processing. From the face images, the square portion of face is extracted. This process removes the background by retaining the central part of the face. Cropped face region is resized to 32×32 for further processing.

3.2 Feature Extraction and Feature Fusion

Gabor filters are extensively used in image processing and pattern recognition. Based on the uncertainty principle, it gives time–frequency location [13]. It is robust to handle texture features even under illumination variation. In our work, feature extraction

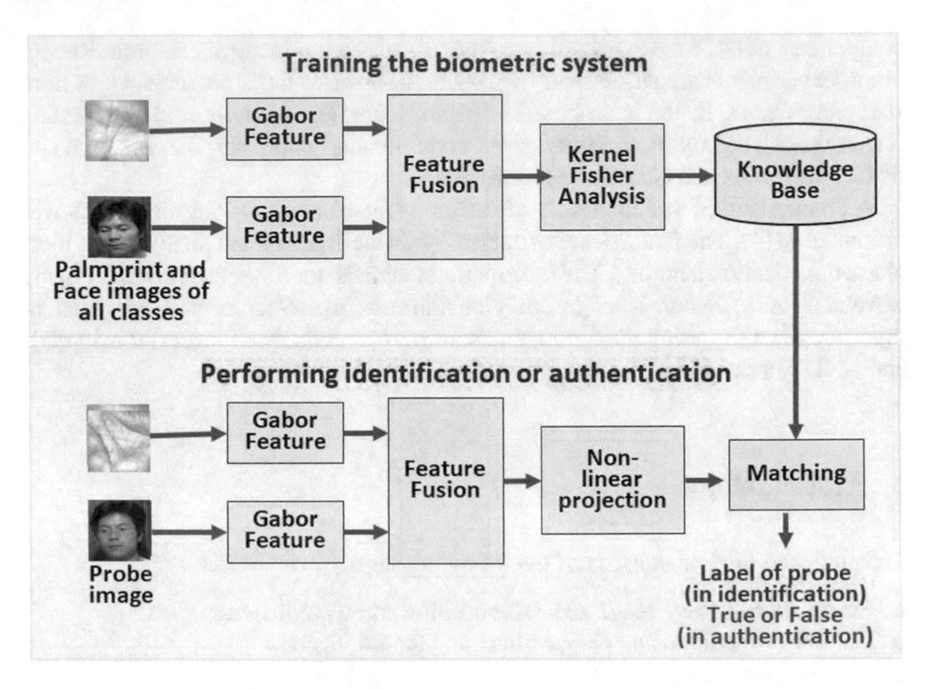

Fig. 1 Block diagram of the proposed bimodal biometric system

is achieved through filtering the images using Gabor filters. The 2D Gabor filter [14] can be designed using Eq. 1.

$$G(x, y, \theta, u, \sigma) = \frac{1}{2\pi\sigma^2} \exp\left\{-\frac{x^2 + y^2}{2\sigma^2}\right\} \times \exp\{2\pi i(ux\cos\theta + uy\sin\theta)\} \quad (1)$$

where

(x, y) is the coordinates of the filter
$i = \sqrt{-1}$
u is the frequency of Gabor wave
θ gives the orientation of the function
σ is the standard deviation of the Gaussian envelope.

In our work, we have designed Gabor filters in five scales and eight orientations.

$$O_{uk}(x, y) = I(x, y) * G(u, k), u = 0, 1, 2, 3, 4;\ \ k = 0, 1, 2, 3, 4, 5, 6, 7. \quad (2)$$

The given image I is convolved by these filters. The parameters are chosen based on the recommendations specified in [13].

The above process is performed for palmprint as well as face images. The feature fusion is achieved by concatenating these feature vectors.

3.3 Kernel Fisher Analysis

The fisher discriminant analysis is used to identify the label of a test sample by associating it with the class of closest mean. It forms the inter-class variance as large as possible and the intra-class variance as small as possible. We have used kernel-based fisher analysis (KFA) using polynomial kernel. The resultant feature vector obtained due to KFA is stored to form the knowledge base.

We have used KFA algorithm presented in [15] for the experimentation. This involves (i) choosing a kernel function, (ii) centering the training data in the feature space, (iii) computing the kernel matrix and block diagonal matrix, (iv) solving the eigenvalue problem and selecting the eigenvectors corresponding to large eigenvalues, and (v) normalizing the eigenvectors to form the eigenspace for feature projections, hence to create knowledge base. More details are presented in [15].

3.4 Matching

In the identification or verification process, the probe is matched against the features stored in the knowledge base. In this step, we extract the Gabor features from both face and palm images of the probe and are concatenated. The difference between grand mean of the training data and test sample is computed. Then nonlinear projection is carried out, so that it is projected on to the feature space. The proximity of the probe with training data is found using the Mahalanobis distance. This measure takes into account of the covariance to compute the distance. It transforms the random vector into a zero mean vector with an identity matrix for covariance.

Mahalanobis distance D_M between two vectors $x = (x_1, x_2, ..., x_n)^t$ and $y = (y_1, y_2, ..., y_n)^t$ is given by,

$$D_M(x, y) = \sqrt{(x - y)^t S^{-1}(x - y)} \quad (3)$$

where S is the covariance matrix computed from x and y.

For identification purpose, experiments are conducted using nearest neighbor classifier to obtain the label of probe. In verification experiments, the probe is validated against the empirically determined threshold.

4 Experimental Results

This section describes datasets used for the experimentation. We also present the results obtained on various datasets with different training and testing configurations.

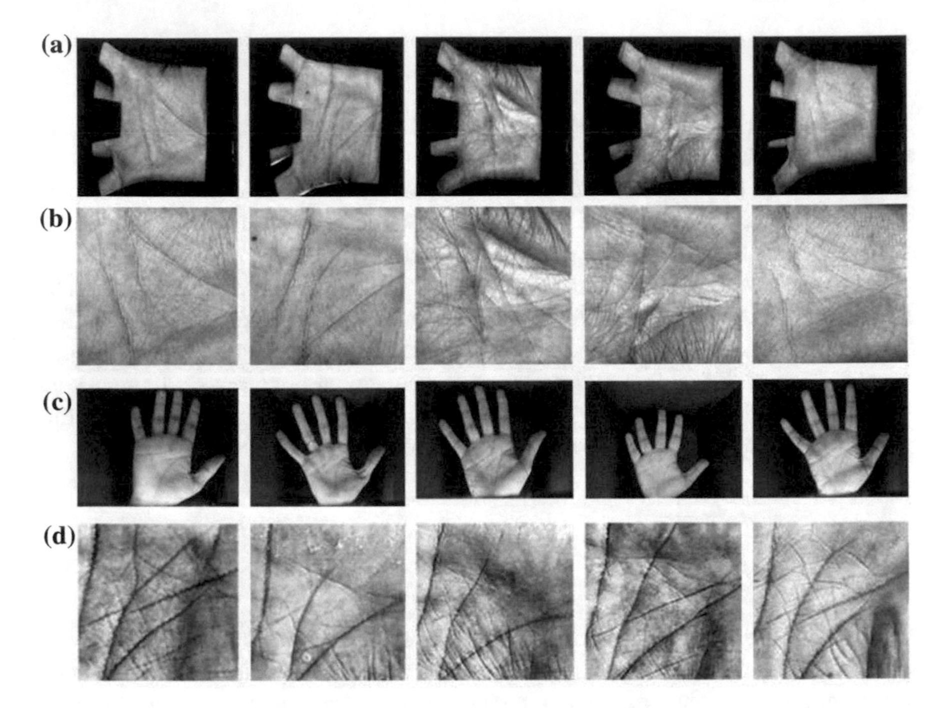

Fig. 2 Sample images from PolyU palmprint and IIT Delhi palmprint database. **a** Sample palmprint images from PolyU palmprint database captured using special acquisition device in the controlled environment. **b** Region of interest extracted from the palmprint. **c** Few palmprint images from IIT Delhi dataset. **d** Segmented images from IIT Delhi palmprint dataset

4.1 Datasets

We have used publicly available palmprint and face databases in our experiments. As there is a difficulty in getting the palm and face image of the same individual, one-to-one mapping is done between the two datasets; i.e., palmprint of an individual selected from the palmprint database and face image of an individual from face database are paired together to form one class.

For palmprint, PolyU 2D palmprint [9] and IIT Delhi palmprint [10] datasets are used. In PolyU dataset, the images are captured using a controlled environment. These images are taken in two sessions with a time difference of around two months. We have used palmprint images from the first session for the training and that of second session for the testing purpose. The total number of classes available here are 374. Figure 2a shows the sample images from PolyU 2D palmprint database. Figure 2b shows the region of interest extracted from the original palmprint image. The IIT Delhi palmprint dataset consists of palm images of 230 persons. The left as well as right-hand palmprint images are captured separately. The total classes available here are 236. Figure 2c shows few images of this dataset. The segmented images are shown in Fig. 2d. For face, CASIA [11] and ORL [12] face image database is used. The

Fig. 3 Sample images from CASIA face and ORL dataset. **a** Sample face images from CASIA face image database. **b** Cropped face images containing only the face region. **c** Few face images from ORL dataset

CASIA dataset includes typical intra-class variations like difference in illumination, expression, pose, spectacles, imaging distance, etc. It contains face images of 500 individuals. Figure 3a shows some of the sample images whereas Fig. 3b shows the face region segmented from the original capture. The ORL face images shown in Fig. 3b include the image of 40 persons with different facial expressions.

4.2 Results

By default, the experiments are conducted considering 200 classes. We have considered 3 samples per class for training. The samples are selected randomly from the dataset. As ORL face database contains only 40 classes, for bimodal experimentation using this dataset, the classes chosen was 40. We have listed experimental results of face and palmprint biometrics in Table 1.

Table 1 Identification results using face and palmprint biometrics and comparative analysis

Type of fusion	Method	Dataset	Recognition rate
Score level [3]	Gabor features	Own dataset	98.25
Feature level [4]	Bit-plane decomposition	PolyU and ORL	91.20
		PolyU and Yale	93.73
Feature level [5]	PCA	PolyU and FERET	95.00
Feature level [8]	Log Gabor feature fusion using PSO	Own Dataset	99.42

Table 2 Performance of the proposed method with unimodal biometrics

Dataset	Verification		Identification rate (in %)
	GAR at 0.1 FAR	GAR at 0.01 FAR	
Face (CASIA) only	94.0	79.0	71.00
Face (ORL) only	100	99.0	95.00
Palmprint (PolyU) only	99.5	96.5	99.25
Palmprint (IITDelhi, Left) only	97.5	96.5	96.00
Palmprint (IITDelhi, Right) only	97.5	97.25	94.50

Table 3 Performance of the proposed method with the bimodality (face and palmprint biometrics)

Sl. no.	Dataset configuration for the proposed method	Verification results		Identification rate (in %)
		GAR at 0.1 FAR	GAR at 0.01 FAR	
1	CASIA Face and PolyU Palmprint	100	99.5	99.50
2	CASIA Face and IITDelhi-left Palmprint	99.5	96.5	98.00
3	CASIA Face and IITDelhi-right Palmprint	99.5	99.0	98.50
4	ORL Face and PolyU Palmprint	100	100	100
5	ORL Face and IITDelhi-left Palmprint	100	100	100
6	ORL Face and IITDelhi-right Palmprint	100	100	100

Identification: The proposed method using bimodal (PolyU Palmprint and CASIA face) biometrics shows identification rate of 99.50, which is better than the unimodal results; 99.25% for palmprint (PolyU) and 71.0% for face (CASIA) modalities. Similarly, identification results on IIT Delhi palmprint and ORL face datasets are also superior to their unimodal results. The results are shown in Tables 2 and 3.

Verification: The proposed method on PolyU Palmprint and CASIA face dataset shows good verification rate (genuine acceptance rate or GAR) of 99.5% at false

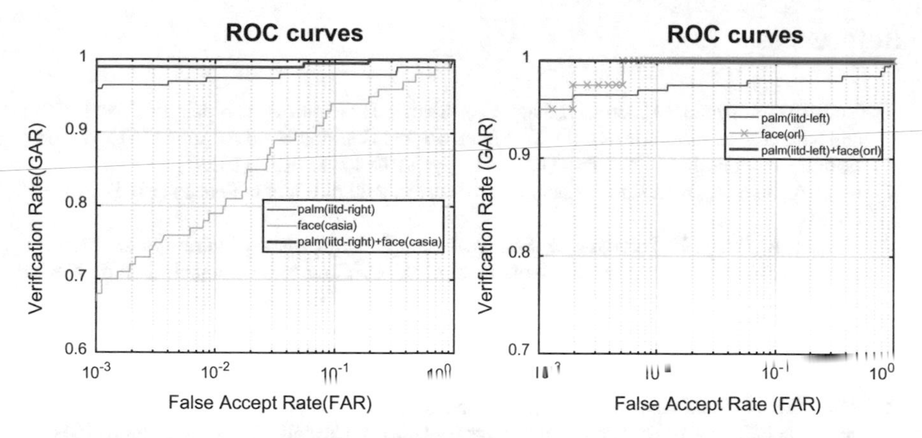

Fig. 4 Comparison of ROC curves of bimodal and unimodal biometrics. ROC curves for configuration 3 and 6 in Table 3 are shown

acceptance rate of 0.01 and 100% GAR at 0.1 FAR. This performance is better than that of the unimodal biometrics. The verification results of unimodal biometric data is shown in Table 2 and that of bimodal are shown in Table 3. The results on IIT Delhi palmprint with CASIA face are also promising. The bimodal verification GAR is 99.5% (at 0.1% FAR) and 99.0% (at 0.01% FAR). Tables 2 and 3 gives results of all experiments.

The receiver operating characteristic(ROC) curves are given in Fig. 4. The ROC curve illustrates the strength of the binary classifier due to the variation of the threshold. Among the six configurations given in Table 3, ROC curves for two configurations are shown in Fig. 4.

5 Conclusion

We have proposed a bimodal biometric identity assurance system that addresses both identification and authentication problems. The method uses palmprint and face modalities for feature fusion. Fusion is achieved by concatenation of the Gabor feature extracted from both palmprint and face images. The kernel Fisher analysis performed to obtain the class information and nonlinear projection is carried out for the matching samples. The comparative analysis of results has been carried out to that of unimodal biometric systems.

References

1. Noushath, S., Imran, M., Jetly, K., Rao, A., Kumar, G.H.: Multimodal biometric fusion of face and palmprint at various levels. In: 2013 International Conference on Advances in Computing, Communications and Informatics (ICACCI), pp. 1793–1798. IEEE (2013)
2. Ross, A., Jain, A.: Information fusion in biometrics. Pattern Recogn. Lett. **24**(13), 2115–2125 (2003)
3. Poinsot, A., Yang, F., Paindavoine, M.: Small sample biometric recognition based on palmprint and face fusion. In: Fourth International Multi-Conference on Computing in the Global Information Technology, 2009. ICCGI'09, pp. 118–122. IEEE (2009)
4. Ribaric, S., Fratric, I., Kis, K.: A biometric verification system based on the fusion of palmprint and face features. In: Proceedings of the 4th International Symposium on Image and Signal Processing and Analysis, 2005, ISPA 2005, pp. 12–17. IEEE (2005)
5. Ahmad, M.I., Mohamad, N., Isa, M.N.M., Ngadiran, R., Darsono, A.M.: Fusion of low frequency coefficients of DCT transform image for face and palmprint multimodal biometrics. In: 2017 3rd IEEE International Conference on Cybernetics (CYBCON), pp. 1–5. IEEE (2017)
6. Ahmad, M.I., Woo, W.L., Dlay, S.: Non-stationary feature fusion of face and palmprint multimodal biometrics. Neurocomputing **177**, 49–61 (2016)
7. Lee, T.Z., Bong, D.B.: Face and palmprint multimodal biometric system based on bit-plane decomposition approach. In: 2016 IEEE International Conference on Consumer Electronics-Taiwan (ICCE-TW), pp. 1–2. IEEE (2016)
8. Raghavendra, R.: PSO based framework for weighted feature level fusion of face and palmprint. In: 2012 Eighth International Conference on Intelligent Information Hiding and Multimedia Signal Processing (IIH-MSP), pp. 506–509. IEEE (2012)
9. PolyU Palmprint Database. http://www.comp.polyu.edu.hk/biometrics/
10. IIT Delhi Touchless Palmprint Database Version 1.0. http://web.iitd.ac.in/ajaykr/database/palm.htm
11. CASIA 2D Face Database v5. http://biometrics.idealtest.org/
12. ORL Face Database. www.cl.cam.ac.uk/research/dtg/attarchive/facedatabase.html
13. Kong, W.K., Zhang, D., Li, W.: Palmprint feature extraction using 2-d gabor filters. Pattern Recogn. **36**(10), 2339–2347 (2003)
14. Štruc, V., Pavešić, N.: The complete gabor-fisher classifier for robust face recognition. EURASIP J. Adv. Signal Process. **2010**, 31 (2010)
15. Liu, C.: Capitalize on dimensionality increasing techniques for improving face recognition grand challenge performance. IEEE Trans. Pattern Anal. Mach. Intell. **28**(5), 725–737 (2006)

Circular Map Pattern Spectrum—An Accurate Descriptor for Shape Representation and Classification

Bharathi Pilar and B. H. Shekar

Abstract In this work, we propose a shape descriptor namely, Circular Map Pattern Spectrum (*CMPS*) for shape representation and classification. The pattern spectrum describes the local figure thickness of the skeleton points with reference to the shape contour. Generally, we say that skeletons are sensitive to contour noise resulting in generation of spurious skeleton branches and hence influences negatively on the performance of the classifier system. Hence, we propose to compute pattern spectrum by using shape contour. These features are invariant to rotation and scale of the object. Further, to make improvement of the accuracy of the classifier, we have explored the 'combined classifier' paradigm where Block-wise binary pattern is combined with CMPS at decision level. The experimentation results on standard shape datasets reveals the performance of the proposed approach. Comparative analysis with some of the existing approaches shows the performance of the proposed approach in terms of the accuracy of the classification.

Keywords Mathematical morphology · Circular map pattern spectrum
Block-wise binary pattern · Combined classifier · Shape classification

1 Introduction

Rapid development in the technology has made extensive usage of automated intelligent systems in a wide range of applications such as industrial image processing, medical applications, defence and biometrics. Design of automated intelligent systems demands development of good machine learning algorithms. The process of machine learning involves object extraction, representation and classification. Shape-based representation scheme is one of the popular object representation schemes available

B. Pilar (✉)
Department of Computer Science, University College, Mangalore, Karnataka, India
e-mail: bharathi.pilar@gmail.com

B. H. Shekar (✉)
Department of Computer Science, Mangalore University, Mangalore, Karnataka, India
e-mail: bhshekar@gmail.com

P. Nagabhushan et al. (eds.), *Data Analytics and Learning*,
Lecture Notes in Networks and Systems 43,
https://doi.org/10.1007/978-981-13-2514-4_18

in the literature. Shape representation generally looks for dominant shape features based on shape contour or shape region. Some methods considers the entire shape to extract global features which are known as global structural methods and some other methods which partitions the shape into primitives, from that extracts local features for shape representation, which is named as structural approaches [30]. The contour-based methods include distance sets [10], elastic matching [2], robust symbolic representation [7], contour points distribution histogram (CPDH) [23], Height Functions [28] etc., whereas region-based methods include Fourier descriptor [3], Zernike Moments [12], Shock graphs [24, 25], Bone graphs [16], Skeletal shape abstraction [8], Amplitude-only log radon transform [11], Hybrid neuro-fuzzy descriptor [6], etc. There are few methods that does the hierarchy of segmentation of shape boundary which represents boundary segments by adopting binary tree. Few examples for such methods include Hierarchical Procrustes Matching (HPM) [18], Shape tree [9], etc. There are techniques which combine different types of features either at feature level or decision level for the purpose of better retrieval accuracy. Learning manifold approach is one such method [5]. It is a fusion of two categories of dissimilarity matrices of shape descriptors resulting in improvement in retrieval results. We are motivated to design a combined classifier which combines Circular Map-Based Pattern Spectrum (CMPS) that captures topological/structural properties and Block-wise Binary Pattern (BBP) that captures regional/local for shape representation and classification. The next sections of this paper is organized as follows. Section 2 contains brief explanation of the proposed approach. Section 3 contains details about experimentations followed by the discussion of results and Sect. 4 is for conclusion.

2 Proposed Shape Descriptor

In our work, we propose a Circular Map-Based Pattern Spectrum which is a variant of Maragoes Pattern Spectrum [17]. The pattern spectrum describes the local figure thickness of the skeleton points with respect to the object boundary. Skeletons are sensitive to contour noise resulting in generation of spurious skeleton branches and hence influences negatively on the performance of the classifier system. In our work, we have used shape contour directly to compute pattern spectrum. To reduce the effect of contour noise, only fixed n points on the contour in the anticlockwise direction are taken. In order to enhance the accuracy of the classifier, we have done decision-level fusion of CMPS with BBP. The BBP captures the local information of the shape. These features are invariant to rotation and scale of the object. The CMPS feature vectors are matched using Euclidean distance. The BBP features are matched using Earth Movers Distance (EMD) metric. In order to improve the classifier accuracy, we have explored the 'combined classifier' paradigm where BBP is combined with CMPS at decision level. The following section presents the proposed approach in detail.

2.1 Circular Map Pattern Spectrum (CMPS)

The basic concept of pattern spectrum was proposed by Maragoas [17] based on Serra's Mathematical morphology. Pattern spectrum concentrates on spatial distribution of the skeleton points with respect to the object boundary. It is a rotation and scale-invariant feature and can be easily represented as histogram. In this method, in order to obtain pattern spectrum, shape is subjected to the process of skeletonization. Later, for every point x on the skeleton, the radius of the maximum circle which can be inscribed inside the shape contour with x as the centre is determined. Each skeleton point associated by radius function is represented in the form of distribution function which forms the feature vectors.

Consider Fig. 1 for illustration. The shape of the given object is shown in Fig. 1a. The skeleton of the shape, shown in Fig. 1b, is obtained by applying distance transform method [14] on Fig. 1a. Figure 1c shows the maximal discs drawn for skeleton points. The thickness map of the skeleton points is represented in the form of 10 bin histogram and is shown in Fig. 1d.

The above method has certain drawbacks. The method computes the skeleton of the shape in order to obtain pattern spectrum. The skeleton is highly sensitive to contour irregularities. The contour noise induces redundant branches in the skeleton which needs to be pruned. The process of skeletonization and skeleton pruning process results in higher time complexity. Also, the skeleton pruning method suffers from the problem of either removal of relevant branches or retaining irrelevant branches that are introduced due to noise. Hence, in our approach, we use directly the contour of the shape instead of skeleton. To reduce the effect of noise, only fixed n points on the contour in the anticlockwise direction are taken. It is found by experimentation that taking $n = 150$ is optimal for standard shape datasets. The spatial

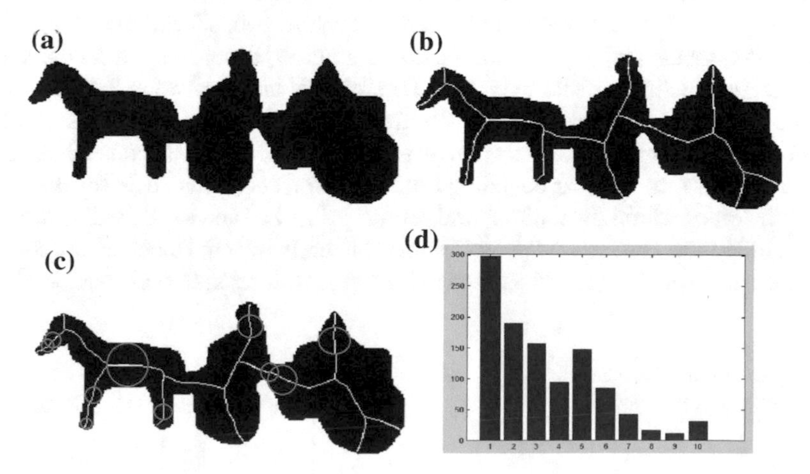

Fig. 1 **a** Given shape; **b** the skeleton obtained due to distance transform method; **c** the maximal disc inscribed on the skeleton points; **d** the histogram representation of the radius of discs with respect to skeleton points

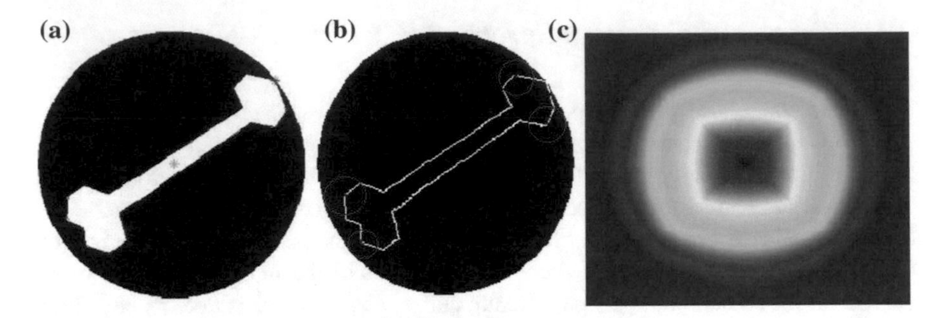

Fig. 2 The steps involved in computing circular pattern spectrum of a shape

distribution of these n contour points with respect to the circular boundary box is used as a feature vector of the shape. For each contour point x, a radius function is associated which is the radius of the maximal disc, with x as the centre and radius are equal to the minimum distance from nearest boundary point of the circular boundary box. The problem with the rectangular bounding box is its sensitivity to the rotation of the shape. So, we have used circular bounding box which is rotational invariant as shown in the Fig. 2.

In Circular Map-Based Pattern Spectrum (CMPS), the feature extraction process is done as follows. The shape is fitted with the circular boundary box. The contour of the shape is traversed in anticlockwise direction and only a fixed n contour points of the shape is taken. To avoid inherent problems with skeleton we have used contour itself to compute the pattern spectrum. The fixed n points are taken in order to avoid the problem of scaling. The radius function of each contour point, with respect to the circular boundary box is obtained. The radius function value for a pixel x is the radius of the maximal disc inscribed inside the object touching the circular boundary box with centre x. The radius value for fixed n contour points forms the *CMPS* of the shape. To compute radius values for the contour points, distance transform map [14] based on the Euclidean distance is used. The forward mask shown in Fig. 3a and 90° rotated backward mask given in Fig. 3b are used as structuring elements. A block which is of the same size as the circular boundary box of the shape is taken. Let r be the radius of the circular bounding box. The block contains all 1s for the points which lie on or within the radius r and all the points lie outside the radius are 0, as shown in Fig. 4a. The forward mask is applied linearly over the block, in the forward direction, i.e. from left to right and top to bottom, resulting in forward pass as shown

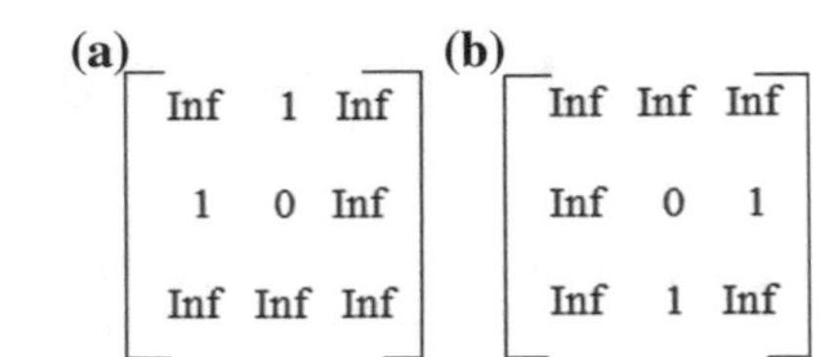

Fig. 3 Structuring elements. **a** Forward mask; **b** backward mask

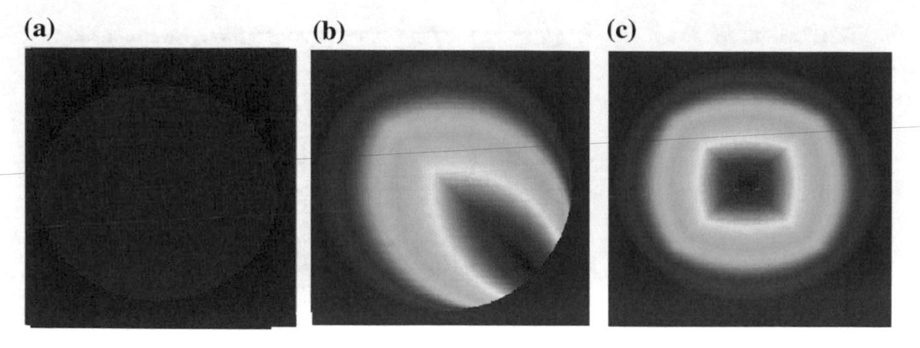

Fig. 4 **a** Block B; **b** result of the forward pass on B; **c** result of the backward pass on B

in the Fig. 4b. The backward mask is applied over the image in reverse direction from right to left and bottom to top resulting in backward pass (Fig. 4c). Now, the block contains largest distance value at the centre with the distance values decremented towards the boundary of the circle. The block is aligned with the shape contour and the values in the distance map, corresponding to the contour point, represents the radius function. This radius vector forms the feature vector for the given shape.

2.2 *Feature Extraction*

The process of computing CMPS is given below.

1. A circular boundary box is fitted to the shape to avoid irrelevant portion of the shape.
2. The shape boundary is extracted and a fixed $N = 150$ countour points are considered for processing.
3. The radius function of each contour point with respect to the circular boundary box is obtained. The radius function value for a pixel x is the radius of the maximal disc inscribed inside the circular boundary box touching the circular boundary box with centre x. The radius value for fixed n contour points forms the CMPS of the shape.
4. To obtain the radius values for the contour points, distance transform map [14] based on the Euclidean distance is used. The forward mask and 90° rotated backward mask is used to obtain the distance map of the circular block. The distance transform map of the block is aligned with the shape contour and the values in the distance map corresponding to the contour points represents the radius function. This radius function forms the feature vector.

The above steps are repeated for all the training set samples to form the CMPS knowledge base for training set.

The process of Block-wise binary pattern-based feature extraction is done as follows. Given the shape of the object, we divide the shape into a set of non overlapping

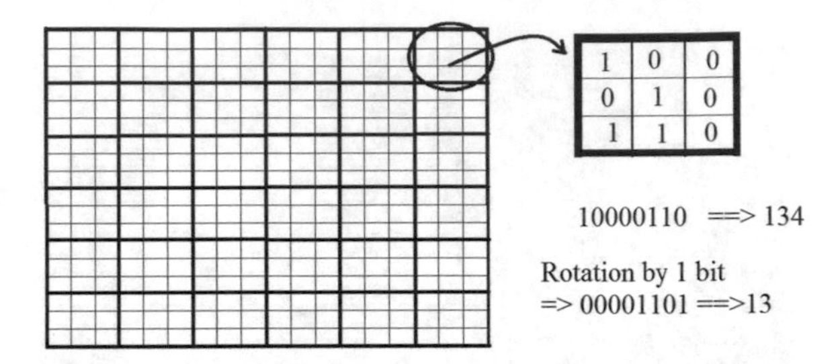

Fig. 5 Block-based binary pattern on a shape

blocks of size 3 × 3. Let x be the central pixel under processing. In each block of 9 pixels, we compute a new value for x based on its surrounding 8 neighbours. The 8 neighbouring pixels are traversed in clockwise direction forming a binary stream of 8 bits as shown in the Fig. 5. The decimal equivalent of the binary stream is computed. This value is rotation variant and position dependent. In order to make it position independent and rotation invariant, one-bit right shift operation is performed on the binary stream, then we get another decimal value equivalent to this binary stream. The one-bit right shift operation is performed and the decimal equivalent of binary stream is recorded. The process is repeated till we get 8 decimal equivalents of binary stream. The minimum value which is rotation invariant will be the computed value for x. The 3 × 3 block is replaced by single pixel x with the computed value. This process is iterated for every 3 × 3 neighbourhood in the training shape. Now, the size of the new shape will be reduced to one-third of the original shape size, thus reducing the number of feature values to be processed. The shape matrix now consists of decimal values. This is converted to a 10 bin histogram and is stored for further processing. Every shape in the dataset undergoes similar procedure in order to yield BBP feature vectors.

2.3 *Classification*

We have used decision-level fusion strategy to combine CMPS and BBP based feature vectors. The Euclidean distance and Earth Movers Distance [20] metric are used to match CMPS and BBP features respectively. Let D_c and D_b be the distance matrices obtained after matching test samples with training samples using CMPS and BBP respectively. The decision-level fusion to compute the resultant matrix is done as follows.

$$D_R = D_c + \beta D_b \tag{1}$$

A set of experiments are performed on given standard dataset to fix the value of β to the corresponding dataset.

To carry out the training process and testing process, we have used leave- one-out strategy. Each shape in the dataset is matched with the remaining $n - 1$ shapes in the dataset to obtain distance matrix. If there are n shapes in the dataset, then we get $n \times n$ distance matrix with main diagonal zero representing self match.

3 Experiments Results and Discussions

The experimental results on MPEG-7, Kmia-216 and Kimia-99 are presented here. The Bull's eye score, Top *N* retrieval and Precision-Recall graphs are used as the evaluation metrics. The Bull's eye score is calculated by considering the top T retrievals, where $T = 2 * N$, N is the number of class samples. The leave- one-out strategy is used for testing.

The experimental results due to the proposed methodology on MPEG-7 dataset is given below. The MPEG-7 dataset consists of 1400 shape from 70 classes with 20 shape samples. Figure 6 shows different class samples of MPEG-7. The class-wise retrieval rate is presented in Fig. 7 and is also given in tabular form in Table 1. The

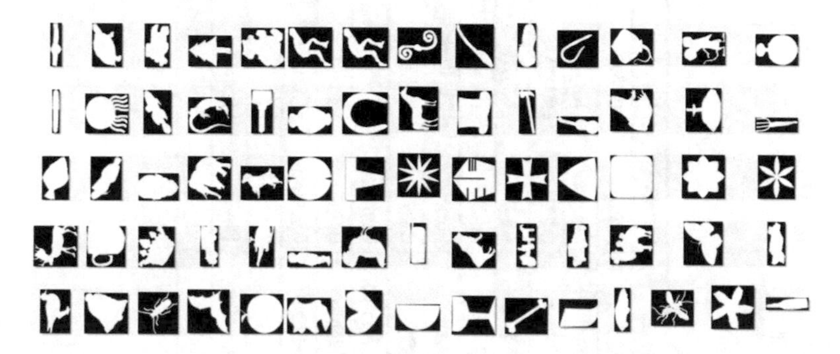

Fig. 6 MPEG-7 dataset: a sample shape from each class

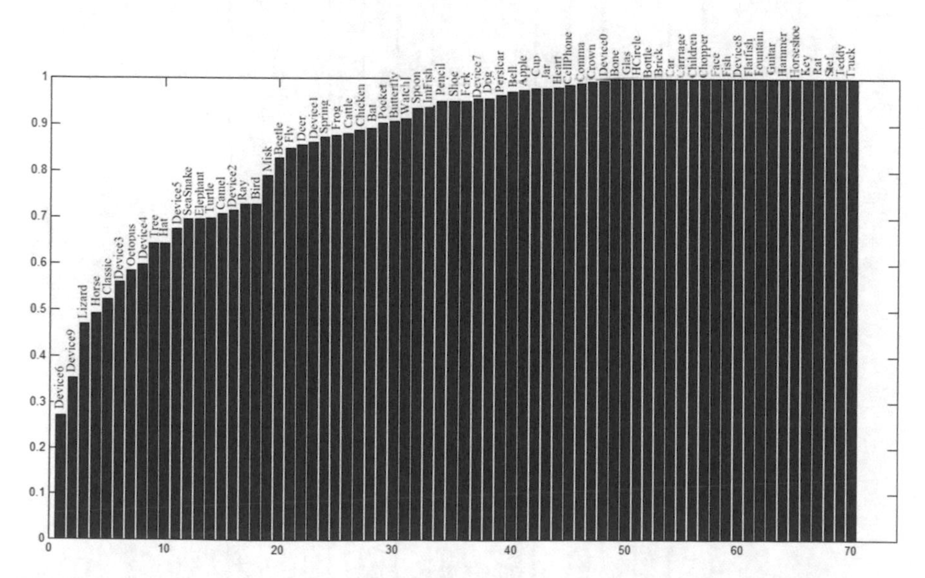

Fig. 7 Class-wise retrieval results for MPEG-7 dataset

Table 1 Class-wise retrieval results for *MPEG-7* dataset

Class	Device6	Device9	Lizard	Horse	Classic	Device3	Octopus	Device4	Tree	Hat
Accuracy	0.2725	0.3525	0.47	0.4925	0.525	0.56	0.585	0.5975	0.6425	0.6425
Class	Device5	SeaSnake	Elephant	Turtle	Camel	Device2	Ray.	Bird	Misk	Beetle
Accuracy	0.675	0.695	0.695	0.6975	0.7075	0.715	0.7275	0.7275	0.79	0.8275
Class	Fly	Deer	Devicel	Spring	Frog	Cattle	Chicken	Bat	Pocket	Butterfly
Accuracy	0.8475	0.855	0.8625	0.8725	0.8775	0.88	0.8875	0.8925	0.9025	0.9075
Class	Watch	Spoon	ImFish	Pencil	Shoe	Fork	Device7	Dog	Perslcar	Bell
Accuracy	0.9125	0.935	0.94	0.95	0.95	0.95	0.955	0.955	0.9625	0.97
Class	Apple	Cup	Jar	Heart	CellPhone	Comma	Crown	Device0	Bone	Glas
Accuracy	0.975	0.9775	0.9775	0.98	0.985	0.99	0.9925	0.995	1	1
Class	HCircle	Bottle	Brick	Car	Carriage	Children	Chopper	Face	Fish	Device8
Accuracy	1	1	1	1	1	1	1	1	1	1
Class	Flatfish	Fountain	Guitar	Hammer	Horseshoe	Key	Rat	Stef	Teddy	Truck
Accuracy	1	1	1	1	1	1	1	1	1	1

Table 2 Retrieval rate (Bull's eye score) of MPEG-7 dataset [13]—a comparative analysis

DataSet	MPEG-7
Proposed approach	**86.48**
HPM [18]	86.35
Symbolic representation [7]	85.92
IDSC + DP [15]	85.40
MCC + shape complexity [1]	84.93
Polygonal multiresolution [2]	84.33
Fixed correspondence [26]	84.05
Optimized CSS [19]	80.54
Generative model [27]	[illegible]
Skeletal contexts [29]	79.92
PS + LBP [22]	79.38
Distance set [10]	78.38
Aligning curves [21]	78.16
CPDH [23]	76.56
SC + TPS [4]	76.51
SC [4]	76.51
CSS [19]	75.44
Visual parts [13]	76.45

retrieval rate in the form of Bull's eye score on MPEG-7 is presented in Table 2. We have also listed the results due to some of the well-known methods in Table 2. With this, we have also made an comparative analysis with *IDSC* [15] and is shown in Fig. 8.

Kimia-99 shape dataset consists of 9 classes with 11 samples in each class. The shapes in Kimia-99 dataset are shown in Fig. 9. The top-*n* where n = 10, matching shapes are presented in Table 3. A comparative analysis with some well- known methods is also performed and is presented in Table 3.

Kimia-216 shape dataset consists of 18 classes with 12 samples in each class. One such sample from each class is shown in Fig. 10. The *Top-11* closest matches obtained due to the proposed methodology is shown in Table 4. A comparative analysis with some well-known methods is also performed and is presented in Table 4.

The performance of the proposed approach in terms of Bull's eye score for various standard shape datasets is shown in Table 5.

	Query	Top 12 retrieval of query shape for MPEG-7		Query	Top 12 retrieval of query shape for MPEG-7
IDSC			IDSC		
Proposed Approach			Proposed Approach		
IDSC			IDSC		
Proposed Approach			Proposed Approach		
IDSC			IDSC		
Proposed Approach			Proposed Approach		
IDSC			IDSC		
Proposed Approach			Proposed Approach		
IDSC			IDSC		
Proposed Approach			Proposed Approach		
IDSC			IDSC		
Proposed Approach			Proposed Approach		
IDSC			IDSC		
Proposed Approach			Proposed Approach		

Fig. 8 Top 12 retrieved shapes of query shape by IDSC approach [15] and proposed approach

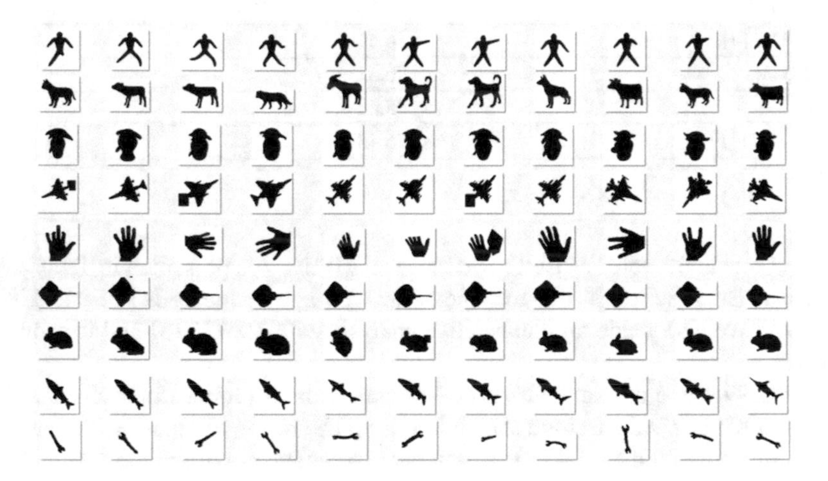

Fig. 9 Kimia 99 dataset: 9 classes with 11 samples in each class

Table 3 Top 10 closest matching shapes on Kimia's 99 dataset—a comparison

Approach	1st	2nd	3rd	4th	5th	6th	7th	8th	9th	10th	Total
SC [4]	97	91	88	85	84	77	75	66	56	37	756
CPDH + EMD (Eucl) [23]	96	94	94	87	88	82	80	70	62	55	808
CPDH + EMD (shift) [23]	98	94	95	92	90	88	85	84	71	52	849
Gen model [27]	99	97	99	98	96	96	94	83	75	48	885
Proposed approach	**99**	**98**	**96**	**96**	**85**	**88**	**82**	**78**	**74**	**69**	**865**

Fig. 10 Kimia 216 dataset : a sample shape from each class

Table 4 Top 11 closest matching shapes on Kimia's 216 dataset—a comparison

Approach	1st	2nd	3rd	4th	5th	6th	7th	8th	9th	10th	11th	Total
SC [4]	214	209	205	197	191	178	161	144	131	101	78	1809
CPDH + EMD(Eucl) [23]	214	215	209	204	200	193	187	180	168	146	114	2030
CPDH + EMD(shift) [23]	215	215	213	205	203	204	190	180	168	154	123	2070
PS + LBP [22]	216	209	205	195	195	197	188	180	179	163	152	2079
Proposed approach	**214**	**212**	**207**	**203**	**200**	**196**	**194**	**184**	**173**	**171**	**134**	**2088**

Table 5 Bull's eye score on MPEG-7, Kimia-216 and Kimia-99 datasets

DataSet	Bull eye score
Kimia-99	94.58
Kimia-216	93.36
MPEG-7	86.48

4 Conclusion

We made an attempt to design a shape descriptor namely, Circular Map Pattern Spectrum (*CMPS*) for shape representation and classification. The skeletal pattern spectrum are highly sensitive to contour noise which influences negatively on the performance of the classifier system. Hence, in our work, we have used shape contour directly to compute pattern spectrum. These features are rotation invariant and scale invariant of the object. Further, in order to improve the classifier accuracy, we have explored the 'combined classifier' paradigm where Block-wise binary pattern is combined with CMPS at decision level. The output of the experiments on standard datasets reveals the performance of the proposed approach. Comparative analysis with some of the well-known approaches demonstrates the classification accuracy of the proposed approach.

References

1. Adamek, T., O'Connor, N.E.: A multiscale representation method for nonrigid shapes with a single closed contour. IEEE Trans. Circuits Syst. Video Technol. **14**(5), 742–753 (2004)
2. Attalla, E., Siy, P.: Robust shape similarity retrieval based on contour segmentation polygonal multiresolution and elastic matching. Pattern Recognit. **38**(12), 2229–2241 (2005)
3. Bartolini, I., Ciaccia, P., Patella, M.: Warp: accurate retrieval of shapes using phase of Fourier descriptors and time warping distance. IEEE Trans. Pattern Anal. Mach. Intell. **27**(1), 142–147 (2005)
4. Belongie, S., Malik, J., Puzicha, J.: Shape matching and object recognition using shape contexts. IEEE Trans. Pattern Anal. Mach. Intell. **24**(4), 509–522 (2002)
5. Chahooki, M.A.Z., Charkari, N.M.: Learning the shape manifold to improve object recognition. Mach. Vis. Appl. **24**(1), 33–46 (2013)
6. Chauhan, P.C., Prajapati, G.I.: 2d basic shape detection and recognition using hybrid neuro-fuzzy techniques: a survey. In: 2015 International Conference on Electrical, Electronics, Signals, Communication and Optimization (EESCO), pp. 1–5. IEEE (2015)
7. Daliri, M.R., Torre, V.: Robust symbolic representation for shape recognition and retrieval. Pattern Recognit. **41**(5), 1782–1798 (2008)
8. Demirci, M.F., Shokoufandeh, A., Dickinson, S.J.: Skeletal shape abstraction from examples. IEEE Transa. Pattern Anal. Mach. Intell. **31**(5), 944–952 (2009)
9. Felzenszwalb, P.F., Schwartz, J.D.: Hierarchical matching of deformable shapes. In: IEEE Conference on Computer Vision and Pattern Recognition, pp. 1–8. IEEE (2007)
10. Grigorescu, C., Petkov, N.: Distance sets for shape filters and shape recognition. IEEE Trans. Image Process. **12**(10), 1274–1286 (2003)
11. Hasegawa, M., Tabbone, S.: Amplitude-only log radon transform for geometric invariant shape descriptor. Pattern Recognit. **47**(2), 643–658 (2014)
12. Kim, W.Y., Kim, Y.S.: A region-based shape descriptor using zernike moments. Signal Process.: Image Commun. **16**(1), 95–102 (2000)
13. Latecki, L.J., Lakamper, R., Eckhardt, T.: Shape descriptors for non-rigid shapes with a single closed contour. In: IEEE Conference on Computer Vision and Pattern Recognition, 2000. Proceedings, vol. 1, pp. 424–429. IEEE (2000)
14. Latecki, L.J., Li, Q.n., Bai, X., Liu, W.Y.: Skeletonization using SSM of the distance transform. In: IEEE International Conference on Image Processing, vol. 5, pp. V–349. IEEE (2007)
15. Ling, H., Jacobs, D.W.: Shape classification using the inner-distance. IEEE Trans. Pattern Anal. Mach. Intell. **29**(2), 286–299 (2007)
16. Macrini, D., Siddiqi, K., Dickinson, S.: From skeletons to bone graphs: medial abstraction for object recognition. In: IEEE Conference on Computer Vision and Pattern Recognition, 2008. CVPR 2008, pp. 1–8. IEEE (2008)
17. Maragos, P.: Pattern spectrum and multiscale shape representation. IEEE Trans. Pattern Anal. Mach. Intell. **11**(7), 701–716 (1989)
18. McNeill, G., Vijayakumar, S.: Hierarchical procrustes matching for shape retrieval. In: IEEE Computer Society Conference on Computer Vision and Pattern Recognition, vol. 1, pp. 885–894. IEEE (2006)
19. Mokhtarian, F., Abbasi, S., Kittler, J.: Efficient and robust retrieval by shape content through curvature scale space. Ser. Softw. Eng. Knowl. Eng. **8**, 51–58 (1997)
20. Rubner, Y., Tomasi, C., Guibas, L.J.: The earth mover's distance as a metric for image retrieval. Int. J. Comput. Vis. **40**(2), 99–121 (2000)
21. Sebastian, T.B., Klein, P.N., Kimia, B.B.: On aligning curves. IEEE Trans. Pattern Anal. Mach. Intell. **25**(1), 116–125 (2003)
22. Shekar, B.H., Pilar, B.: Shape representation and classification through pattern spectrum and local binary pattern–a decision level fusion approach. In: Fifth International Conference on Signal and Image Processing (ICSIP), pp. 218–224. IEEE (2014)
23. Shu, X., Wu, X.J.: A novel contour descriptor for 2d shape matching and its application to image retrieval. Image Vis. Comput. **29**(4), 286–294 (2011)

24. Siddiqi, K., Kimia, B.B., Tannenbaum, A., Zucker, S.W.: Shapes, shocks and wiggles. Image Vis. Comput. **17**(5), 365–373 (1999)
25. Siddiqi, K., Shokoufandeh, A., Dickinson, S.J., Zucker, S.W.: Shock graphs and shape matching. Int. J. Comput. Vis. **35**(1), 13–32 (1999)
26. Super, B.J.: Retrieval from shape databases using chance probability functions and fixed correspondence. Int. J. Pattern Recognit. Artif. Intell. **20**(08), 1117–1137 (2006)
27. Tu, Z., Yuille, A.L.: Shape matching and recognition–using generative models and informative features. In: Computer Vision-ECCV 2004, pp. 195–209. Springer, Berlin (2004)
28. Wang, J., Bai, X., You, X., Liu, W., Latecki, L.J.: Shape matching and classification using height functions. Pattern Recognit. Lett. **33**(2), 134–143 (2012)
29. Xie, J., Heng, P.A., Shah, M.: Shape matching and modeling using skeletal context. Pattern Recognit. **41**(5), 1756–1767 (2008)
30. Zhang, D., Lu, G.: Shape-based image retrieval using generic fourier descriptor. Signal Process.: Image Commun. **17**(10), 825–848 (2002)

Features Fusion for Retrieval of Flower Videos

D. S. Guru, V. K. Jyothi and Y. H. Sharath Kumar

Abstract This paper presents a Flower Video Retrieval System (FVRS). An algorithmic model is proposed for the retrieval of natural flower videos using Local Binary Pattern (LBP) and Gray-Level Co-occurrence Matrix (GLCM) as texture features and Scale-Invariant Feature Transform (SIFT) features. For a given query flower video, the system retrieves similar videos from the database using Multi-class Support Vector Machine (MSVM). Euclidean distance is used as a proximity measure. The proposed model has been verified on keyframes selected from cluster-based approaches from natural flower videos. Our own dataset is used for the experimentation, which consists of 1919 videos belonging to 20 classes of flowers. It has been observed that the proposed model generates good retrieval results from the fusion of the features SIFT, GLCM, and LBP.

Keywords Multi-class support vector machine · Gray-level co-occurrence matrix Local binary pattern · Scale-invariant feature transform · Flower videos retrieval

1 Introduction

Developing a systematic video retrieval system is a dynamic field of research. Due to the ease of availability of recent video capturing devices such as cameras and mobiles, users could capture a huge number of videos with the large storage

D. S. Guru · V. K. Jyothi (✉)
Department of Studies in Computer Science, University of Mysore,
Manasagangotri, Mysore 570006, India
e-mail: jyothivk.mca@gmail.com

D. S. Guru
e-mail: dsg@compsci.uni-mysore.ac.in

Y. H. Sharath Kumar
Department of Information Science, Maharaja Institute of Technology,
Mysore (MITM), Mandya 571438, India
e-mail: sharathyhk@gmail.com

P. Nagabhushan et al. (eds.), *Data Analytics and Learning*,
Lecture Notes in Networks and Systems 43,
https://doi.org/10.1007/978-981-13-2514-4_19

volume. Developing a flower video retrieval system is a specific field with a number of applications.

Retrieval of flower videos has tremendous applications such as searching patent flower videos in the database [1], interest in users for knowing the flower names for medicinal use, decoration and for cosmetics, etc. Floriculture has become one of the important commercial traders in agriculture [2] due to the increase in the demand for flowers. Floriculture trades, nursery, marketing flowers and flower plants, seed production, and extracting oils from flowers [2]. In such cases, automation of retrieval of flower video plays a vital role.

Designing a system for retrieval of flower videos is a challenging task when the capturing of flower video is in natural environment with the movement of the camera or mobile devices. Flowers in the videos pose a number of challenges such as illumination vary with the weather conditions, viewpoint variation, with large variation in scaling, occluding of flowers, and multiple instances of the flowers in the videos [3]. Due to the wind, there is a movement of flowers in videos causes a small amount of blurriness and the clarity may also reduce.

2 Related Works

There is an immense increase in the quantity of video information. Method to search such a huge volume of data should be effective and simple [4]. There exists a various form of video information such as audio, text, and visual objects, etc. Flower video is also one such video where we consider flowers as objects. In this section, we discuss the literature review of some of the video retrieval works.

Video can be retrieved based on query types where query can be an example, a sketch, and an object. From a given video or image, query-by-example extracts low-level features. Query-by-sketch extracts features from sketches and an image of the object is used as a query in query by objects, to find, and retrieve the similar videos [5]. In [4], the authors proposed an approach to retrieve similar clips from a video using the query by example with the feature Local Binary Pattern Variance and Euclidean distance to find the similar clips in the video. In [6], the authors proposed a system to retrieve videos using local features and fuzzy k-Nearest Neighbor with two distance measures such as Euclidean and Manhattan to find the similarity. In [7], the authors proposed content-based video retrieval system to retrieve videos from the database based on query input from the user using support vector machine classifier, it supports for two-class problem. In [8], Scale-Invariant Feature Transform features are encoded with temporal information to generate temporal concentration SIFT for large-scale video retrieval. To find the highest similar video, an ensemble similarity-based method for video retrieval has been proposed in [9] using Gaussian model. In [10], the authors proposed a framework for retrieval of videos using SVMs and relevance feedback based on the Fisher Kernel. In [11], after shot boundary detection, color structure descriptors and the edge histogram descriptor are used to characterize the keyframes and the similarity between keyframes is calculated using dynamic-weighted feature similarity.

From the literature, we found that there is no attempt, on retrieval of flower videos, hence we are motivated to develop a system to retrieve flower videos. Flowers in videos pose some challenges such as occlusions, a large variation in lighting effects, viewpoint variations, multiple flowers, cluttered environment, etc. Videos are collected with different seasons such as summer, winter, and rainy.

In this work, the videos are retrieved based on the query by video using the texture and Scale-Invariant Feature Transform features and the Multiple Support Vector Machine (Multi-class SVM) classifier. The proposed system is presented in the following sections.

3 Proposed Model

The proposed model retrieves similar videos using query-by-video framework among various video retrieval systems. The proposed method involves four stages, namely keyframe selection, segmentation, feature extraction, generation of support vectors and finally similarity matching to retrieve the similar videos. Figure 1 shows the block diagram of the proposed model and the stages are explained in the subsequent sections.

3.1 Keyframe Selection

A video contains redundant information which increases computational complexity. To reduce the maximum amount of computational burden, it needs to select most

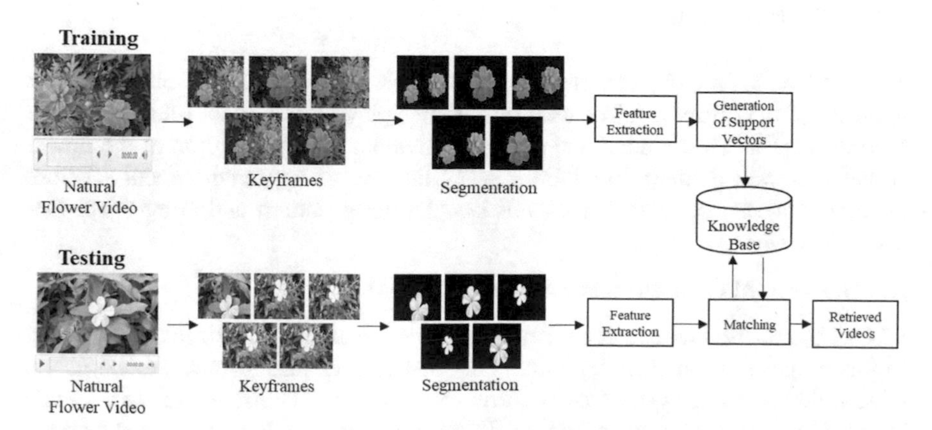

Fig. 1 Block diagram of the proposed Flower Video Retrieval System (FVRS)

representative frames to represent a video, which contains necessary information of the video. This helps the user to retrieve videos of their interest. In our proposed work, we represent the flower videos with a set of keyframes. In [12], we have proposed an algorithmic model to select a set of keyframes in flower videos using cluster based approaches. The model [12], extracts color features such as red, blue, and HSV, a texture feature GLCM and a gradient feature Histogram of Oriented Gradients and entropy feature. The similar frames are grouped using Gaussian Mixture Model and k-means clustering methods. The centroids are selected as keyframes. Finally, for a video, the set of keyframes with high-fidelity value is selected as the final set of keyframes.

3.2 Segmentation

After keyframe selection, the next step in the proposed video retrieval system is the flower region segmentation. Isolating an object of interest from other objects or from its background is called segmentation. Algorithm should stop segmentation process when the object of interest is isolated [13]. We segment the flower region using statistical region merging proposed in [14].

3.3 Feature Extraction

Feature extraction is an important and essential task in video retrieval to match the similar videos. Different features viz., textural features and scale-invariant features are extracted from keyframes of the video and are explained in the following sections.

3.3.1 Textural Features

Textural features are object surface visual features that are independent of color or intensity of the object [5]. The textural features play an important role to describe a flower region. They contain the information about the organization of the flower surfaces as well as their correlations with the surrounding environment. Texture features that we use in this work are Local Binary Pattern and Gray-Level Co-occurrence Matrix.

A. **Gray-Level Co-occurrence Matrix (GLCM)**

GLCM can be used to extract statistical information about the texture of the flower. In this proposal, from each keyframe, GLCM-based texture features are extracted. GLCM defines the gray-level co-occurrence of values in keyframe. In [15], 14 different statistical features are proposed. These are used for extracting texture features from segmented flower region of each keyframe. We represent these 14 features as a feature vector.

B. **Local Binary Pattern (LBP)**

LBP is one popularized local feature extraction technique for texture description. In [16], the authors presented the LBP texture feature of 3×3 neighborhood. By the value of center pixel, the pixels of eight neighbors are thresholded. To obtain the LBP value of the center pixel, thresholded binary values are weighted by powers of two and summed up. In [16], the authors originally proposed Local Binary Patterns, the idea behind LBP concatenates the binary gradient directions that are generated. It extracts information that is invariant to local grayscale variations in the keyframe. It is computed at each pixel location, considering the values of a small circular neighborhood with radius r pixels around the value of a central pixel *Cp*. Formally, the LBP operator is defined as follows [17].

$$LBP(P_{kfs}, r) = \sum_{pl=0}^{pl-1} S(q_{pl}, C_p)2^{pl} \tag{1}$$

where pl is the number of pixels in the neighborhood, r is the radius and

$$s(x) = \begin{cases} 1, & if\, x \geq 0 \\ 0 & otherwise \end{cases}.$$

3.3.2 Scale-Invariant Feature Transform (SIFT)

SIFT is an efficient technique to extract local features. It detects local points in a segmented flower region of the keyframe with different angles and it provides description of detecting points which are invariant to geometric transformations namely, rotation, scaling, and translation. In [17], the author proposed SIFT which has a feature extraction method. SIFT detects local patterns which exist in various views of keyframes and gives a description of these patterns which are invariant to geometric transformations. Then, SIFT assigns orientation and direction property on detecting positions of the key point locations in the keyframe. Then, it generates unique descriptor, it is constructed with a kernel of the 4×4 histogram of 8 bins [17]. These histograms compute the magnitude and.

Direction of the gradient is present around the feature point in the region of 16×16 pixels. The results of the histograms represented in the form of descriptors. These feature descriptors are normalized using Discrete Cosine Transform (DCT). The normalized features are used to generate a feature vector.

3.4 Multi-class Support Vector Machine

Support vector machine is a computationally significant tool for supervised learning [18], is extensively used in classification and retrieval problems. The fundamental idea of SVM classifier is to obtain the optimal separating hyperplane between two classes. The optimal hyperplane is defined as the maximum margin between training samples that are closest to the hyperplane. Initially, SVM is a two-class problem, it generates the support vectors to separate two classes that are shown in Fig. 2a. In a meticulous problem, it takes more time and it is difficult to generate support vectors to separate the multiple classes. Generating support vectors for each iteration takes more time. The idea we have used in the proposed work is initially we train the model by generating support vectors for all "*n*" classes shown in Fig. 2b. These are the vectors separates one class for all other classes, then predict the query matches to which class videos using trained models.

Let $C_1, C_2, \ldots, C_n$ be n number of classes.

Let $C_1 = s_1, s_2, \ldots, s_m,\ C_2 = s_1, s_2, \ldots, s_m,\ C_n = s_1, s_2, \ldots, s_m$, are the support vectors of the above classes.

In general,

$$C_i = \sum_{k=1}^{n-1} \sum_{j=1}^{m} c_k s_j \tag{2}$$

where C_i consists a set of support vectors s_j, separates nth class from all other $(n-1)$ classes.

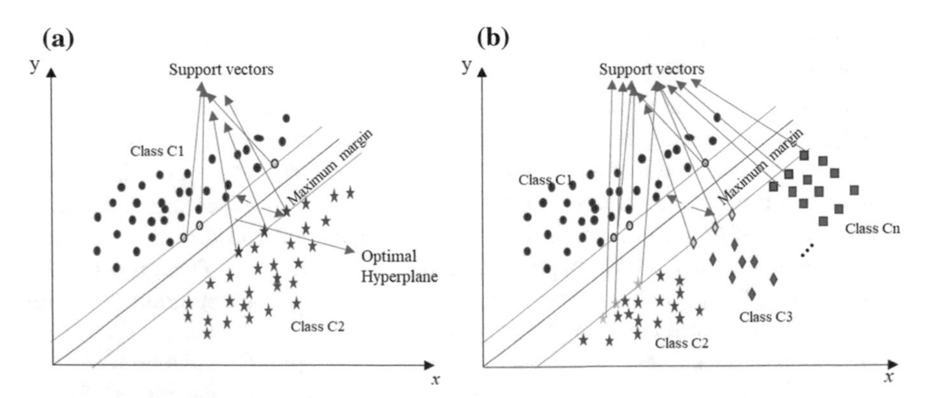

Fig. 2 **a** Two-class SVM. **b** Multi-class SVM

3.5 Algorithm of the Proposed Model

Step 1: Convert video V_i into frames and preprocess the frames to resize and remove noise

$$V_i = \{f_1, f_2, f_3, \ldots, f_n\} \tag{3}$$

Step 2: Select keyframes

$$Kf_{c1}, Kf_{c2}, Kf_{c3}, \ldots, Kf_{cm} \tag{4}$$

Step 3: Segment the flower regions from keyframes.
Step 4: Generate the feature vectors for segmented keyframes.

$$FM = [\partial_1, \partial_2, \partial_3, \ldots, \partial_p]$$

Step 5: Generation of Support Vectors.
Step 6: Retrieve the videos matched to query videos.

$$sim(TS_{KF_i}, TN_{KF_j}) = dist(TS_{KF_i}, TN_{KF_j}) \tag{5}$$

where

TS_{KF_i} is the feature vector of keyframes of a query video,
TN_{KF_j} is the feature vector of keyframes of all videos in the database.

4 Dataset

Standard flower video dataset is not available, we have collected our own flower video dataset which contains 1919 videos with 20 different classes in which each class is of 35–160 videos. The videos are of 4–60 s. These flower videos are captured in different environments which poses to the challenges such as variation in viewpoint, illumination, multiple and partial instances and cluttered background. The devices used for capturing videos are Samsung Galaxy Grand Prime, Canon, and Sony Cybershot camera with different environmental conditions. The samples of flower videos with large intraclass variation from 20 classes are shown in Fig. 4. The videos captured through canon camera are subjected for experimentation.

5 Experimental Results

The efficiency of retrieval of flower videos is analyzed in this section. Our own dataset is used for experimentation. Initially keyframes are selected to represent a video. Then statistical region merging segmentation has been applied to select the flower regions from keyframes. Then, SIFT, LBP and GLCM features are extracted from segmented keyframes. Generate support vectors for training features using multi-class Support Vector Machine classifier.

The efficiency of the proposed retrieval method is estimated using precision, recall, F-measure, and accuracy. Tables 1, 2, 3 and 4 show that the average retrieval results of the features SIFT, GLCM, and LBP with combining the two features such as SIFT+GLCM, SIFT+LBP, GLCM+LBP, and fusion of all the features SIFT+GLCM+LBP. From the results, we can notice that the good accuracy can be achieved with the combination of all features. Figure 3 shows that the average accuracy of each feature and the combination of features. Figure 4 shows the samples of flower videos with a large intraclass variation from 20 classes. Figure 5 shows the videos retrieved. The precision, recall, F-measure, and accuracy are given below,

$$\text{Precision} = \frac{NRV}{(NRV + NNRV)} \tag{6}$$

Table 1 60% training–40% testing

Features	Accuracy	Precision	Recall	F-measure
LBP	17.24	0	0	0
GLCM	30.13	52.63	45.45	48.78
SIFT	59.6	64	36.36	46.38
LBP+GLCM	58.95	81.59	70.45	75.61
SIFT+GLCM	65.26	69.23	61.23	65.06
SIFT+LBP	68.03	67.44	65.91	66.67
SIFT+GLCM+LBP	70.92	72.09	70.45	71.26

Table 2 70% training–30% testing

Features	Accuracy	Precision	Recall	F-measure
LBP	19.12	0	0	0
GLCM	33.51	48.28	42.42	45.16
SIFT	61.93	72.73	48.48	58.18
LBP+GLCM	63.16	75	72.73	73.85
SIFT+GLCM	68.07	85.71	72.73	78.69
SIFT+LBP	71.75	67.65	69.7	68.66
SIFT+GLCM+LBP	75.44	80.65	75.76	78.12

Table 3 80% training–20% testing

Features	Accuracy	Precision	Recall	F-measure
LBP	19.95	0	0	0
GLCM	36.97	33.33	40.91	36.73
SIFT	64.89	78.95	68.18	64.89
LBP+GLCM	70.48	69.57	72.73	71.11
SIFT+GLCM	69.15	81.82	81.82	81.82
SIFT+LBP	70.47	71.43	90.91	80
SIFT+GLCM+ LBP	70.47	73.08	86.36	79.17

Table 4 90% training–10% testing

Features	Accuracy	Precision	Recall	F-measure
LBP	23.78	0	0	0
GLCM	41.62	46.15	54.55	50
SIFT	69.73	77.78	63.64	70
LBP+GLCM	74.05	88.89	72.73	80
SIFT+GLCM	75.13	90.91	90.91	90.91
SIFT+LBP	79.46	90.91	90.91	90.91
SIFT+GLCM+ LBP	**83.24**	75	81.82	78.26

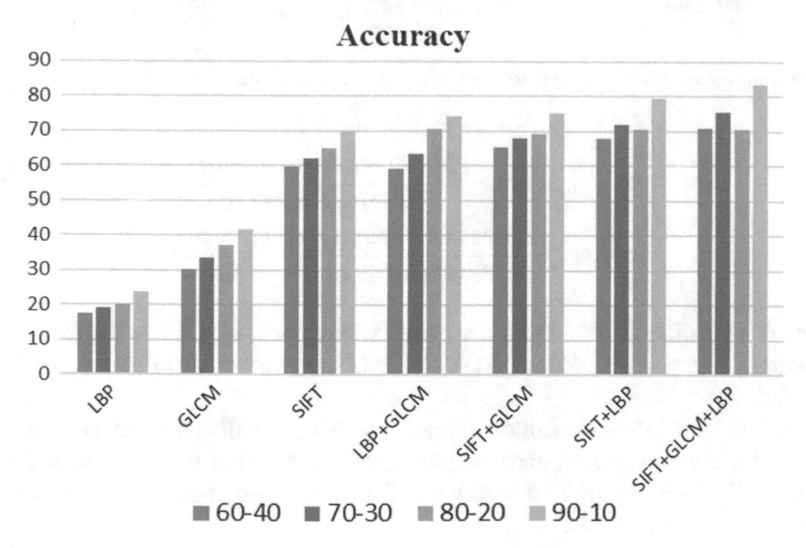

Fig. 3 Average retrieval accuracy based on LBP, GLCM, SIFT, and the combination of these features along with the varying intervals

$$\text{Recall} = \frac{NRV}{(NRV + NRNR)} \tag{7}$$

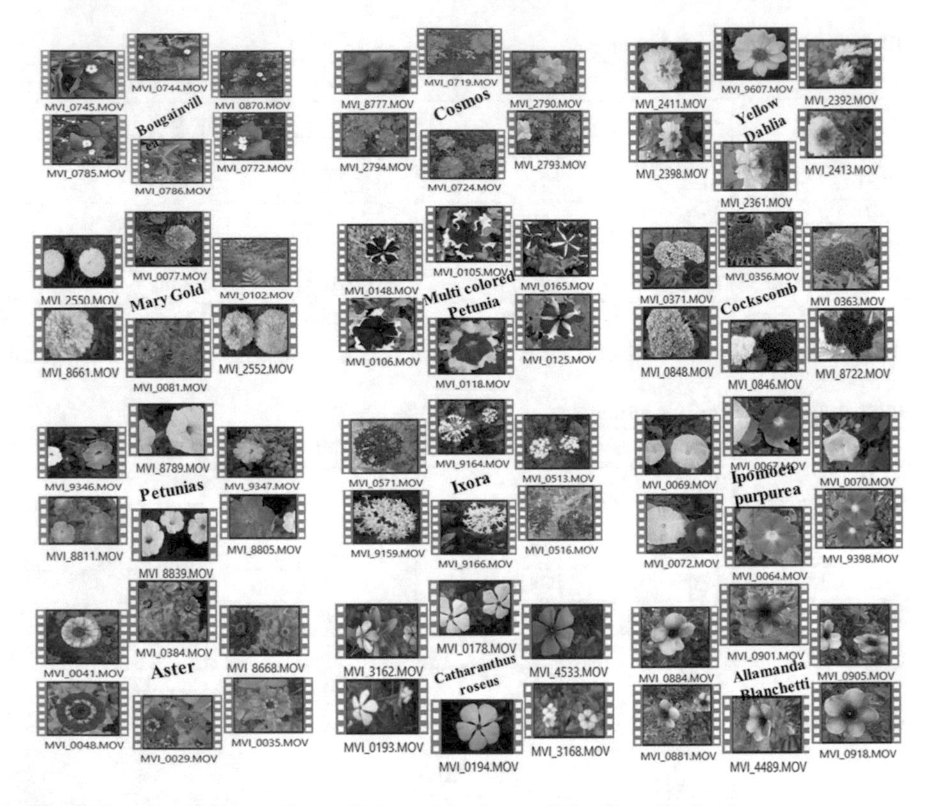

Fig. 4 Samples of flower videos with large intraclass variation from 20 classes

$$\text{F} - \text{measure} = \frac{2 * Precision * Recall)}{(Precison + Recall)} \tag{8}$$

$$\text{Accuracy} = \frac{\text{sum of videos correctly retrieved}}{\text{Total number of query videos}} \tag{9}$$

where NRV is the retrieved number of relevant videos, $NNRV$ is the retrieval of number of nonrelevant videos and $NRNR$ is the number of relevant videos not retrieved.

The following tables illustrate accuracy, precision, recall and F-measure for each feature, for combination of each features and combination of all features for the analysis of 20 classes of flower videos. The results are shown with varying the training and testing videos.

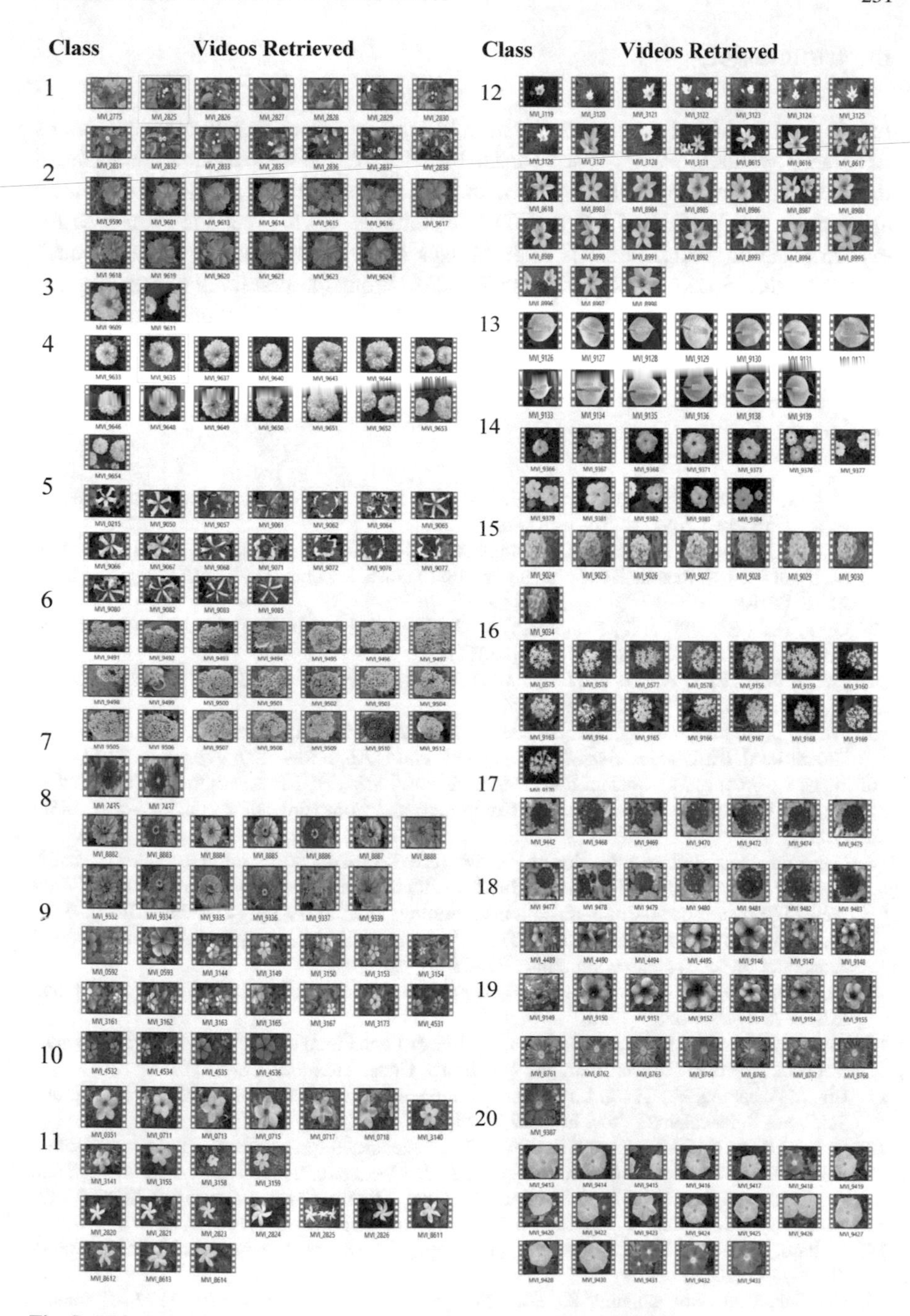

Fig. 5 Videos retrieved from the cluster-based keyframe selection method

6 Conclusion

In this work, we have developed a system to retrieve videos to support applications like query by video. We have used cluster-based keyframe selection method to select the keyframes to represent the video. In the proposed model, segmented flowers are represented by SIFT, LBP, and GLCM features. Own dataset is examined for experimentation, which consists of 20 classes of flower videos. We represent each flower by descriptors SIFT, LBP, and GLCM. From the results obtained, we can observe that good accuracy can be achieved through the fusion of all features.

References

1. Das, M., Manmatha, R., Riseman, E.M.: Indexing flower patent images using domain knowledge. IEEE Intell. Syst. **14**(5), 24–33 (1999)
2. Guru, D.S., Sharath Kumar, Y.H., Manjunath, S.: Texture Features and KNN in classification of flower images. Recent Trends Image Process. Pattern Recogn. RTIPPR IJCA Special Issue 21–29 (2010)
3. Guru, D.S., Sharath, Y.H., Manjunath, S.: Textural features in flower classification. Math. Comput. Model. **54**(2011), 1030–1036 (2011)
4. Shekar, B.H., Uma, K.P., Holla, Raghurama: Video clip retrieval based on LBP variance. Proc. Comput. Sci. **89**(2016), 828–835 (2016)
5. Weiming, H., Xie, N., Zeng, L.X., Stephen, M.: A survey on visual content-based video indexing and retrieval. IEEE Trans. Syst. Man Cybernet. Part C Appl. Rev. **41**(6), 797–819 (2011)
6. Asha, S., Sreeraj, M.: Content based video retrieval using SURF descriptor. In: Third International Conference on Advances in Computing and Communications, pp. 212–215. IEEE (2013)
7. Nagaraja, G.S., Rajashekara Murthy, S., Deepak, T.S.: Content based video retrieval using support vector machine classification. In: 2015 International Conference on Applied and Theoretical Computing and Communication Technology (iCATccT), pp. 821–827. IEEE (2015)
8. Yingying, Z., Xiaoyan, H., Qiang, H., Tian, Qi: Large scale video copy retrieval with temporal concentration SIFT. Neurocomputing **187**, 83–91 (2016)
9. Li, D., Li-Zuo, J.: A video retrieval algorithm based on ensemble similarity. IEEE 638–642 ISBN 978-1-4244-6585 (2010)
10. Ionut, M., Bogdan, I., Jasper, U., Nicu, S.: Fisher kernel temporal variation-based relevance feedback for video retrieval. Comput. Vis. Image Understand. **143**, 38–51 (2016)
11. Bin, L., Wenbing, X., Xiang, L.: Design of video retrieval system using MPEG-7 descriptors. Sci. Verse ScienceDirect Proc. Eng. **29**, 2578–2582 (2012)
12. Guru, D.S., Jyothi, V.K., Sharath Kumar, Y.H.: Cluster based approaches for keyframe selection in natural flower videos. In: 17th International Conference on Intelligent Systems Design and Applications, ISDA 2017, 14–16th, Dec. 2017, vol. 736, pp. 474–484. Springer, AISC (2017)
13. Gonzales, R.C, Woods, R.E, Eddins, S.L.: Digital Image Processing, 3rd ed. (2008)
14. Richard, N., Frank, N.: Statistical region merging. IEEE Trans. Pattern Anal. Mach. Intell. **26**(11), 1–7 (2004)
15. Haralick, R.M., Shanmugam, K., Dinstein, I.H.: Textural features for image classification. IEEE Trans. Syst. Man Cybermat. **SMC-3**(6), 610–621 (1973)
16. Ojala, T., Pietikainen, M., Maenpaa, T.: Multiresolution gray scale and rotation invariant texture classification with local binary patterns. IEEE Trans. Pattern Anal. Mach. Intell. **24**(7), 971–987 (2002)

17. Loris, N., Alessandra, L., Sheyl, B.: Survey on LBP based texture descriptors for image classification. Expert Syst. Appl. **39**, 3634–3641 (2012)
18. Arun Kumar, M., Gopal, M.: A hybrid SVM based decision tree. Pattern Recogn. **43**, 3977–3987 (2010)
19. Asnath, Y., Amutha, R.: Discrete cosine transform based fusion of multi-focus images for visual sensor networks. Signal Process. **95**, 161–170 (2014)
20. Bagher, B.A., Haghighat, A.A., Sevedarabi, H.: Multi-focus image fusion for visual sensor networks in DCT domain. Comput. Electr. Eng. **37**(5), 789–797 (2011)
21. Lowe, D.G.: Distinctive image features from scale-invariant keypoint. Int. J. Comput. Vis. **60**, 91–110 (2004)

Offline Signature Verification: An Approach Based on User-Dependent Features and Classifiers

K. S. Manjunatha, H. Annapurna and D. S. Guru

Abstract This work aims at proposing an approach for verification of offline signature based on user-dependent features and classifiers. We have used 50 global features of shape, geometric, and texture category. The features suitable for each user are selected by means of a computationally efficient filter-based feature selection method. In addition, user dependency has been considered at classifier level also. Based on lowest equal error rate obtained with training samples, the decision is made on the features and classifier to be used for a user. In the first stage, we conducted experiments without any feature selection but with user-dependent classifier. Further, experiments have been carried out under varying number of features for all users with user-dependent classifiers. To evaluate the performance of the proposed model, experimentation has been conducted on MCYT offline signature dataset which is one of the standard benchmark datasets. The EER that we obtained indicates the effectiveness of the usage of writer-dependent characteristics.

Keywords Offline signature · User-dependent features · User-dependent classifier
Frequency of classifier · Feature selection

1 Introduction

The technology used for automatic recognition of individuals based on their biological or behavioral characteristics is referred to as biometrics [1]. Biological characteristics include face, hand vein, knuckle, iris, etc., and behavioral characteristics

K. S. Manjunatha (✉)
Maharani's Science College for Women, Mysuru 570001, Karnataka, India
e-mail: kowshik.manjunath@gmail.com

H. Annapurna · D. S. Guru
Department of Studies in Computer Science, University of Mysore, Manasagangothri, Mysuru 570006, Karnataka, India
e-mail: annapurnavmurthy@yahoo.co.in

D. S. Guru
e-mail: dsg@compsci.uni-mysore.ac.in

P. Nagabhushan et al. (eds.), *Data Analytics and Learning*,
Lecture Notes in Networks and Systems 43,
https://doi.org/10.1007/978-981-13-2514-4_20

include voice, gait, and handwritten signature [2]. Signature is being used as one of the most popular behavioral biometric traits to identify a person in many applications such as banking transactions, security, and document verification. Signature verification can be done either offline or online based on the acquisition [3]. In offline signature verification, static features are extracted from the signature images. Due to lack of availability of dynamic features, designing an offline signature verification system is more complex when compared to online signature verification system. Due to its wide applications, significant research has been taking place in the field of offline signature verification during the last three decades.

Offline signature verification can be categorized as user-dependent approach or user-independent approach. In a user-dependent approach, system uses a different set of features, a different thresholds, and also different classifiers for each user. In case of user-independent approach, a common set of features, a common threshold, and also a same classifier have been used for all the users.

Many models have been proposed for offline signature verification system using different types of feature with different comparison techniques during the last few decades. As reported in the literature survey [4], different types of features used in offline signature verification system are global features of different categories, namely geometric features, direction features, graphometric features, structure features, shape features, and grid-based features. Further, for comparison purpose classifiers such as Neural Network (NN), Support Vector Machines (SVM), Hidden Markov Models (HMMs), clustering techniques, Fuzzy Modeling Approach, Feature matching technique, and contourlet-based methods are used. From the literature survey, it is observed that most of the existing models are writer-independent as the features and the classifier used for verification are common across all writers. But while verifying the signature manually, a human expert considers different set discriminating features for different users. Even the matching technique adopted is also different for different user. Based on this observation, in this work, an approach has been proposed for offline signature verification utilizing user-dependent characteristics. User dependency has been considered at both feature level and classifier level.

The organization of this paper is as follows. In Sect. 2, we discuss the different stages of our proposed method. In Sect. 3, details of experimental setup with results obtained are discussed. We presented the comparative study of our model with other existing verification models in Sect. 4. Finally conclusions are drawn in Sect. 5.

2 Proposed Method

The four different stages of the proposed model

1. Computation of features
2. Selection of user-dependent features
3. Selection of user-dependent classifier
4. Verification.

2.1 Computation of Features

In this approach, after applying, the necessary preprocessing steps such as binarization, noise removal, and thinning, altogether 50 global features are computed from each signature, the details of which are shown in Table 1.

2.2 Selection of User-Dependent Features

After extracting features, features for each user are selected by means of a feature selection technique proposed in [5]. As this approach is suitable for feature selection in a multi-clustered data, we have adopted this in our work because generally genuine signatures of a writer form some natural groupings. Here, initially for every feature, a score is estimated which shows how much relevant the features in preserving the cluster structure of signatures of a particular writer. For instance, Let N be the number of writers and n be the number of signatures available for training purpose from each writer. Let P be the number of features computed for each writer. Hence, for each writer, we have a data matrix of dimension $n\ X\ P$. For each of the P features, a relevancy score is estimated as suggested in [5]. The features are sorted based on the

Table 1 50 computed global features

F#	Details	F#	Details	F#	Details
1	No. of connected components	12	Baseline shift [10]	35	Slope of the off-diagonal points of the bounding box
2	Center feature	13	No. of end points [11]	36 and 37	Global centroid
3	Outline feature [12]	14–19	Six fold surface features	38	Fine ink distribution [12]
4	Height [10]	20	Max. horizontal histogram [11]	39	Coarse ink distribution [12]
5	Width [10]	21	Max. vertical histogram [11]	40–43	Four area
6	Aspect ratio [13]	22	Normalized area of the signature [13]	44	Wrinkleless
7	Orientation	23	Core feature [12]	45	Perimeter
8	Area [14]	24	Slope	46–48	Tri-surface features
9	Slant angle [10]	25	Ratio	49	Major axis length
10	Kurtosis [14]	26–33	Directional frontiers [12]	50	Minor axis length
11	Skewness [14]	34	High pressure region [12]	–	–

computed score and finally select only d features with higher scores for each user. It is interesting to note that *d* features that we select vary from a user to a user. In order to use only those features during verification, we store the indices of each of the features selected for each writer in knowledge base,

2.3 Selection of User-Dependent Classifier

Decision on the classifier selection for a user arrives as follows [6]. Let n_g be the number of available genuine signature for training purpose which is further divided into training set and testing set. For validation, we have considered 50% of training samples. During validation in order to estimate FAR equal number of random forgeries is used. Based on the samples used for validation, EER is estimated from each of the classifier C_{Lj} ($j = 1, 2, 3, \ldots, m$), where m is the number of classifiers used. The experimentation is conducted for 20 trials by considering randomly chosen training and validation samples in each trial. In each trial, we select the classifier with lowest EER for a writer. The best classifier for each user is decided by computing the frequency of a particular writer [6]. To calculate the frequency of classifier, first, we count the number times each classifier is selected for a particular user for all trials, then it is divided by the number of trials conducted. The classifier with highest frequency of selection will be selected as the classifier for that particular user. The selected classifiers along with the features selected are stored in the knowledge base.

2.4 Verification

To test whether the claimed signature is genuine or not, we retrieve features and classifier selected for the respective user during training. The test signature is fed to the respective classifier selected for the claimed writer which decides the genuinity of the test signature

3 Experimentation and Results

The efficacy of the proposed model is established by conducting experimentations on MCYT-75 [7]. The MCYT-75 dataset consists of signatures of 15 genuine signatures and 15 skilled forgery signatures of 75 writers. In our experimentation, we have used 50 global features and 6 different classifiers namely, Probabilistic Neural Network (PNN), Naïve Bayesian (NB), Nearest Neighbor (NN), Support Vector Machine (SVM), Principal Component Analysis (PCA), and Linear Discriminant Analysis (LDA). For more details, the reader can refer [8].

To train the model, we have considered 5 and 10 genuine signatures from each user. In this phase, we validate the system by using 50% of the training samples and the same number of random forgeries. We conducted the experimentation for 20 trials with random selection of training and validation samples in each trial. User-dependent features and user-dependent classifier are fixed in this stage. Using the remaining genuine and skilled forgeries testing is done. We have also considered random forgeries for testing.

Initially, we conducted experiments by using a common set of features irrespective of the writer but for verification, user-dependent classifier is used. In the next stage, we conducted experiments using user-dependent features and also user-dependent classifier. The EER obtained in both the cases is presented in Table 2.

From the above table it is very clear that, obtained EER is lower when both features and classifier is writer dependent compared to EER obtained with only user-dependent classifier.

Further, we conducted experiments by means of each of the six classifiers as a common classifier across all writers. To represent the classifiers, Naïve Bayesian (NB), Nearest Neighbor (NN), Support Vector Machine (SVM), Probabilistic Neural Network (PNN), Linear Discriminant Analysis (LDA), and Principal Component Analysis (PCA), we have used the symbols C_{L1},C_{L2},C_{L3},C_{L4},C_{L5}, and C_{L6}, respectively.

The obtained EER with the usage of common classifier as well as user-dependent classifier is shown in Table 3. The last row in Table 3 is the EER obtained with the proposed model.

From the result shown in Table 3, it clearly indicates that when a common classifier is used for all writers, the EER obtained is higher when compared to the EER obtained with the usage of user-dependent classifier.

We also conducted experimentation with user-dependent classifier for varying feature dimension from 5 to 50 in steps of 5. The obtained EER is as shown in Table 4.

Table 2 Error rate obtained without user-dependent features and with user-dependent features

Method	Skilled		Random	
	5	10	5	10
Without user-dependent features but with user-dependent classifier	7.86	5.97	3.27	2.96
With user-dependent features and user-dependent classifier	5.06	3.67	1.56	0.94

Table 3 EER obtained with common classifier (writer-independent classifier)

Classifier used	Skilled		Random	
	5	10	5	10
C_{L1}	11.51	10.94	8.13	5.24
C_{L2}	10.87	9.77	6.22	6.13
C_{L3}	8.38	6.93	4.73	3.04
C_{L4}	18.67	17.39	12.94	10.22
C_{L5}	15.40	11.63	9.21	4.06
C_{L6}	20.94	18.41	14.50	11.01
Proposed model (with user-dependent features and user-dependent classifier)	5.06	3.67	1.56	0.94

Table 4 Performance of the proposed model with varying feature dimension

Number of features	Skilled		Random	
	5	10	5	10
5	7.97	6.98	3.04	2.99
10	7.91	6.95	2.12	2.22
15	7.24	5.88	2.33	1.05
20	6.03	5.59	2.01	1.44
25	6.76	5.34	1.56	0.98
30	5.44	4.87	1.59	1.62
35	5.06	4.09	1.60	1.04
40	5.94	3.67	1.73	1.36
45	5.62	3.99	1.74	0.94
50	5.49	3.72	1.62	0.96
Without user-dependent features and with user-dependent classifier	7.86	5.97	3.27	2.96

4 Comparative Analysis

In this work, we have compared the EER of the proposed model with other existing models. In order to compare the performance of our model with other models EER has been used as a measure. We considered the models evaluated on MCYT dataset and the results are tabulated in Table 5. Even though, other models taken into consideration for comparative analysis are writer independent, we have used dataset as a parameter for comparative study.

From the Table 5, it can be observed that the proposed approach result is lowest EER when compared to most of the existing models except [9] which is a writer-independent model while ours is a user-dependent model. Further, when compared to outer models, our model works with lower feature dimension.

Table 5 EER/AER of the conteporary models on MCYT-75 dataset

Model	Training samples size	EER/AER
Alonso et al. [15]	05	22.40
	10	20.00
Wen et al. [16]	05	15.02
Prakash et al. [17]	09	18.26
	09	17.83
Ooi et al. [18]	05	13.86
	10	09.87
Manjunatha et al. [19]	05	13.67
	10	09.53
Vargas et al. [20]	05	12.02
	10	08.80
Ferrer et al. [21]	05	12.02
Gilperez et al. [22]	05	10.18
	10	06.44
Soleimani et al. [23]	05	13.44
	10	09.86
Hafemann et al. [9]	05	03.70
	10	02.87
Proposed model (*Without user-dependent features with only user-dependent classifier*)	05	07.86
	10	05.97
Proposed model *(With user-dependent features and user-dependent classifier)*	05	05.06
	10	03.67

5 Conclusion

A new approach has been proposed in this paper for offline signature verification which is based on user-dependent features and classifiers. We arrive at the decision on features and classifiers suitable for each user based on minimum EER criteria. To bring out the superiority of the proposed model, considerable experimentation has been carried out on MCYT dataset. The results obtained clearly shows the effectiveness of the proposed approach.

References

1. Jain, A.K., Ross, A., Nandakumar, K.: Introduction to Biometrics, Springer. https://doi.org/10.1007/978-0-387-77326-1 (2011)
2. Plamondon, R., Lorette, A.: Automatic signature verification and writer identification—the state of the art. Pattern Recogn. **2**, 107–131 (1989)
3. Jain, A.K., Griess, F.D., Connell, S.D.: On-line signature verification. Pattern Recognit. **35**, 2963–2972 (2002)
4. Pal, S., Blumenstein, M., Pal, U.: Automatic off-line signature verification systems: a review. In: International Conference and Workshop on Emerging Trends in Technology, Mumbai, India, pp. 20–27 (2011)
5. Cai, D., Zhang, C., He, X.: Unsupervised feature selection for multi-cluster data. In: International Conference on Knowledge Discovery and Data Mining, pp. 333–342 (2010)
6. Manjunatha, K.S., Manjunath, S., Guru, D.S., Somashekara, M.T.: Online signature verification based on writer dependent features and classifiers. Pattern Recogn. Lett. **80**, 129–136 (2016)
7. Ortega-Garcia, J., Fierrez-Aguilar, J., Simon, D., Gonzalez, J., Faundez-Zanuy, M., Espinosa, V., Satue, A., Hernaez, I., Igarza, J.-J., Vivaracho, C. et al.: MCYT baseline corpus: a bimodal biometric database. In: IEE Proceedings Vision, Image and Signal Processing, vol. 150(6), pp. 395–401 (2003)
8. Duda, R.O., Hart, P.E., Stork, D.G.: Pattern Classification, 2nd edn. Wiley, USA (2001)
9. Luiz Hafemann, G., Sabourin, R., Oliveira, L.S.: Learning features for offline handwritten signature verification using deep convolutional neural networks. Pattern Recogn. **70**, 163–176 (2017)
10. Qi, Ying Yong, Hunt, Bobby R.: Signature verification using global and grid features. Pattern Recognit. **27**(12), 1621–1629 (1994)
11. Ramachandra, A.C., Rao, J.S., Raja, K.B., Venugopla, K.R., Patnaik, L.M.: Robust offline signature verification based on global features. In: IEEE International Advance Computing Conference, pp. 1173–1178 (2009)
12. Huang, K., Yan, H.: Off-line signature verification based on geometric feature extraction and neural network classification. Pattern Recognit. **30**, 9–17 (1997)
13. Kruthi, C., Deepika, C.S.: Offline signature verification using support vector machine, IEEE Trans. https://doi.org/10.1109/icsip.2014.5(2014)
14. Karouni, A., Daya, B., Bahlak, S.: Offline signature recognition using neural networks approach. Procedia Comput. Sci. **3**, 155–161 (2011)
15. Alonso-Fernandeza, F., Fairhurst, M.C., Fierrez, J., Ortega-Garciaa, J.: Automatic measures for predicting performance in off-line signature. In: IEEE Proceedings of the International Conference on Image Processing, ICIP, vol. 1, pp. 369–372 (2007)
16. Wen, J., Fang, B., Tang, Y., Zhang, T.: Model-based signature verification with rotation invariant features. Pattern Recogn. **42**(7), 1458–1466 (2009)
17. Prakash, H.N., Guru, D.S.: Offline signature verification: an approach based on score level fusion. Int. J. Comput. Appl. **1**(18), 0975–8887 (2010)
18. Ooi, S.Y., Teoh, A.B.J., Pang, Y.H., Hiew, B.Y.: Image-based handwritten signature verification using hybrid methods of discrete Radon transform, principal component analysis and probabilistic neural network. Appl. Soft Comput. **40**, 274–282 (2016)
19. Manjunatha, K.S., Guru, D.S., Annapurna, H.: Interval-valued writer-dependent global features for off-line signature verification. In: proceedings of MIKE-2017, LNAI 10682, pp. 133–143 https://doi.org/10.1007/978-3-319-71928-3_14 (2017)
20. Vargas, J.F., Ferrer, M.A., Travieso, C.M., Alonso, J.B.: Off-line signature verification based on grey level information using texture features. Pattern Recogn. **44**(2), 375–385 (2011)

21. Ferrer, M.A., Vargas, J.F., Morales, A., Ordóñez, A.: Robustness of offline signature verification based on gray level features. IEEE Trans. Inf. Forensic Secur. **7**(3), 966–977 (2012)
22. Gilperez, A., Alonso-Fernandez, F., Pecharroman, S., Fierrez, J., Ortega-Garcia, J.: Off-line Signature Verification Using Contour features. In: Proceedings of the International Conference on Frontiers in Handwriting Recognition, ICFHR (2008)
23. Soleimani, A., Araabi, B.N., Fouladi, K.: Deep multitask metric learning for offline signature verification. Pattern Recogn. Lett. **80**, 84–90 (2016)

Simple DFA Construction Algorithm Using Divide-and-Conquer Approach

Darshan D. Ruikar and Ravindra S. Hegadi

Abstract In real-world applications like network software design, pattern recognition, compiler construction moreover in some of the applications of formal and natural language like accepter, spell checker and advisor, language dictionary, the Deterministic Finite Automata (DFA) plays an important role. For such applications, specification rules to construct DFA are more complex in nature. More generally, the rules consist of the intricate words like 'and', 'or', 'not having', 'followed by'. Constructing DFA for such rules is time-consuming and tedious process. To make the construction process easier, simple DFA construction algorithm is proposed. The proposed algorithm is based on divide-and-conquer (D&C) algorithm design strategy. To apply D&C, the first step is to divide given language specification rule into a number of manageable pieces of sub-language specification rules, second construct DFA for each sub-language by any efficient method and at last, combine the DFA using closure properties of the regular set to obtain resultant DFA. To verify the correctness of the proposed algorithm, the obtained resultant DFA is compared with DFA constructed by a conventional method.

Keywords Deterministic finite automata · Regular expression · Closure properties · Divide and conquer · Validation rules · Natural language processing Formal language

1 Introduction

According to the theory of formal languages, language is a collection of strings which satisfy the given validation rules [1]. Based on validation rules the formal

D. D. Ruikar (✉) · R. S. Hegadi
School of Computational Sciences, Solapur University,
Solapur 413255, India
e-mail: ddruikar@sus.ac.in

R. S. Hegadi
e-mail: rshegadi@gmail.com

P. Nagabhushan et al. (eds.), *Data Analytics and Learning*,
Lecture Notes in Networks and Systems 43,
https://doi.org/10.1007/978-981-13-2514-4_21

languages can be categorized into two classes: regular language and not a regular language. A regular language is a more restricted language and validation rules are bit simpler because any sort of dependencies does not exist between occurrences of input symbols in the sting. Whereas, some sorts of dependencies exist in the not regular language.

A Finite Automaton (FA) is constructed to identify belongingness of the supplied string (membership problem). FA constructed for regular language considers only current state and current input symbol to decide next transition. Contrary to this automaton constructed for not regular languages require some memory (for instance, stack) to remember the dependency between the occurrences and they consider memory status along with current state and input symbol to decide the next transition. Hence, the automata construction for regular language is a bit simpler than of automata construction for not regular language.

A regular expression is an algebraic representation of regular language whereas DFA is used to solve membership problem [2]. Moreover, string matching, pattern reorganization, ROBOT behaviour, state charts, digital circuit design, compiler design (lexical analyzer) and string processing are few application areas where DFA plays an important role. Apart from this, DFA find its extensive applications in knowledge engineering, game theory, computer graphics and linguistics. In natural or formal languages, large set of vocabulary can be described by using DFA. These DFA can be used as a spell checker, spell advisor or language dictionary. More often, DFA specification rules contain elementary words like ‘starts with’, ‘end with’ or ‘having sub-string’. But in real-life applications, specification rules may contain intricate constraints such as ‘and’, ‘or’, ‘not’, ‘followed by’ or ‘reversal’. To construct a DFA with such constraints is a bit difficult. The proposed algorithm will solve this difficulty by applying divide-and- conquer strategy to construct the DFA.

This paper presents a simple method for DFA construction based on the D&C approach. The paper is organized as follows: Sect. 2 briefly describes related preliminaries and definitions. The literature survey is discussed in Sect. 3. Section 4 contains information about D&C strategy and details about implementation procedure. Section 5 presents the proposed algorithm and its detailed explanation. Section 6 evaluates proposed algorithm with the help of counterexample. Conclusion is presented in Sect. 7.

2 Background

This section briefly introduces some preliminaries and definitions related to finite automata. Notations and definitions are followed from [1, 3].

2.1 Preliminaries

The symbol is an abstract entity that can not formally define. The alphabet is a finite set of symbols for a particular language (denoted by $\sum$). The string is a sequence of symbols chosen from a specific alphabet set. Language is set to string over alphabet (represented by $\sum^*$). According to Chomskys hierarchy, there are various types of formal languages: regular language, context-free language, context-sensitive language, recursive language and recursively enumerable language [4]. In all, the regular language is a most intersected and restricted language having very specific acceptance specifications. These specifications can be represented nicely in terms algebraic notations called as a regular expression. As like an expression regular expression is also a finite sequence of operators and operands. The operators are '+' (union), '.' (concatenation), '*' (Kleen closure) and operands are the input symbol chosen from $\sum$ [5]. Expansion of regular expression results in set of strings, i.e. regular language set. Closure operation, a specific operation is closed when the result of that operation belongs to some set. For instance, the set of integers are closed under the operations: addition, subtraction, multiplication and modulo division but not closed under division. Similarly, regular sets are closed under the operations: union, concatenation, Kleene closure, set deference, intersection, complementation, substitution, homomorphism, inverse homomorphism and quotient with the arbitrary set [6, 7].

2.2 Definitions

Definition 1 *Regular Expression*

Formally, the regular expression is recursively defined as follows:

1. Φ is a regular expression denoting an empty language.
2. ϵ is regular expression indicating the language containing an empty string.
3. a $\in \sum$ is regular expression denoting finite language a.
4. If R and S be the regular expressions denoting the languages L_R and L_S respectively, then R + S, R.S, R^* are regular expressions corresponding to languages $L_R \cup L_S$, $L_R.L_S$, L_R^* respectively.
5. A regular expression obtained by applying any of the above rules is also a regular expression.

Definition 2 *Deterministic Finite Automata*

DFA is 5 tuple M = (Q, $\sum$, δ, q_0, A) where

1. Q = finite non-empty set of states
2. $\sum$ = finite non-empty set of input symbols (alphabet set)

3. δ = state transition function ($\delta : Q \times \sum \rightarrow Q$)
4. $q_0 \in Q$ = an initial state
5. $A \subseteq Q$ = finite non-empty set of final states.

3 Literature Survey

In the literature, various successful attempts are made to construct the DFA. The most common way is to convert a regular expression to DFA with or without using intermediate NFA [8, 9]. A new DFA construction method based on a set of string is presented in [10]. In [11], recurrent neural networks are trained to behave like DFA. Learning from example approach [12] is used for construction and minimization of DFA is discussed in [13]. Hill climbing with the heuristically guided approach is used for construction, minimization and to implement regular set recognizer (RR). To avoid sub-optimal results, adoptive search is applied. In total, 14 counterexamples are demonstrated (7 simple examples and 7 its reversal) to show construct validity of the proposed approach. Constructed DFA correctly accept strings in right list and rejects the stings in the wrong list. Multi-start, random hill climber evolutionary algorithm that uses smart state labelling is implemented in [14] to learn DFA construction through examples. A novel DFA construction method is presented [15]. The presented method effectively uses CLOSURE ( ) and GOTO ( ) functions of LR parser to construct minimal DFA from regular grammar moreover two applications are also discussed. First, ambiguity test for given grammar and second to check whether the string is part of language generated by the grammar. They developed a second-degree polynomial algorithm to construct DFA from regular grammar and polynomial time DFA minimization algorithm.

All above-discussed methods restricted to construction of DFA whose validation rules are elementary and not much complicated in nature. The proposed algorithm deals DFA construction problem with complicated language specification rules.

4 Methodology

The general idea about divide-and-conquer strategy and detailed explanation of how D&C strategy is applied to construct the DFA for complex specification rules are described in this section.

Divide and conquer is one of most useful and widely used algorithm design strategies to solve problems where the problem is more complex that it cannot be solved directly in single attempt. To solve such problem, divide-and-conquer strategy suggest that divide problem P into smaller instances of subproblems $P_1, P_2, P_3, \ldots, P_k$, of the same type that of the original problem. As a second step, solve each subproblem separately; finally, devise the method to combine the solution as a whole. Recursive application of D&C is necessary if subproblems are relatively larger [16]. There are basically three tasks in D&C strategy: divide, conquer, and combine.

1. Divide: the problem is divided into smaller instances until they are manageable, i.e. the instance must be small enough that it can be solved directly without splitting is produced.
2. Conquer: solve smaller instances by the efficient method.
3. combine: combine the solutions of all smaller instances to get the final result.

The same strategy can be useful to construct DFA for a complex set of validation rules.

1. Divide: split the language specification rules into sub language specification rules surrounded by intricate words 'or', 'and', 'not having'.
2. Conquer: construct DFA for each smaller sub-instance by using an efficient technique.
3. Combine: to obtain the result as a whole, combine constructed DFA using appropriate closure properties of the regular set.

If language specification rule contains the intricate words 'or', 'and', 'not having' then to obtain combined DFA, as a result, use the closure properties of a regular set like regular sets are closed under the operations union, intersection and set difference respectively. The resultant combined DFA may contain few numbers of extra states, due to consideration of all possibilities of the combination. To discard these extra states along with its transitions, the DFA minimization technique (for instance, Myhill Nerode theorem [17]) can be applied to obtain the minimized DFA. The brief application procedure of the Myhill Nerode theorem is demonstrated in [18].

5 Proposed Algorithm

This section facilitates detailed explanation of a proposed algorithm to construct DFA using divide-and-conquer approach.

The proposed algorithm ConstructDFA ( ) accepts a sequence of language specification rule (P) over alphabet set $\sum$ and constructs a DFA M as an output. Algorithm ConstructDFA(), first check whether the given specification rule is too small, i.e. elementary validation rule. If so it constructs a DFA for the same. If the specification rule contains one or more intricate words, then in the next step, ConstructDFA() split the specification rule into multiple manageable subproblems (P_1, P_2, P_3, ..., P_k,) based on the surrounding part of intricate words. For each subproblem, it recursively calls itself to construct DFA and call to algorithm Closure with appropriate operation is used to combine the DFA constructed for each subproblem. Combining process is continued till the algorithm outputs a single resultant DFA. The implementation procedure is described in detail as follows.

Algorithm ConstructDFA ( ) give a call to algorithm Closure ( ) with three parameters. First parameter is the name of the operator, second and third parameters are call to algorithm ConstructDFA () with sub-specification rules P_1 and P_2, respectively. These two recursive function calls intern returns two DFAs $M_1(Q_1, \sum, \delta_1, q_{01}, A_1)$ and $M_2(Q_2, \sum, \delta_2, q_{02}, A_2)$, respectively. Then, algorithm Closure ( ) combines the accepted DFAs based on closure property of regular set and returns combined DFA as result. The combined DFA is passed to algorithm Minimize ( ) to obtain minimized DFA.

6 Evaluation

A counterexample is solved with both the approaches: first by the proposed algorithm as well as second by the conventional approach. The obtained DFAs are compared, to comment on the correctness of the proposed algorithm. Example: Construct a DFA over $\sum = \{0, 1\}$ where all strings start with 0 and having 00 as a substring.

Algorithm 1 Construct DFA (P)

Require: P // set of specification rules
Ensure: DFA M (Q,$\sum$, δ, q_0,A)
1: **if** small(P) **then**
2: // P is elementary validation rule
3: //construct DFA by any efficient technique
4: Construct DFA M (Q,$\sum$, δ, q_0,A) for P
5: **else**
6: // split problem based on surrounding part of intricate word
7: **switch** (IntricateWord)
8: **case** 'or'**:**
9: Closure ('Union', ConstructDFA (P_1), ConstructDFA (P_2))
10: break
11: **case** 'and'**:**
12: Closure ('Intersection', ConstructDFA (P_1), ConstructDFA (P_2))
13: break
14: **case** 'notHaving'**:**
15: Closure ('SetDifference', ConstructDFA (P_1), ConstructDFA (P_2))
16: break
17: **end switch**
18: **end if**
19: Minimize(M)
20: Return (M (Q,$\sum$, δ, q_0,A))

Algorithm 2 Closure (IntricateWord, $M_1(Q_1, \sum, \delta_1, q_{01}, A_1)$, $M_2(Q_2, \sum, \delta_2, q_{02}, A_2)$)

Require: IntricateWord
Ensure: Combined DFA M (Q,$\sum$, δ, q_0,A)
1: //Let $M_1(Q_1, \sum, \delta_1, q_{01}, A_1)$ and $M_2(Q_2, \sum, \delta_2, q_{02}, A_2)$
2: //be the DFAs for validation rules P_1 and P_2 respectively
3: // P_1 validation rule before IntricateWord
4: // P_2 validation rule after IntricateWord
5: Let $M(Q, \sum, \delta, q_0, A)$ be the combined DFA, where
6: Q = $Q_1 \times Q_2$
7: $\delta([q_1, q_2], a) = [\delta_1(q_1, a), \delta_2(q_2, a)]$ where ($q_1 \in Q_1$ and $q_2 \in Q_2$)
8: $q_0 = [q_{01}, q_{02}]$
9: **switch** (IntricateWord)
10: **case** 'Union':
11: $A_{P_1 \cup P_2} = \{[q_1, q_2] | q_1 \in A_1$ OR $q_2 \in A_2\}$
12: break
13: **case** 'Intersection':
14: $A_{P_1 \cap P_2} = \{[q_1, q_2] | q_1 \in A_1$ AND $q_2 \in A_2\}$
15: break
16: **case** 'SetDifference':
17: $A_{P_1 - P_2} = \{[q_1, q_2] | q_1 \in A_1$ AND $q_2 \notin A_2\}$
18: break
19: **end switch**
20: Return (M (Q,$\sum$, δ, q_0,A))

Algorithm 3 Minimize (M)

Require: DFA M (Q,$\sum$, δ, q_0,A)
Ensure: Minimized DFA M (Q,$\sum$, δ, q_0,A)
1: // Myhill Nerode theorem says; states along with its transitions can be excluded,
2: //those do not lie on a path which leads to accepting state from initial state.
3: //Rest of the DFA is minimal.
4: //LetM (Q,$\sum$, δ, q_0,A) be any DFA accepting L
5: //We define relations $D_0, D_1, \ldots$ on Q as follows.
6: //State q is distinguishable from state p by a string of length i is written as
7: $q D_i p$ iff either $q \in A$ and $p \notin A$ or $q \notin A$ and $p \in A$
8: For $i \geq 0$ we define State q is distinguishable from state p by a string of length $\leq$ i iff $q D_{i-1} p$ or there exists $a \in \sum$ such that $\delta(q, a) D_{i-1} \delta(p, a)$
9: State q is distinguishable from state p iff q D p
10: Return (M (Q,$\sum$, δ, q_0,A))

6.1 DFA Construction by Applying of Proposed Algorithm

At first, algorithm ConstructDFA ( ) accepts a sequence of language specification rule P = starts with 0 and having substring 00. In the specified sequence 'and' intricate word is present. Hence, given P is divided into subproblems: $P_1 =$ starts with 0 and $P_2 =$ having substring 00. Because these are surrounded by intricate word 'and'. In the next step, ConstructDFA ( ) calls to algorithm Closure ( ) with parameters 'and', M_1 and M_2. Algorithm closure will return the combined DFA M. That DFA is passed to algorithm Minimize ( ) to obtain minimized resultant DFA.

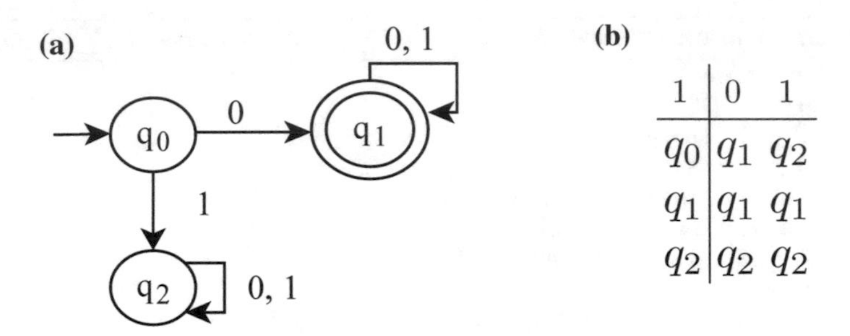

1	0	1
q_0	q_1	q_2
q_1	q_1	q_1
q_2	q_2	q_2

Fig. 1 **a** Transition diagram for M1 and **b** Transition table for M1

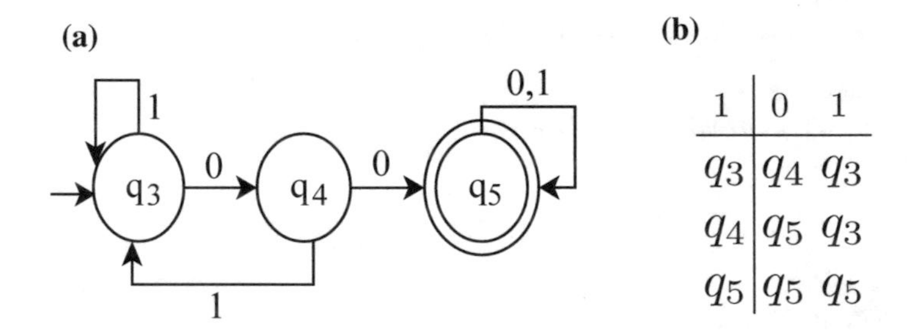

1	0	1
q_3	q_4	q_3
q_4	q_5	q_3
q_5	q_5	q_5

Fig. 2 **a** Transition Diagram for M2 and **b** Transition table for M2

Let $M_1(Q_1, \sum, \delta_1, q_{01}, A_1)$ and $M_2(Q_2, \sum, \delta_2, q_{02}, A_2)$ be the DFA for P_1 and P_2 shown in Figs. 1a and 2a respectively.

$Q_1 = \{q_0, q_1, q_2\}, \sum = \{0, 1\}, q_{01} = q_0, A_1 = \{q_1\}, \delta_1$ is shown in Fig. 1b.

$Q_2 = \{q_3, q_4, q_5\}, \sum = \{0, 1\}, q_{02} = q_3, A_2 = \{q_5\}, \delta_2$ is shown in Fig. 2b.

The intricate word 'and', so the DFA M_1 and M_2 are combined using closure property: regular sets are closed under intersection. The accepting state is $[q_1, q_5]$ because $q_1 \in A_1$ and $q_5 \in A_2$.

Let $M_3(Q_3, \sum, \delta_3, q_{03}, A_3)$ be the DFA for language L= $L_1 \cap L_2$ constructed by combining the DFAs M_1 and M_2 using algorithm Closure(). The combined DFA M_3 is shown in Fig. 3a.

$Q_3 = \{[q_0, q_3], [q_1, a_4], [q_2, q_3], [q_1, a_5], [q_1, q_3], [q_2, a_4], [q_2, q_5]\}, \sum = \{0, 1\}$, $q_{03} = [q_0, q_3]$, $A_3 = \{[q_1, q_5]\}$, δ_3 is shown in Fig. 3b. Now to optimize the result, Myhill Nerode theorem, the DFA minimization technique is applied on DFA M_3. The execution of the algorithm Minimize() is shown in Fig. 4a. Let $M_4(Q_4, \sum, \delta_4, q_{04}, A_4)$ be the minimized DFA shown in Fig. 4b.

For the sake of simplicity, consider $p_0 = [q_0, q_3]$, $p_1 = [q_1, q_4]$, $p_2 = [q_1, q_5]$, $p_3 = \{[q_2, q_3], [q_2, q_4], [q_2, q_5]\}$, $p_4 = [q_1, q_3]$, and redraw the transition diagram in Fig. 4b. Let $M_5(Q_5, \sum, \delta_5, q_{05}, A_5)$ be the redrawn DFA shown in Fig. 5a.

$Q_5 = \{p_0, p_1, p_2, p_3, p_4\}, \sum = \{0, 1\}, q_{05} = p_0, A_5 = \{p_2\}, \delta_5$ is shown in Fig. 5b.

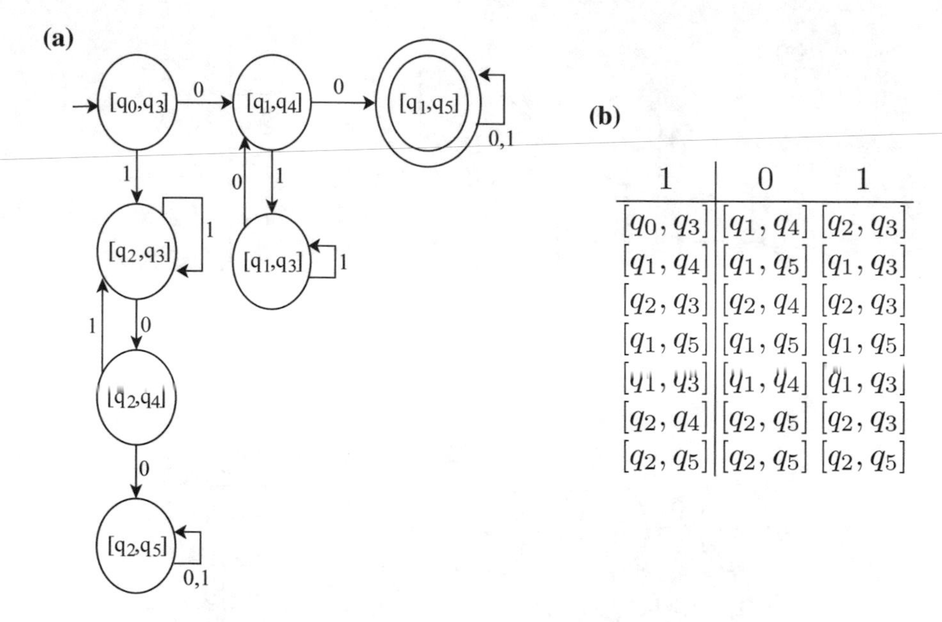

1	0	1
$[q_0, q_3]$	$[q_1, q_4]$	$[q_2, q_3]$
$[q_1, q_4]$	$[q_1, q_5]$	$[q_1, q_3]$
$[q_2, q_3]$	$[q_2, q_4]$	$[q_2, q_3]$
$[q_1, q_5]$	$[q_1, q_5]$	$[q_1, q_5]$
$[q_1, q_3]$	$[q_1, q_4]$	$[q_1, q_3]$
$[q_2, q_4]$	$[q_2, q_5]$	$[q_2, q_3]$
$[q_2, q_5]$	$[q_2, q_5]$	$[q_2, q_5]$

Fig. 3 **a** Transition diagram for M_3 and **b** Transition table for M_3

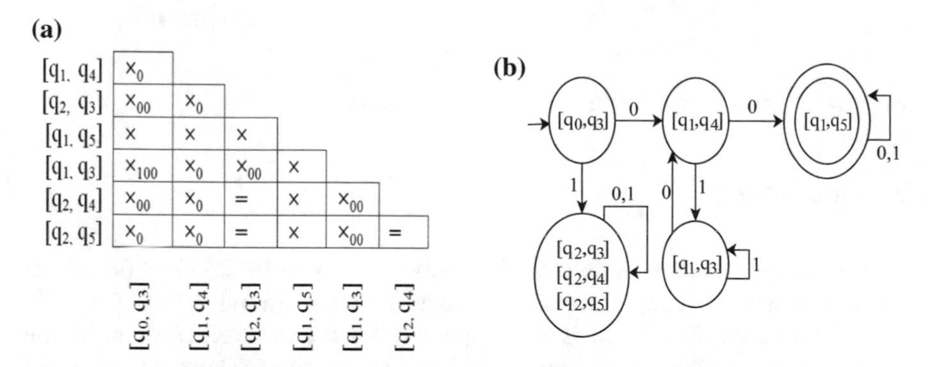

	$[q_0, q_3]$	$[q_1, q_4]$	$[q_2, q_3]$	$[q_1, q_5]$	$[q_1, q_3]$	$[q_2, q_4]$
$[q_1, q_4]$	X_0					
$[q_2, q_3]$	X_{00}	X_0				
$[q_1, q_5]$	X	X	X			
$[q_1, q_3]$	X_{100}	X_0	X_{00}	X		
$[q_2, q_4]$	X_{00}	X_0	=	X	X_{00}	
$[q_2, q_5]$	X_0	X_0	=	X	X_{00}	=

Fig. 4 **a** Execution of Myhill Nerode Theorem and **b** Transition Diagram for M_4

6.2 *DFA Construction by Conventional Approach*

To check the correctness of proposed algorithm, DFA is constructed for same validation rule using conventional method. Let $M_6(Q_6, \sum, \delta_6, q_{06}, A_6)$ be the required DFA. The transition diagram is shown in Fig. 6a.

$Q_6 = \{q_0, q_1, q_2, q_3, q_4\}$, $\sum = \{0, 1\}$, $q_{06} = q_0$, $A_6 = \{q_2\}$, δ_6 is shown in Fig. 6b.

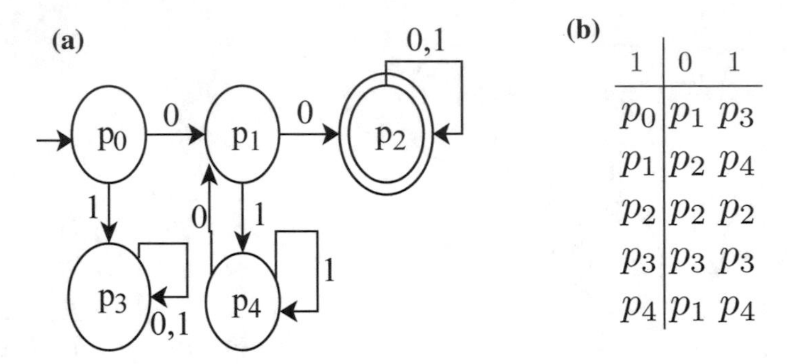

1	0	1
p_0	p_1	p_3
p_1	p_2	p_4
p_2	p_2	p_2
p_3	p_3	p_3
p_4	p_1	p_4

Fig. 5 **a** Transition diagram for M_5 and **b** Transition table for M_5

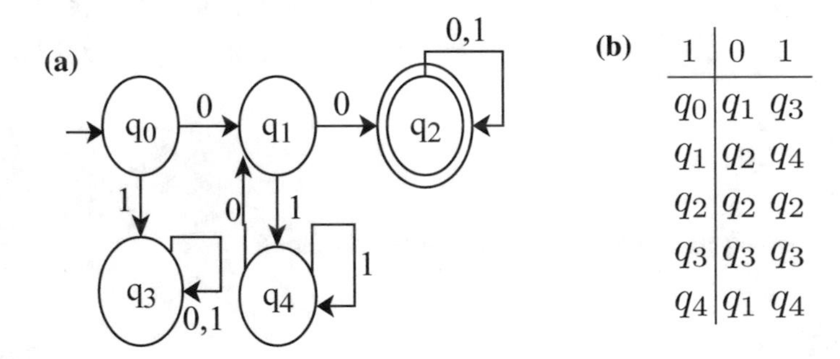

1	0	1
q_0	q_1	q_3
q_1	q_2	q_4
q_2	q_2	q_2
q_3	q_3	q_3
q_4	q_1	q_4

Fig. 6 **a** Transition diagram for M_6 and **b** Transition table for M_6

7 Conclusion

By comparing the entries in the transition tables shown in Figs. 5b and 6b, all the entries are same. Hence, it is concluded that the proposed algorithm effectively construct the same DFA for complicated specification rules with effective utilization of D&C strategy. This comparison also shows the accuracy of proposed algorithm as the result obtained by both (conventional and proposed) methods are the same. The proposed algorithm has four major steps: (1) divide given specification rule into small pieces surrounded by intricate words, (2) construct DFA for each subdivided rules, (3) decide appropriate closure property to combine the DFAs by considering given intricate word, and (4) apply DFA minimization algorithm to optimize the result.

Acknowledgements The first author thanks the Ministry of Electronics and Information Technology (MeitY), New Delhi for granting Visvesvaraya Ph.D. fellowship through file no. PhD-MLA\4(34)\201-1 Dated: 05/11/2015.

References

1. Martin, J.C.: Introduction to Languages and the Theory of Computation, vol. 4. McGraw-Hill NY, USA (1991)
2. Rabin, M.O., Scott, D.: Finite automata and their decision problems. IBM J. Res. Dev. **3**(2), 114–125 (1959)
3. Hopcroft, J.E.: Introduction to Automata Theory, Languages, and Computation. Pearson Education India (2008)
4. Chomsky, N.: On certain formal properties of grammars. Inf. Control **2**(2), 137–167 (1959)
5. Myers, G.: A four Russians algorithm for regular expression pattern matching. J. ACM (JACM) **39**(2), 432–448 (1992)
6. Lochan, M., Garhwal, S., Kumar, A.: Closure properties of prefix-free regular languages. Int. J. Comput. Appl. **52**(8) (2012)
7. Stearns, R.E., Hartmanis, J.: Regularity preserving modifications of regular expressions. Inf. Control **6**(1), 55–69 (1963)
8. Ben-David, S., Fisman, D., Ruah, S.: Embedding finite automata within regular expressions. Theor. Comput. Sci. **404**(3), 202–218 (2008)
9. Berry, G., Sethi, R.: From regular expressions to deterministic automata. Theor. Comput. Sci. **48**, 117–126 (1986)
10. Daciuk, J., Mihov, S., Watson, B.W., Watson, R.E.: Incremental construction of minimal acyclic finite-state automata. Comput. Linguist. **26**(1), 3–16 (2000)
11. Omlin, C.W., Giles, C.L.: Constructing deterministic finite-state automata in recurrent neural networks. J. ACM (JACM) **43**(6), 937–972 (1996)
12. Parekh, R., Honavar, V.: Learning DFA from simple examples. In: International Workshop on Algorithmic Learning Theory, pp. 116–131. Springer (1997)
13. Tomita, M.: Learning of construction of finite automata from examples using hill-climbing: RR: Regular Set Recognizer (1982)
14. Lucas, S.M., Reynolds, T.J.: Learning deterministic finite automata with a smart state labeling evolutionary algorithm. IEEE Trans. Pattern Anal. Mach. Intell. **27**(7), 1063–1074 (2005)
15. Kumar KS, M.D.: A novel method to construct deterministic finite automata from a given regular grammar. Int. J. Sci. Eng. Res. **6**(3), 106–111 (2015)
16. Horowitz, E., Sahni, S.: Fundamentals of Computer Algorithms. Google Scholar (1978)
17. Nerode, A.: Linear automaton transformations. Proc. Am. Math. Soc. **9**(4), 541–544 (1958)
18. Högberg, J., Larsson, L.: DFA minimisation using the Myhill-Nerode theorem. a a **1**(q2) q3

Estimating the Rating of the Reviews Based on the Text

Mohammadamir Kavousi and Sepehr Saadatmand

Abstract User-generated texts such as reviews and social media are valuable sources of information. Online reviews are important assets for users to buy a product, see a movie, or choose a restaurant. Therefore, rating of a review is one of the reliable factors for all the users to read and trust the reviews. This paper employs the texts of the reviews to evaluate and predict the ratings. Moreover, we study the effect of lexical features generated from text as well as sentimental words on the accuracy of rating prediction. We show that words with high information gain score are more efficient compared to TF–IDF. In addition, we explore the best number of features for predicting the ratings of the reviews.

Keywords Review mining · Natural language processing · Machine learning Big data

1 Introduction

With a rapid growth of Internet and online shopping systems, customers share an opinion on online platforms to help others in making wiser choices about the products that they liked or disliked. This contribution has resulted in active communities which are known as valuable sources for both scholars and industry owners. Online reviews are important assets for users to buy a product, see a movie, or choose a restaurant. For instance, the ratings of the reviews highlight the quality of the products and, and as a result, these features are among the reliable factors for all users to read and trust the reviews or purchase a product [1]. Moreover, ratings are used in recommender systems where users receive automatic suggestions based on similar choices and attributes with other users. Based on the importance of these attributes, business

M. Kavousi (✉) · S. Saadatmand
Southern Illinois University, Carbondale, USA
e-mail: mohammadamir.kavousi@siu.edu

S. Saadatmand
e-mail: sepehr.saadatmand@siu.edu

P. Nagabhushan et al. (eds.), *Data Analytics and Learning*,
Lecture Notes in Networks and Systems 43,
https://doi.org/10.1007/978-981-13-2514-4_22

owners, and academic scholars have studied reviews to find an efficient technique to estimate the rating based on the content of the text [2–6]. The research in this area is very wide and is not just limited to finding the rating. Other areas such as opinion extraction and sentiment analysis, building recommendation systems, and summarizing the texts are among domains that people explore in this area [7–12].

There are a huge number of studies on predicting ratings of different products on Amazon (please refer to Sect. 2 for more information). In this paper, we focus on predicting the ratings of films (documentary and nondocumentary) based on the text of the reviews. Documentary films have reviews with richer texts focusing on the themes of the films, while nondocumentary films consist of attributes such as famous casts. We believe that a combination of these two types of films and using the textual features will provide good insights about the underlying characteristics of the texts.

The rest of the paper is organized as follows: Sect. 2 discusses the related work. Section 3 discusses the data collection. In the method Sect. 4, we first explain the feature selection and discuss the classifiers that we chose for this study, and then report the results of the classifier. Section 5 discusses the conclusion and future directions for improving the paper.

2 Related Works

As mentioned earlier, prior work on review mining is very vast. Researchers in this area have tried to find efficient algorithms for predicting rating, helpfulness, and sentiment of the reviews [11, 13–16]. In this section, we explore the most related work in the area of review mining and analyze the papers that used text of the reviews to extract information.

Bing and Zhang (2012) reviewed the most efficient algorithms in opinion mining and sentiment analysis of the reviews. Supervised and unsupervised approaches are employed to extract the sentiment of the reviews on sentence or document level [17]. Since a positive or negative opinion about a product (e.g., Camera) does not show the feeling of the opinion holder about every specific feature of that product (e.g., picture quality of a camera), aspect-based sentiment analysis is introduced to find the opinions related to each aspect of a product on a sentence level.

In a research, De Albornoz et al. [4] used both topic and sentiment of the reviews on sentence level to assess the impact of text-driven information in predicting the rating of the review in recommendation system. This article aims to predict the overall rating of a product review based on the user's opinion about the different product features that are evaluated in the review. After identifying the features that are relevant to consumers, they extract the user's opinions about the different product features. The salience of different product features and the values that quantify the user opinions are used to construct a vector of feature intensities to represents the review. This vector was used as the input of the machine learning model that classifies the review into different rating categories. As a result, they got 84% for Logistic, 83% for LibSVM, and 81.9% for FT. The errors are assumed to happen because of:

(1) mislabeled instances in the training set, (2) very frequent spelling errors in the reviews, and (3) the presence of neutral sentences that do not express any opinion but are necessarily classified as positives or negatives.

In another research [18] conducted a study to create a panel dataset of products from Amazon.com. The dataset consists of product-specific characteristics and the details of the product review. They generated a training set with two classes of documents; (a) a set of "objective" documents that contain the product descriptions of each of the 1000 products and (b) a set of "subjective" documents that contain randomly retrieved reviews. Based on their findings, they identified the reviews that are expected to be helpful to the users, and display them first, which can improve the usefulness of the reviewing mechanism for the users of the electronic marketplace [19–22].

Kim et al. [5] have suggested an automated assessment for the review helpfulness by considering that the length of the review is more useful than the features. The main contribution of this paper is: (a) they implemented a system for automatically ranking reviews according to helpfulness by using state of the art SVM regression and (b) analyzing different classes of features to capture review helpfulness; including structural (e.g., HTML tags, punctuation, and review length), lexical (e.g., ngrams), syntactic (e.g., percentage of verbs and nouns), semantic (e.g., product feature mentions), and metadata (e.g., star rating). Their best system achieved Spearman correlation coefficient scores of 0.656–0.604 against a gold standard for two products as MP3 players and digital cameras. They also performed a detailed analysis of different features to study the importance of several feature classes in capturing helpfulness. They found that the most useful features were review length, unigrams, and product rating. While semantic features like mentions of product features and sentiment words seemed to be subsumed by the simple unigram features, structural features, and syntactic features had no significant impact [23].

Rezapour and Diesner [24] studied the movie reviews from a new perspective that, as claimed by the authors, is new in the area of review and opinion mining. In this work, the authors captured the impact of movies from the reviews by first creating a novel dictionary of impact and then annotating the reviews on sentence level with various types of impact as a change in behavior, change in cognition, etc. They used three different classifiers with three sets of features to classify the impact of each film on the review writers. The results showed that SVM classifier and the combination of all features is the best predictor of impact in reviews. In another work, the amount of alignment of social media (Facebook and reviews), news articles and the transcript of the film were studied [25]. It was found that social media are more aligned with the main subject of the films. The results of these two studies show that film is important sources that are capable of influencing people's behavior and cognition.

The study presented in this paper builds upon the previous research. We study the impact of lexical feature as well as the impact of feature size on the classification task. Moreover, our analysis will show the importance of using the correct size for rating classification. Note that this study is a small-scale rating prediction. Using the insights of this paper, for data parallel processing of big data, more sophisticated

Table 1 Number of reviews for each movie

Name	#Reviews	5 Star	4 Star	3 Star	2 Star	1 Star
Food inc.	2462	1949	357	78	30	48
Boyhood	2253	872	342	318	296	425
Fed up	1401	1069	185	82	29	36
Blackfish	955	750	117	42	15	31
The imitation game	829	577	158	54	14	26
Super-size me	670	324	152	71	41	82
Inside job	437	308	54	21	10	44
Citizenfour	199	168	17	10	1	3

algorithms based on MapReduce can be used for speeding up the processing time, e.g., look at [26–31].

3 Data

We used the reviews of eight films from Amazon.[1] Table 1 shows the names and the number of the reviews for each film. Around 65% of the reviews are 5 star and 20% of them are 1, 2, and 3 stars. To have more distinct classes, 4-star reviews were excluded from the dataset. For preprocessing, first, all the reviews were divided into two categories/classes as High (5 stars) and Low (1, 2, and 3 stars). We then removed the stop words and tokenized the sentences to words. All the preprocessing was done using python NLTK [32] and custom programs. The resulted dataset consisted of around 39,802 sentences and 307,138 words. Words in a collection of documents mostly follow the zip's law, where the rare and common words are scattered on the end of two tails of the graph. To address this problem, we removed the words with less than 10 counts. The words with the high counts will be downscaled in feature selection, using TF–IDF or information gain. More details can be found in the following section.

4 Methodology

Reviews as users generated texts entail feelings and personal ideas of the users [24]. Unlike usual opinion mining and sentiment analysis in the field of review mining, movies such as documentary films do not benefit from special attributes such as famous cast or directors. Since the aim of a documentary movie is to raise the awareness about a social issue or introduce a new topic, individuals try to focus on

[1]The reviews of the films were randomly selected in 2014.

these areas in their reviews as well [24]. To find the ratings of the reviews, some prior studies leveraged helpfulness level as one of the features in rating classification. The helpfulness level is also a great help in creating the recommendation systems. We did not consider this feature in our study for two reasons: (1) documentary films are not among popular genres of films, and therefore the number of reviews written and viewed by customers is limited. In some cases, this lack of interest results in limited number of helpfulness rate. Moreover, the number of reviews which were viewed more than one time and also had helpfulness rate were around 1400, which is a very small input data for the classifiers. (2) In this study, we just focus on textual features with no external input.

4.1 Feature Selection

TF–IDF: Term frequency–inverse document frequency is a numerical statistic that is intended to reflect how important a word is to a document in a collection or corpus. It is often used as a weighting factor in information retrieval and text mining. The TF–IDF value increases proportionally to the number of times a word appears in the document but is offset by the frequency of the word in the corpus, which helps to adjust for the fact that some words appear more frequently in general. As mentioned earlier, using this feature will also help in downscaling the words with high frequency (such as stop words).

We considered top 500 and 900 words with the highest TF–IDF to train the classifiers.

Information Gain: Information gain measures the number of bits of information obtained for category prediction by knowing the presence and absence of a term in a document. To leverage this feature, we calculated the information gain of all words and choose top 200, 600, 900, and 1000 words as features. We will compare the results of the classifiers using different sets of features to find the ones that help the best in predicting the ratings.

Sentiment: One of the popular features in estimating the reviews is using sentiment of the words. In sentiment analysis, each word is tagged with a polarity as positive, negative, or neutral. There are two well-known approaches for analyzing the sentiment of the texts. In this paper, we considered a lexicon-based approach which leverages a predefined lexicon consisting words as well as their polarity (as a tag or ratio). To get the sentimental words, we used MPQA subjectivity lexicon [6], and extracted and tagged words from the reviews. In total, 2055 words were extracted.

After extracting the words of each feature set, we created the feature vectors using Python scikit-learn library [33]. Table 2 shows the list of features that were used in this study. In addition, top 10 words of each feature set, as Info-Gain, TF–IDF, and sentiment are listed in Table 3.

Table 2 List of the chosen features

TF–IDF	Top 500 words
TF–IDF	Top 900 words
Info-Gain	Top 200 words
Info-Gain	Top 600 words
Info-Gain	Top 900 words
Info-Gain	Top 1000 words
Sentiment word	Top sentiment words in more with a count of 5 or more

Table 3 List of the top words in selected features

TF–IDF	Info-Gain	Sentiment
Unethical	SeaWorld	Heavily
Origin	Boyhood	Praise
Slim	Concept	Amazing
Oversight	Interesting	Hard
Marijuana	Great	Cruel
Burned	Documentary	Attraction
Juices	Arquette	Innocent
Sharks	McDonalds	Wired
ingest	Sea	Hard
investigate	Opening	Pure

4.2 Classifiers

We used two well-known classifiers, Support Vector Machines (SVMs) and Naïve Bayes (NB) to classify the ratings [12, 34]. These two algorithms are among the most common ones in this area of research. Support vector machines are universal learners and are based on the structural risk minimization principle from computational learning theory. In general, SVMs are highly accurate, and with an appropriate kernel, they can work well even if the data is not linearly separable in the base feature space. They are also especially, popular in text classification problems where very high-dimensional spaces are the norm. Therefore, based on theoretical evidence, SVMs should perform well for text categorization. Naïve Bayes is one of the simple classifiers. They perform well with semi-supervised learning or fully supervised classifications.

After creating the feature vectors, we randomly selected 90% of the data for training the classifiers. The rest of the data will be used for testing the classifiers with the highest accuracy and the most efficient feature sets. We used WEKA [35] to implement the two algorithms. Table 4 shows the results of the selected features and the values of average accuracy, precision, recall, and F-score of two classifiers, SVM and NB.

Table 4 Results of classifiers

Features	SVM (%)				NB (%)			
	Acc	P	R	F	Acc	P	R	F
500 TF–IDF	50.58	95	77	84	50.43	99	77	86
900 TF–IDF	51.17	94	77	84	50.48	99	77	86
200 Info-Gain	78.48	91	89	88	**71.86**	**86**	**87**	**85**
600 Info-Gain	**81.9**	**90**	**92**	**90**	71.05	87	86	89
900 Info-Gain	81.61	90	91	90	70.65	87	86	89
1000 Info-Gain	77.84	90	89	89	71.39	88	86	87
Sentiment	75.60	87	89	87	71.04	84	87	85

Table 5 Prediction confidence of the classifiers

Features	Confidence (%)	Confidence (%)
Top 500 TF–IDF	1.17	0.8
Top 900 TF–IDF	2.3	0.96
Top 200 Info-Gain	62.1	44.6
Top 600 Info-Gain	63.9	42.11
Top 900 Info-Gain	63.3	42
Top 1000 Info-Gain	55	43
Sentiment	52	43

Based on the results in Table 4 the best highest accuracy was resulted using top 600 Info-Gain words and top 900 Info-Gain words for SVM and top 200 Info-Gain for NB. The average accuracy value, 82.0%, and the F-score value, 90% are the highest among all other features. SVM classifier is performing much better than NB.

We can also see that using information gain is more efficient than sentiment and TF–IDF. The precision value of TF–IDF is very high, but unfortunately, it is just predicting the high-rank reviews which are the larger class. The average accuracy of both top 500 and 900 TF–IDF is the lowest among all others and as Table 5 shows the classifiers are not showing enough confidence in predicting the classes. Based on the Average accuracy, prediction confidence and F-score values in Tables 4 and 5 both top 600 and top 900 information gain words obviously are the best features. Unfortunately, the sentiment of the words did not help us in rating as we expected, but compared to TF–IDF are still among the top features.

Comparing the result of this paper with other papers, and especially with De Albornoz's work [4] (83% for SVM), we showed that the result of this research is almost comparable with the other works in this area. One important note here is that reviewed works leveraged various types of features as well sentiment. The result presented in this work is solely based on lexical features.

To take a step further, we tested the trained classifier on the 10% test set data (as explained before). The results are slightly different from what we got before. Table 6 is showing the features and the result of the classifiers. Since we have a small data set,

Table 6 Result of a test set

	SVM (Overall accuracy %)	NB (Overall accuracy %)
Info-Gain_Top 200	78	72
Info-Gain_Top 600	77	70

we decided to choose (1) top 200 words to avoid overfitting, (2) and top 600 words as one of the best features. Same as the training, SVM resulted in higher accuracy compared to NB. However, the 200 info-Gain words worked better.

5 Conclusion

Based on the results in Tables 4 and 5, increasing the number of attributes in features was not helpful in enhancing the prediction of the ratings. One assumption is that the features proportional to the size of the data worked better than the high or low number of the attributes. The higher number may result in overfitting of the classifier and the low number may not be able to extract the necessary information from the content to classify the reviews. The best features were top 600 and 900 information gain words with the highest average accuracy, 82%, confidence of the classifier, 64%, and F-score, 90%. Running the test data showed that sometimes the words of the largest group or domain can dominate the selected features like info-Gain and may result in biased classifiers.

In addition, we found that TF–IDF is not always the best metric for extracting the most salient words from the document. We showed that words with the highest information scores perform better. As noted in the methodology and results, we did not combine the features to increase the accuracy. We found that the words in different feature sets highly overlap, which would result in overfitting the classifiers.

We plan to explore other algorithms and approaches in the future. With the popularity of the deep learning algorithms, we can test the same approach using word embedding and LSTM or CNN. In addition, we all considered adding other features such as syntactic features, the length of the reviews and n-gram words to our analysis.

Finally, in our future work, we plan to add social media texts as new features to the rating prediction. Same as reviews, social media such as Twitter, consists of user-generated texts. We believe that this new feature can tremendously help in rating prediction of the reviews, especially in sparse matrix situations. Tweets are great sources of user-specific features such as sentiments, hashtags, location, and texts. We plan to extract the text and hashtags related to films and add them as the sentimental and/or topical word to our features to expand and improve the prediction models. Hashtags were used in the previous study in social media analysis for topic modeling [36], sentiment analysis [37], and opinion mining [38]. Numbers of research leveraged social media information to predict a movie's success [39]. However, the research on combining these two user-generated texts is not well explored in the area of rating prediction.

References

1. Zhang, R., Yu, W., Sha, C., He, X., Zhou, A.: Product-oriented review summarization and scoring. Front. Comput. Sci. **9**(2), 210–223 (2015)
2. Bagherzadeh, R., Bayat, R.: Investigating online consumer behavior in iran based on the theory of planned behavior. Mod. Appl. Sci. **10**(4), 21 (2016)
3. Danescu-Niculescu-Mizil, C., Kossinets, G., Kleinberg, J., Lee, L.: How opinions are received by online communities: a case study on Amazon. com helpfulness votes. In: Proceedings of the 18th International Conference on World Wide Web, pp. 141–150. ACM (2009)
4. De Albornoz, J.C., Plaza, L., Gervás, P., Díaz, A.: A joint model of feature mining and sentiment analysis for product review rating. Adv. Inf. Retr. Springer 55–66. (2011)
5. Kim, S.M., Pantel, P., Chklovski, T., Pennacchiotti, M.: Automatically assessing review helpfulness. In: Proceedings of the 2006 Conference on Empirical Methods in Natural Language Processing, pp. 423–430. Association for Computational Linguistics (2006)
6. Wilson, T., Wiebe, J., Hoffmann, P.: Recognizing contextual polarity in phrase-level sentiment analysis. In: Proceedings of the Conference on Human Language Technology and Empirical Methods in Natural Language Processing, (HLT'05). Association for Computational Linguistics, pp. 347–354 (2005)
7. Cui, H., Mittal, V., Datar, M.: Comparative experiments on sentiment classification for online product reviews. In: Proceedings of the 21st International Conference on Artificial intelligence, pp. 1265–1270 (2006)
8. Hu, M., Liu, B.: Mining and summarizing customer reviews. In: Proceedings of the Tenth ACM SIGKDD International Conference on Knowledge Discovery and Data Mining, pp. 168–177. ACM (2004)
9. Jindal, N., Liu, B.: Opinion spam and analysis. In: Proceedings of the 2008 International Conference on Web Search and Data Mining, pp. 219–230. ACM (2008)
10. Liu, B., Zhang, L.: A Survey of Opinion Mining and Sentiment Analysis. Mining Text Data, pp. 415–463. Springer (2012)
11. Liu, C.-L., Hsaio, W.H., Lee, C.H., Lu, G.C., Jou, E.: Movie rating and review summarization in mobile environment. IEEE Trans. Syst. Man Cybern. Part C Appl. Rev. **42**(3), 397–407 (2012)
12. Pang, B. Lee, L.: Seeing stars: exploiting class relationships for sentiment categorization with respect to rating scales. In: Proceedings of the 43rd Annual Meeting on Association for Computational Linguistics (2005)
13. Haddadpour, F., Siavoshani, M.J., Noshad, M.: Low-complexity stochastic generalized belief propagation. In: 2016 IEEE International Symposium on Information Theory (ISIT), pp. 785–789. IEEE (2016)
14. Hong, Y., Lu, J., Yao, J., Zhu, Q., Zhou, G.: What reviews are satisfactory: novel features for automatic helpfulness voting. In: Proceedings of the 35th International ACM SIGIR Conference on Research and Development in Information Retrieval, pp. 495–504. ACM (2012)
15. Hongning, W., Yue, L., Chengxiang, Z.: Latent aspect rating analysis on review text data: a rating regression approach. KDD **10** (2010)
16. Lu, Y., Tsaparas, P., Ntoulas, A., Polanyi, L.:. Exploiting social context for review quality prediction. In: Proceedings of the 19th International Conference on World Wide Web (WWW) (2010)
17. Wilson, T., Wiebe, J., Hoffmann, P.: Recognizing contextual polarity in phrase-level sentiment analysis. In: Proceedings of the Conference on Human Language Technology and Empirical Methods in Natural Language Processing. ACL (2005)
18. Ghose, A., Ipeirotis, P.G.: Designing novel review ranking systems: predicting the usefulness and impact of reviews. In: Proceedings of the Ninth International Conference on Electronic Commerce, pp. 303–310. ACM (2007)
19. Chehardeh, M.I., Almalki, M.M., Hatziadoniu, C.J.: Remote feeder transfer between out-of-phase sources using STS. In: Power and Energy Conference at Illinois (PECI), 2016 IEEE pp. 1–5. IEEE (2016)

20. Ghose, A., Ipeirotis, P.G.: Estimating the helpfulness and economic impact of product reviews: mining text and reviewer characteristics. IEEE Trans. Knowl. Data Eng. **23**(10), 1498–1512 (2011)
21. Mozaffari, S.N., Tragoudas, S., Haniotakis, T.: A generalized approach to implement efficient cmos-based threshold logic functions. IEEE Trans. Circuits Syst. I Regul. Pap. (2017)
22. Mozaffari, S.N., Tragoudas, S., Haniotakis, T.: A new method to identify threshold logic functions. In: 2017 Design, Automation and Test in Europe Conference and Exhibition (DATE), pp. 934–937. IEEE (2017)
23. Zhuang, L., Jing, F., Zhu, X.Y.: Movie review mining and summarization. In: Proceedings of the 15th ACM International Conference on Information and Knowledge Management, pp. 43–50. ACM (2006)
24. Rezapour, R., Diesner, J.: Classification and detection of micro-level impact of issue-focused documentary films based on reviews. In: Proceedings of the 2017 ACM Conference on Computer Supported Cooperative Work and Social Computing, pp. 1419–1431. ACM (2017)
25. Diesner, J., Rezapour, R., Jiang, M.: Assessing public awareness of social justice documentary films based on news coverage versus social media. In: 2016 Proceedings, Conference (2016)
26. Daghighi, A., Kavousi, M.: Scheduling for data centers with multi-level data locality. In: 2017 Iranian Conference on Electrical Engineering (ICEE), pp. 927–936. IEEE (2017)
27. Kavousi, M.: Affinity Scheduling and the Applications on Data Center Scheduling with Data Localit. arXiv:1705.03125 (2017)
28. Wang, W., Zhu, K., Ying, L., Tan, J., Zhang, L.: Maptask scheduling in mapreduce with data locality: throughput and heavy-traffic optimality. IEEE/ACM Trans. Netw. (TON) **24**(1), 190–203 (2016)
29. Xie, Q., Lu, Y.: Priority algorithm for near-data scheduling: throughput and heavy-traffic optimality. In: 2015 IEEE Conference on Computer Communications (INFOCOM), pp. 963–972. IEEE (2015)
30. Xie, Q., Yekkehkhany, A., & Lu, Y.: Scheduling with multi-level data locality: Throughput and heavy-traffic optimality. In: 2016-the 35th Annual IEEE International Conference on Computer Communications, INFOCOM IEEE, pp. 1–9. IEEE (2016)
31. Yekkehkhany, A., Hojjati, A., Hajiesmaili, M.H.: GB-PANDAS: throughput and heavy-traffic optimality analysis for affinity scheduling. ACM SIGMETRICS Perform. Eval. Rev. **45**(2), 2–14 (2018)
32. Bird, S., Klein, E., Loper, E.: Natural Language Processing with Python: Analyzing Text with the Natural Language Toolkit: O'Reilly Media, Inc (2009)
33. Pedregosa, F., Varoquaux, G., Gramfort, A., Michel, V., Thirion, B., Grisel, O., Vanderplas, J.: Scikit-learn: machine learning in python. J. Mach. Learn. Res. **12**, 2825–2830 (2011)
34. Ganu, G., Elhadad, N., Marian, A.: Beyond the stars: improving rating predictions using review text content. In: Proceedings of the 12th International Workshop on the Web and Databases, pp. 1–6 (2009)
35. Hall, M., Frank, E., Holmes, G., Pfahringer, B., Reutemann, P., Witten, I.H.: The WEKA data mining software: An update. ACM SIGKDD Explor. Newsl. **11**(1), 10–18 (2009)
36. Yang, S. H., Kolcz, A., Schlaikjer, A., Gupta, P.: Large-scale high-precision topic modeling on twitter. In: Proceedings of the 20th ACM SIGKDD International Conference on Knowledge Discovery and Data Mining, pp. 1907–1916. ACM (2014)
37. Rezapour, R., Wang, L., Abdar, O., Diesner, J.: Identifying the overlap between election result and candidates ranking based on hashtag-enhanced, lexicon-based sentiment analysis. In: 2017 IEEE 11th International Conference on Semantic Computing (ICSC), pp. 93–96. IEEE (2017)
38. Lim, K. W., Buntine, W.: Twitter opinion topic model: extracting product opinions from tweets by leveraging hashtags and sentiment lexicon. In: Proceedings of the 23rd ACM International Conference on Information and Knowledge Management, pp. 1319–1328. ACM (2014)
39. Lehrer, S. Xie, T.: Box office buzz: does social media data steal the show from model uncertainty when forecasting for Hollywood? Technical Report. National Bureau of Economic Research (2016)

40. Li, F., Han, C., Huang, M., Zhu, X., Xia, Y.J., Zhang, S., Yu, H.: Structure-aware review mining and summarization. In: Proceedings of the 23rd International Conference on Computational Linguistics, pp. 653–661. Association for Computational Linguistics (2010)
41. Mudambi, S., M., Schuff, D.: What makes a helpful online review? A study of customer reviews on amazon.com. MIS Q. **34**(1), 185–200 (2010)

Multimodal Biometric Recognition System Based on Nonparametric Classifiers

H. D. Supreetha Gowda, G. Hemantha Kumar and Mohammad Imran

Abstract The paper addresses the unimodal and multimodal (fusion prior to matching) biometric recognition system from the promising traits face and iris which uniquely identify humans. Performance measures such as precision, recall, and f-measure and also the training time in building up the compact model, prediction speed of the observations are tabulated which gives the comparison between unimodal and multimodal biometric recognition system. LPQ features are extracted for both the modalities and LDA is employed for dimensionality reduction, KNN (linear and weighted), and SVM (linear and nonlinear) classifiers are adopted for classification. Our empirical evaluation shows our proposed method is potential with 99.13% of recognition accuracy under feature level fusion and computationally efficient.

Keywords K-NN · SVM · LDA · Fusion · Biometric

1 Introduction

The automated biometric system analysis the physiological and behavioral patterns stored in the database and perform either recognition (1:N comparison) or verification (1:1 comparison) in authenticating persons identity. Identification of humans globally is very much integral in todays society and biometric system has that potential in providing tighter security in various fields and finds application widely in law enforcement, business sector, border control, banking transactions, access control, surveillance, etc. A typical biometric system comprises of the following modules,

H. D. Supreetha Gowda (✉) · G. Hemantha Kumar · M. Imran
Department of Computer Science, University of Mysore, Mysore 560007, India
e-mail: supreethad3832@gmail.com

G. Hemantha Kumar
e-mail: ghk.2007@yahoo.com

P. Nagabhushan et al. (eds.), *Data Analytics and Learning*,
Lecture Notes in Networks and Systems 43,
https://doi.org/10.1007/978-981-13-2514-4_23

(a) Preprocessing stage: In this stage, distortions that arises while image acquisition are addressed by normalizing data by feature standardization, mean subtraction ways. (b) Region of Interest (ROI): Highlighting the interesting region by template matching approach, appearance based approach. Usually, ROI methods depend on the choice of database. (c) Feature extraction stage: this stage is very much crucial and selection of dominant features from the entire image by suitable feature reduction techniques and finally which governs the reliability of deploying system. (d) Matching module: here, the new feature set is compared with the templates stored through the best suitable classification algorithms. (e) Decision: identity of the user is revealed to accept or reject based on the score generated by matching threshold [1]. Unimodal biometric system relies on single source of trait either physiological or behavioral in developing system, but it is more likely subjected to noisy sensor data, spoofing, non-universality and other undesirable suspects that makes the system less secure. To address these issues, multimodal system of more than one modality which meets the tighter security systems as spoofing all the traits simultaneously and cracking into the system all at once is highly impossible, failure to enroll rate is also reduced here. Face being a physiological trait comprises of several features such as texture features, geometrical, size and shape feature, etc., and widely accepted trait. Iris is most reliable trait which develops in embryonic stages which is guarded by scleara and it holds different pattern for the twins also. Multimodal biometric fusion [2] can be done in two ways, first, fusion prior to classification which comprises of two types, namely sensor level and feature level and fusion after matching, which can be done by score level and decision level fusion. In sensor level fusion, traits obtained from same or different sensors are combined in obtaining a single fused image which is more accurate and complete than the data got from single sensor [3]. In feature level fusion, the feature vectors obtained from different feature extractors are augmented and before that the vectors are checked for compatibility for merging and curse of dimensionality should be taken care of. The fusion performs in two stages, feature normalization and dominant feature selection to decrease the dimensionality.

In our work, we are exploring facial and iris traits extracting LPQ features and LDA for dimensionality reduction and performing both the fusion schemes of pre-classification strategies, finally the recognition is performed through the classifiers KNN (linear and weighted) and SVM (linear and nonlinear).

The skeleton of the paper is as follows: Section 2 gives literature review to multiple multimodal biometric systems. Section 3 presents the methods and material used in this research. Section 4 discusses experimental results. Conclusion and future work are drawn in Sect. 5.

2 Review of Literature

Poh et al. [2] have designed score level fusion for evaluating score level fusion-based multibiometric systems adopting quality-dependent (acquired samples quality $\sum_i w_i y_i$ where w_i is the weight and y_i is the output), client-specific (individual specific features), cost-sensitive (specific cost of the system) fusion algorithms. Score level fusion is considered as more practical problem, training the multibiometric baseline system with efficient and suitable algorithms which are dependable on data and finally the features being input to fusion classifier. The authors also have given the key points on database contents which may affect the performance when dealing with Biased intra-session performance, Degraded performance when the device is changed for template and query sample collection, importance of acquisition device on performance of system. Toh et al. [4] developed a biometric verification system and addressed the score level fusion problem based on novel and fast error rate minimization, experimented on face and iris modalities. Authors have opted for EER as it is compact and optimal instead of the performance measures such as FAR, FRR.

Nanni et al. [1] based on score level fusion they have proposed a novel method in combining matchers and tested various classifiers like SVM, AdaBoost, and random subspace methods. Score level fusion is generally categorized in three levels namely, Transformation based (normalizing the scores to common domain), classifier based (training the classifier to judge imposter and genuine samples), and density based (likelihood ratio) fusion strategies. Gaussian model is usually not reliable in matching scores and hence mixture of Gaussians is used here which needs lot of training data. Random subspace of AdaBoost neural network achieved better performance than other employed methods. Performance of Fingerprint based biometric system degrades due to its surface like cuts, sweat and poor ridges, facial trait adopted system is very sensitive to emotions, iris modality contracts, and dilates in the outdoor environment and hence the authors choose finger vein trait as it is robust to vasodilatation. Kang et al. [5] proposed a biometric recognition system combining finger vein and finger geometry features. The system is deployed constructing small device, the finger geometry model is based on fourier descriptors and it is insensitive to translation and rotation variations and obtained good recognition results based on MAX, MIN, and SUM rules and attained a very low equal error rate.

Sim et al. [6] proposed a multimodal biometric recognition system employing face and iris (iris formed in embryonic stages) traits based on weighted score level fusion technique using exhaustive search criteria in attaining good accuracy. Face recognition is challenging in uncontrolled environments and also in dealing with partial occlusions. The authors have localized the off-angle iris boundary of iris images and have increased the discriminating ability lowering the FAR and FRR. Haar Wavelet and Neural Network are used to extract iris features. Rodrigues et al. [7] proposed a secure system by conducting experiments when the system is targeted to be spoofed by using likelihood ratio (lessens the probability error)-based fusion and fuzzy logic schemes (describes the Heuristics) and finally, the experimentation indicated trade-off between recognition accuracy and robustness against spoof attack.

Lakshmanan [8] proposed a geometric approach on Ear modality considering the middle portion of ear and extracted the features that is invariant to both scale and rotation. Particle swarm optimization was used in optimization of matching score weights, $S_ear = w_1 D_1 + w_2 2$, where D_1, D_2 are the distances and w_1, w_2 are the respective weights assigned. Kabir et al. [9] developed score reliability-based weighting (SRBW) technique by estimating weights and in turn resulting in elevation of recognition accuracy by score level fusion. Matcher or the classifier providing the lowest EER (FAR and FRR is same) is allocated the highest weight and vice versa. Mezai et al. [10] adopted face and voice traits and proposed an efficient particle swarm optimization where the weights of belief assignments were done. Score level fusion was done as it presents a tradeoff between ease of use and information. If the classifiers outputs the similar accuracy, then transformation and learning methods perform well but if it results in different accuracy, then handling such disagreement may not be possible, but PSO chooses the best fusion level and its parameters in attaining greater accuracy.

3 Methods and Materials

In this section, we have explored different methods which used in this research.

3.1 LPQ

The observed image $b(x)$ is given by convolution, $b(x) = (o * h)(x)$. Where $o(x)$ is the original image and $h(x)$ is point spread function, in many cases point spread function is centrally symmetric. In Fourier domain the above expression is written as $B(u) = O(u)H(u)$, where $B(u)$, $O(u)$, $H(u)$ are the discrete Fourier transforms of blurred image $b(x)$, original image $o(x)$ and PSF $h(x)$. For centrally symmetric blur $\angle H(u) = \begin{cases} 0 & if H(u) \geq 0 \\ \prod & if H(u) < 0 \end{cases}$ Low-frequency components contain most of the blur invariant information and it is better to select low frequencies F(k, u) computed for all pixels at four points $u_i = [a, 0]^T$, $[0, a]^T$ and $[a, -a]^T$. The coefficients are quantized using $B(x) = [ReF(x),\ imF(x)]$.

3.2 LDA

LDA is a data classification and dimensionality reduction algorithm which maximizes the within class variance and between class variances and draws the decision with maximum separation between classes. The mean vector of classes a and b is given by $\mu_i = \frac{1}{N_i} \sum_{a \in w_i} a$ and $\mu_i = \frac{1}{N_i} \sum_{b \in w_i} b$ and now distance between the projected means, i.e., objective function is given by, $\left| W^T = \mu_1 - \mu_2 \right|$. The co-variance matrix of class w_i is given by, $S_i = \sum_{b \in w_i} (a - \mu_i)(a - \mu_i)^T$. The within- class scatter matrix and between-class scatter matrix are given by, $W^T S_w W$ and $W^T S_b W$.

3.3 K-Nearest-Neighbors

KNN is a nonparametric technique which places a new observation to the class for which it belongs by comparing the new observation to the learning set with respect to the covariates. By using various similarity measures such as Euclidean, Minkowski, etc. $L = (c_i, x_i)\, i = 1 \ldots n_L$ is the learning set of a new observation and $c_i \in c_1 \ldots c_n$ is class membership. The nearest neighbor is determined by distance function $d(x, x_{(1)}) = min_i(d(x, x_i))$. By user-specific choice, the parameter k can be selected in determining the new observation to the class of closest data samples. For $k = 1$ is the simplest nearest neighbor method and for n_L, majority vote is considered globally from learning dataset results which gives classification of a new observation.

3.4 Weighted K-Nearest-Neighbors

The idea is the extension of KNN, distance closer to a new observation (c, x) is assigned the highest weight than to the data sample that is far from the observation. $L = (c_i, x_i)\, i = 1 \ldots n_L$ is the feature set that is learnt from set of observations x_i and the class labels c_i when x is a new observation and class c has to be predicted. We find the nearest neighbors of new samples by $d(x, x_i)$, then the normalized distances D_i by using any kernel function to weights $w(i) = k(D_{(i)})$ on prediction of class label c weighted majority voting from the nearest k neighbors $\hat{c} = max_r \sum_{i=1}^{k} w_i I(c(i) = r)$.

3.5 Support Vector Machine and Polynomial SVM

Mapping the low dimensional original feature set of vectors to higher dimensional space makes the nonlinearly separable data to linearly separable. $g(x) = W^t s_i + b$ being the linear function, W^t is the weight vector perpendicular to hyperplane that determines the orientation of hyperplane and b is the bias that represents the position of hyperplane in d dimensional space. Let s_i be the feature vector, in the two class SVM problem it is classified to the class for which it belongs according to this rule, $g(s_i) = \begin{cases} W^t s_i + b > 0\; if\; s_i \in c_1 \\ W^t s_i + b < 0\; if\; s_i \in c_2 \end{cases}$. If s_i lies on the hyperplane then it is $W^t s_i + b = 0$. Let y_i be the category of labels given by (x_i, c) where $y_i = \pm 1$, once the data to the classifier is fed in the form, if $s_i \in c_1$ where $y_i = -1$ and $s_i \in c_2$ where $y_i = +1$. Position of the hyperplane depends on the support vectors. In the equation, $\frac{W \cdot s + b}{\|W\|} \geq \gamma$ the margin γ has to be maximized by minimizing $\|W\|$ and maximizing the bias b. By using the Lagrangian Multipliers $L(W, b) = \frac{1}{2}(w \cdot w) - \sum \alpha_i [y_i[W \cdot s_i + b] - 1]$ the derivatives of W and b are found in identifying the unknown data feature vector to be classified.

SVM polynomial kernel mapping is given by: $\Re^2 \rightarrow \Re^3$, $(x_1, x_2) \rightarrow (z_1, z_2, z_3) = (x_1^2, \sqrt{2x_1x_2} + x_2^2)$. When d is large kernel needs to perform n multiplications.

4 Experimental Results

In this section, on the selection of two powerful and reliable psychological modalities such as face and iris, we have extracted LPQ features and on top of that we have applied dimensionality reduction technique named LDA. Once the dimensionality reduction features are extracted, we have passed the feature set to the well-practiced classifiers such as KNN (linear and weighted) and SVM (linear and nonlinear) for classification accuracy, and the block diagram our model is as shown in Fig. 1. The recognition results is tabulated in terms of accuracy, $Precision = \frac{TruePositives}{TruePositives + FalsePositives}$, $Recall = \frac{TruePositives}{TruePositives + FalseNegatives}$ and $F - measure = \frac{2 \times Precision \times Recall}{Precision + Recall}$. Instead of creating virtual dataset, we have used the multimodal dataset named MEPCO Biometric Database [11], MEPCO multimodal biometric image database is formed by combining three modalities such as face, iris, and fingerprint, collected from 115 uses, into 1 dataset. We have chosen only iris and facial traits to our experimentation.

Tables 1 and 2 shows iris and face features being subjected to classifiers as said above depicts the classification accuracy of unimodal recognition system with the performance measures. Iris modality is performing outstanding than the facial trait with the highest accuracy on both the types of the classifiers with its variants. Iris with the recognition accuracy of 97.39% on k-nn linear and 97.50% on SVM nonlinear classifier and is upholding performance than the face trait.

Table 3 shows the experimentation on multibiometric recognition results obtained on performing both the kinds of fusion prior to matching techniques, namely sensor level and feature level fusion. Both the levels of fusion are performed and have undergone the classifiers such as K-nn (linear), K-nn(weighted), SVM linear, and SVM nonlinear resulting in the performance of measure entries. Sensor level fusion usually distorts the images when fused from different kinds of sensors and handling

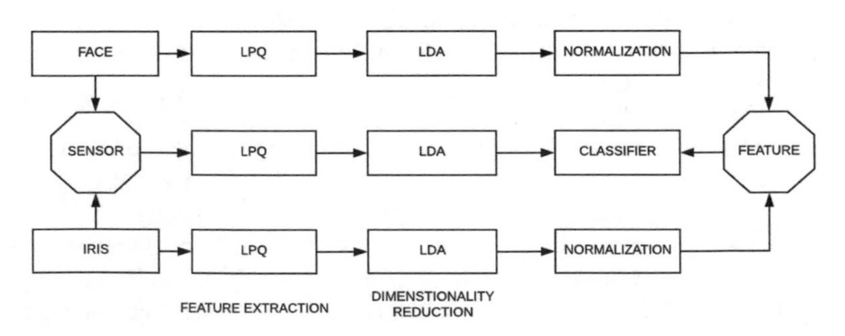

Fig. 1 Block diagram on fusion of face and iris

Table 1 Recognition accuracy of iris unimodal biometric system

Classifiers	Iris			
	Accuracy	Precision	Recall	F-measure
K-nn (Linear)	97.39	0.9754	0.9739	0.9746
K-nn (Weighted)	96.95	0.9710	0.9696	0.9703
SVM (linear)	97.39	0.9754	0.9739	0.9746
SVM (Weighted)	97.50	0.9784	0.9750	0.9786

Table 2 Recognition accuracy of face unimodal biometric system

Classifiers	Face			
	Accuracy	Precision	Recall	F-measure
K-nn (Linear)	74.78	0.7510	0.7478	0.7494
K-nn (Weighted)	74.35	0.7627	0.7435	0.7530
SVM (linear)	63.91	0.6429	0.6391	0.6410
SVM (Weighted)	61.30	0.5856	0.6130	0.5990

Table 3 Recognition accuracy of multimodal biometric system

Classifiers	Fusion of iris and face				
	Levels of fusion	Accuracy	Precision	Recall	F-measure
K-nn (Linear)	Sensor level	67.33	0.6572	0.6733	0.6655
K-nn (Weighted)	Sensor level	67.39	0.6889	0.6739	0.6814
SVM (linear)	Sensor level	53.04	0.5304	0.5223	0.5263
SVM (Weighted)	Sensor level	52.17	0.4984	0.5217	0.5098
K-nn (Linear)	Feature level	99.13	0.9855	0.9913	0.9883
K-nn (Weighted)	Feature level	99.13	0.9855	0.9913	0.9883
SVM (linear)	Feature level	98.69	0.9826	0.9869	0.9848
SVM (Weighted)	Feature level	97.83	0.9768	0.9782	0.9775

the high dimensional nature of hyperspectral sensor data is challenging. From the table, we can see that feature level fusion contains the raw data which is rich has performed better and achieved 99.13% of accuracy and seems to be an reliable fusion technique for the tight security systems when compared with the sensor level fusion scheme.

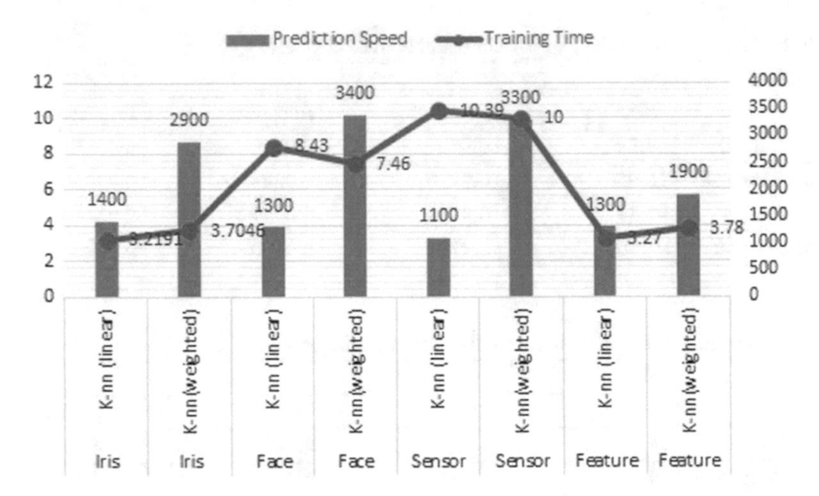

Fig. 2 Bar chart for K-NN classifier for training time in sec and prediction speed observations per sec

Face and Iris Multimodal Biometric Recognition System based on preclassification fusion schemes. Table 3 shows the Training time consumed in building the unimodal and multimodal recognition model of face and iris traits and also both the tables consists of the attribute prediction speed of observations per second, i.e., the number of samples the model developed is capable of going under observation per second, which gives the time complexity of the model. The entries in the tables are graphically depicted in the bar chart Figs. 2 and 3. SVMs are taking more

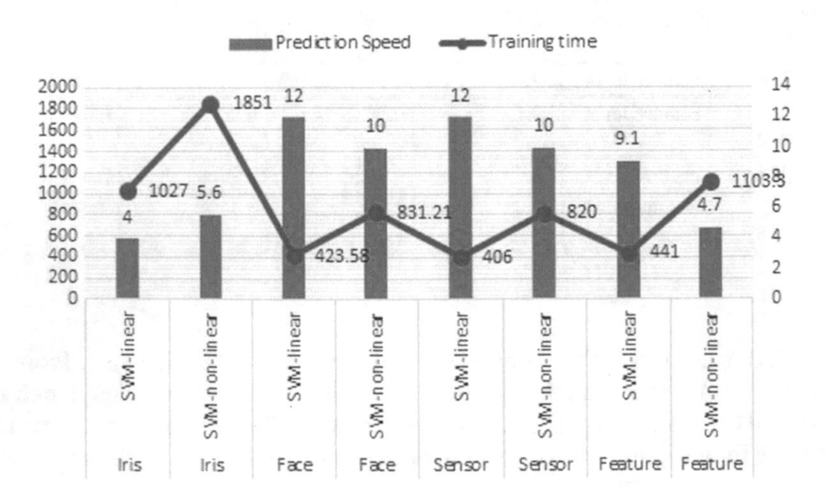

Fig. 3 Bar chart for SVM classifier for training time in sec and prediction speed observations per sec

training time and prediction time than K-NNs classifiers which can be seen in the bar chart. Hence, SVMs need more time to tune its hyperparameters for the data. When there is limited set of features in more than one dimensions, SVMs tends to perform very well. However, our biometric data in both unimodal and multimodal case has more features sets in high dimension, SVMs consume more training and prediction time. KNN usually detect linear or nonlinear distributed multidimensional data, it takes less time on both training and testing phase.

5 Conclusion

On choosing physiological traits (face, iris), we have evaluated the system by extracting LPQ features and for efficient computation, we have taken LDA for dimensionality reduction, then the well-known classifiers such as KNN (linear and weighted) and SVM (linear and nonlinear) are opted for classification. Our results give the accuracy in terms of performance measures such as precision, recall, and f-measure for both unimodal and multimodal (fusion prior to matching schemes) biometric system. The results also shows which trait is reliable on the employed feature extraction techniques and classifiers. Naturally, the multimodal biometric system would yield high accuracy. In our case, feature level fusion with 99.13% accuracy indicates the outstanding performance of our developed system. We have also tabulated training time needed to build the compact model and also the prediction speed for the samples per second time, which is shown graphically in gaining the computation time idea of the deployed system. Our future work will address on behavioral modalities and analyzing results on various classifiers.

References

1. Nanni, L., Lumini, A., Brahnam, S.: Likelihood ratio based features for a trained biometric score fusion. Expert Syst. Appl. **38**(1), 58–63 (2011)
2. Poh, N., Bourlai, T., Kittler, J.: A multimodal biometric test bed for quality-dependent, cost-sensitive and client-specific score-level fusion algorithms. Pattern Recognit. **43**(3), 1094–1105 (2010)
3. Zhang, Q., Liu, Y., Blum, R.S., Han, J., Tao, D.: Sparse representation based multi-sensor image fusion for multi-focus and multi-modality images: a review. Inf. Fusion **40**, 57–75 (2018)
4. Toh, K.-A., Kim, J., Lee, S.: Biometric scores fusion based on total error rate minimization. Pattern Recognit. **41**(3), 1066–1082 (2008)
5. Kang, B.J., Park, K.R.: Multimodal biometric method based on vein and geometry of a single finger. IET Comput. Vis. **4**(3), 209–217 (2010)
6. Sim, H.M., Asmuni, H., Hassan, R., Othman, R.M.: Multimodal biometrics: weighted score level fusion based on non-ideal iris and face images. Expert Syst. Appl. **41**(11), 5390–5404 (2014)
7. Rodrigues, R.N., Ling, L.L., Govindaraju, V.: Robustness of multimodal biometric fusion methods against spoof attacks. J. Vis. Lang. Comput. **20**(3), 169–179 (2009)

8. Lakshmanan, L.: Efficient person authentication based on multi-level fusion of ear scores. IET Biom. **2**(3), 97–106 (2013)
9. Kabir, W., Ahmad, M.O., Swamy, M.N.S.: Score reliability based weighting technique for score-level fusion in multi-biometric systems. In: 2016 IEEE Winter Conference on Applications of Computer Vision (WACV), pp. 1–7, Lake Placid, NY (2016)
10. Mezai, L., Hachouf, F.: Score-level fusion of face and voice using particle swarm optimization and belief functions. IEEE Trans. Hum.-Mach. Syst. **45**(6), 761–772 (2015)
11. http://115.249.38.8/mepcobiodb/HomePage.html

Classification of Multi-class Microarray Cancer Data Using Ensemble Learning Method

B. H. Shekar and Guesh Dagnew

Abstract Nowadays, microarray cancer analysis is one of the top research areas in the field of machine learning, computational biology, and pattern recognition. Classifying cancer data into their respective class and its analysis plays a key role in diagnosis, identifying negative and positive cases as well as treatment in the case of binary classes. In the case of multi-class classification, the aim is to identify the type of cancer. The main challenge in microarray cancer datasets is the curse of dimensionality and lack of sufficient sample data. To overcome this problem, feature selection and dimensionality reduction are explored in identifying relevant features. In this work, we propose an ensemble learning method for multi-class cancer data classification. The Information Gain (IG) is used for feature selection which works by ranking attributes according to their relevance with respect to the class label. Three classifiers are used, namely k-Nearest Neighbor, Logistic Regression, and Random Forest. tenfold cross validation is applied to train and test the model. Experiments are conducted on the standard multi-class cancer datasets, namely Leukemia 3 class, Leukemia 4 class, Harvard Lung cancer 5 class, and MLL 3 class. To evaluate the performance of the model, various performance measures such as Classification Accuracy, F1-measure, and Area Under the Curve (AUC) are used. Confusion matrix is used to show whether or not samples are correctly classified. Comparison of each classifier's performance is presented on the basis of performance evaluation criteria. Significant performance improvement is observed in the results due to feature selection for three of the classifiers with the exception of random forest's performance on MLL Leukemia whose result is found to be good on the original dataset compared to the selected features. For the rest of the datasets, all classifiers registered better result due to feature selection.

Keywords Feature selection · Dimensionality reduction · Ensemble learning Microarray cancer data classifier

B. H. Shekar · G. Dagnew (✉)
Department of Computer Science, Mangalore University, Karnataka, Mangalore, India
e-mail: guesh.nanit@gmail.com

B. H. Shekar
e-mail: bhshekar@gmail.com

P. Nagabhushan et al. (eds.), *Data Analytics and Learning*,
Lecture Notes in Networks and Systems 43,
https://doi.org/10.1007/978-981-13-2514-4_24

1 Introduction

Cancer is a group of diseases characterized by the growth of cells abnormally and uncontrollably invades the body to destroy normal tissues [1]. Classification of microarray cancer data plays an important role in identification, diagnosis, prognosis, and treatment of cancer cell [12]. Classification of multi-class microarray cancer data remains as one of the most challenging research topics in the field of Bioinformatics, Machine Learning, and Pattern Classification. Nowadays, survival rates from cancer are improving due to enormous research achievements aiming towards enhancement in cancer screening and treatment. The main difficulty with microarray dataset classification arises from lack of enough samples and high dimensional features that lead to complexity in analysis. Microarray cancer data lacks sufficient sample size and contains redundant and irrelevant features which cause curse of dimensionality [11, 13, 19]. Many research works have been carried out related to binary class microarray cancer data classification. Classifying multi-class microarray data is still a hot research topic as a result of challenges in class imbalance whereby classes with a small number of samples are usually neglected due to biases of most of the models towards classes with more number of elements [4, 14]. More challenges arise when the problem is related to multi-class classification [18].

Classification task requires designing and development of a model, computation of input data which represent objects and foresees the class label with respect to the object under verification [11]. In [6, 8], an ensemble learning method is proposed using bi-objective genetic algorithm and hesitant fuzzy set approach respectively. An integrated particle Swarm optimization (PSO algorithm) with the C4.5 classifiers called PSOC4.5 was proposed by [5] to address gene selection problem. They evaluate the method with average performance using fivefold cross validation method on various microarray cancer datasets. A modification of analytic hierarchy process (MAHP) for gene selection was proposed in [16]. Adaptive rule-based classifier was proposed by [9] for big biological datasets. They use Decision Tree (DT) and k-Nearest-Neighbor (KNN) to construct their model. We have seen a comparative research work done by [15] to classify cancer data using machine learning methods. It is a well known fact that the feature selection plays a great role in minimizing complexity arising from the curse of dimensionality by selecting informative features and neglecting insignificant ones. In our work, we propose to explore to the Information Gain (IG) task for feature selection. It filters all features in the dataset in ascending order of importance with respect to the class. It assigns zero ranks to all irrelevant features, which is considered to be an adaptive threshold value of the filter.

The rest of the paper is outlined as follows, Sect. 2 introduces the proposed methodology, Sect. 3 deals with the evaluation methods, Sect. 4 deals with dataset description. Section 5 presents the experimental results and discussion. Finally, conclusion is drawn in Sect. 6.

2 Proposed Methodology

In this work, an ensemble learning method for multi-class microarray cancer data classification based on three established classifiers is proposed. In a classification problem, the classifier is usually given a known dataset for training and unknown data to test the model. 10fold cross validation method was used to train and test the classifier. Cross validation defines a dataset to test the model in the training phase so as to overcome the problem of overfitting. As shown in Fig. 1, given a microarray cancer dataset, three classifiers; k-Nearest Neighbor (KNN), Logistic Regression (LR), and Random Forest (RF) trained using tenfold cross validation. It works by iteratively trains each sample nine times and tests at the 10th iteration. The workflow of the proposed model is shown in Fig. 1, which clearly depicts the data entry point in the data file,the data view table, learning of the classifiers from the training data, and evaluation of the method on the test samples.

In the case of the KNN classifier, it is a parameter-free model which uses the number of neighbors k to overcome the complexity of the model. In this work we use $k = 5$ neighbors. To predict the class label y of a new data point, the model looks for the minimum distance d data points in the training set x_i as shown in Eq. (1).

$$y(x_{new}) = d(x_i, x_{new}) \tag{1}$$

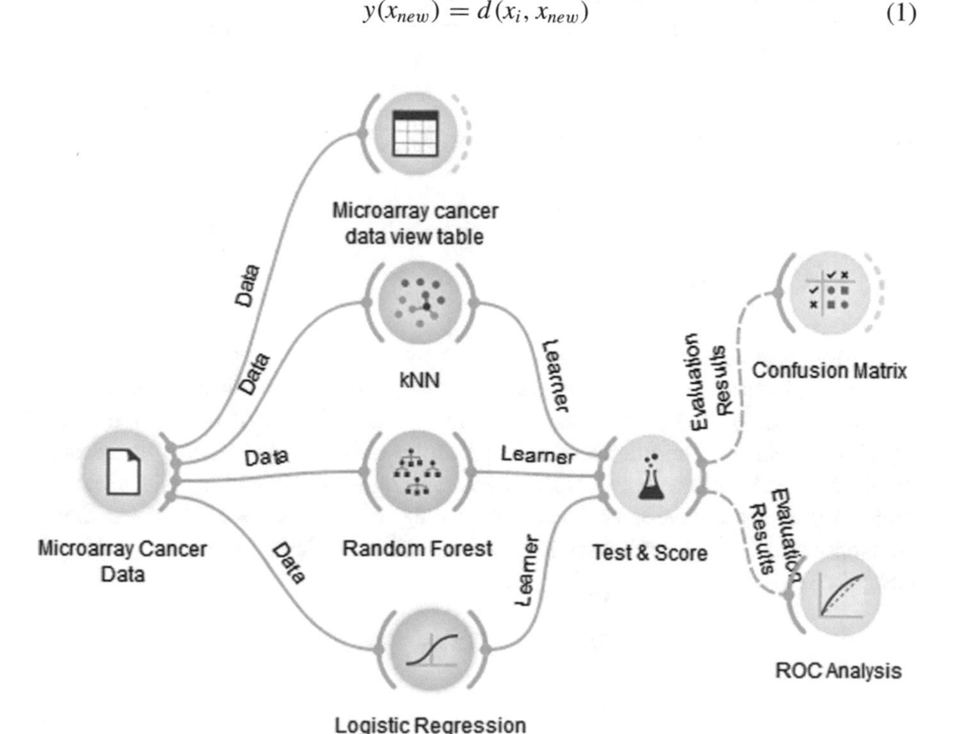

Fig. 1 Conceptual framework of the proposed model

Given number of classes C, if $C \geq 2$, then C number of two-class logistic regression classifiers are fitted where one class is compared with the rest of the classes combined. When a new data point X_i is arrived, it is classified according to where the classifier value is the largest for this data point. Logistic regression model uses a ridge regression or $L2$ and λ parameters to control the complexity of the model. It adds a penalty term $L2$ to the sum of squared errors of linear regression model as shown in Eq. (2):

$$E_{ij} = \sum_{i=1}^{n} \left(y_i - \sum_{j=1}^{p} x_{ij} w_j \right)^2 + \lambda \sum_{j=1}^{p} w^2 \tag{2}$$

where E is the error, X is a matrix of n by P (n samples and p features), y is a label, and w are coefficients, λ is a parameter which controls the shrinkage of the coefficients. The larger the value of λ, the greater the shrinkage and the coefficients become more robust to collinearity [17].

On the other hand, the Random forest model is an ensemble method which works by fitting a number of decision tree classifiers on the subset of the data set and then it uses average of those decision trees which improve the predictive accuracy and handles over-fitting. The probability of a sample to be member of a certain class is expressed by the probability of each features f of that sample towards class c as shown in Eq. (3).

$$p(c|f) = p_1(c|f) + p_2(c|f) + \ldots + p_n(c|f) = \sum_{i=1}^{n} p_i(c|f) \tag{3}$$

2.1 Feature Selection

Feature selection plays a key role in microarray-related research works in order to handle problems such as overfitting, running time and space complexity. Information gain was used for feature reduction. Information gain is one of the filter methods which evaluate the importance of the feature by ranking with respect to the class label. Important features get a higher rank and higher value and usually irrelevant features will get a rank of zero and hence they will not be considered for model development [3, 7]. Information gain is a measure of change in entropy which is computed based on Eq. (4).

$$IG(S, X) = E(S) - E(S, X) \tag{4}$$

$$IG(S, X) = Entropy(S) - \sum_{v \in Values(X)} \frac{|S_v|}{|S|} \cdot Entropy(S_v) \tag{5}$$

Here, S is the set of instances and X is a feature, $|S|$ is the size of S, S_v stands for a subset of S with $X_v = v$ and $Values(X)$ is the set of all possible values of X. Each feature's information gain will be calculated according to Eq. (4). Information gain is the amount of information obtained from each feature. The focus of information gain is to find an attribute that gives the largest information gain by ranking the features in accordance to their relevance.

$$E(S, X) = \sum_{n=1} -p_i Log_2 P_i \tag{6}$$

where E stands for entropy, S the sample size, X is a feature and p is the probability. The entropy of each feature is computed as given by Eq. (6). Entropy is a measure used to compute how pure or mixed a given attribute is in the distribution.

3 Evaluation Metrics

To evaluate the results of the proposed model, performance measures such as Classification Accuracy (CA), recall, precision, F1-measure, Receiver Operating Characteristic (ROC), and Area Under the Curve (AUV) are used. As shown in Eq. (7), classification accuracy computes correctly classified to the total sample size in the test data. Since accuracy alone is not sufficient to measure the performance of a model, other performance metrics are also considered.

$$Accuracy = \frac{TP + TN}{TP + TN + FP + FN} \tag{7}$$

Recall also known as sensitivity is a true positive rate, which is the ratio of true positive (TP) to the sum of true positive (TP) and false negative(FN). It measures the number of true positives (TP) to the ratio of total TP and FN as shown in Eq. (8).

$$Recall = \frac{TP}{TP + FN} \tag{8}$$

Another performance measure used in this work is the *precision* which is also referred to as *positive predictive value* (PPV). Equation 9 shows the precision which computes the ratio of correctly classified samples (true positives) to the sum of true positives and false positives.

$$Precision = \frac{TP}{TP + FP} \tag{9}$$

Moreover, as shown in Eq. (10), F1-measure is also used as a performance metric. This metric is applied to neutralize the biases in Precision and Recall. F1-measure considers the harmonic mean of Precision and Recall [17].

$$F_1 = 2 \cdot \frac{Precision \times Recall}{Precision + Recall} \tag{10}$$

Moreover, confusion matrix is also used as evaluation criteria. Confusion matrix is essential in identifying correctly classified and wrongly classified samples in the test data. Samples along the diagonal are correctly classified and off the diagonal elements are wrongly classified.

4 Dataset Description

Four publicly available microarray cancer datasets which are downloaded from Shenzhen University website [21] were used in this work and their description is shown in Table 1. Moreover, name of each datasets, sample size, number of features, number of classes, and selected number of features are presented. Three of the datasets are from Leukemia cancer family and the remaining one is lung cancer data with five classes. The number of selected features for MLL 3 class dataset are about 7% of the total number of features and about 93% features were discarded. This implies that the values of MLL 3 class dataset have significant variance and only a few relevant ones are selected. The same is true for all the datasets regarding the size of selected features with respect to the original number of features.

Table 1 List of datasets and their description

Dataset name	Sample size	Original number of features	Number and names of classes	Selected number of features
MLL_3C	72	12583	3 (ALL, MLL, AML)	916
Leukemia_3C	72	7130	3 (AML, T-cell, B-cell)	4612
Lung cancer Harvard1_5C	202	12600	5 (ADEN, COID, NORMAL, SCLC, SQUAL)	9662
Leukemia_4C	72	7130	4 (B-cell, BM, PB, T-cell)	800

5 Experimental Results and Discussion

In this section, present experimental results of the proposed model on four standard microarray multi-class cancer datasets. The results obtained due to the proposed methodology is discussed below.

Since the proposed method is an ensemble learning method, three established classifiers are used to train and test the model, these are k-Nearest Neighbor, Logistic Regression, and Random Forest tree. The results from each classifier before and after feature selection is presented in Table 2. In terms of classification accuracy, better results are found after feature selection with the exception of LR which shows relatively the same result on lung cancer and MLL Leukemia. Significant results are registered by RF and KNN on all datasets after feature selection. In the case of LR classifier, it shows improvement on Leukemia 3 class dataset. In the case of F1-Measure, all classifiers show improvement with the exception of LR which still shows the same performance regardless of feature selection on lung cancer and MLL leukemia.

As can be seen from Table 2, the performance of the classifiers have shown improvement in terms of classification accuracy (CA), area under the curve (AUC) precision and recall and this is due to the feature selection task carried out using information gain. Three classifiers namely k-Nearest Neighbor (KNN), Logistic Regression and Random Forest shows significant performance improvement on leukemia 3 class dataset after feature selection. Moreover, the performance of three of the classifiers is improved on leukemia 4 class and leukemia 3 class datasets for all the metrics as shown in Table 2. Three of the classifiers KNN, LR and RF shows improvement in classification accuracy on Leukemia 3 class data as a result of feature selection. The performance of KNN and RF is improved on Leukemia 4 class, Lung cancer, and MLL datasets but little improvement is observed for LR classifier on the same datasets.

An optimal number of features with respect to the sample size highly depends on the type of classifier to be used. As a result of this, a different number of features have been found after feature selection. The classifiers produce better results after feature selection, but with some exceptions. This is observed in the performance of random forest (RF) that decreases in MLL Leukemia see Table 2.

Tables 3, 4, 5, 6, 7, and 8 shows confusion matrix on leukemia 3 class and leukemia 4 class datasets using the three classifiers. Table 3 shows confusion matrix using KNN classifier on leukemia 3 class dataset. The number of correctly classified samples is improved after feature selection and the improvements are shown along the diagonal in boldface. Similarly, Table 4 shows confusion matrix of leukemia 3 class dataset using LR classifier. As this classifier is shown little improvement, only the B-cell class is shown improvement by one element and the other two classes register the same regardless of feature selection. The same behavior is reflected for LR classifier on Leukemia 4 class dataset as shown in Table 7 which shows little improvement.

Table 5 shows confusion matrix of Leukemia 3 class dataset using RF classifier and it shows improvement after feature selection. The improvement is shown in bold

Table 2 Classification report using KNN, LR, and RF before and after feature selection

Dataset	Classification report before feature selection						Classification report after feature selection				
Leukemia_3C	Method	AUC	CA	F1	Precision	Recall	AUC	CA	F1	Precision	Recall
	KNN	0.99	0.87	0.87	0.89	0.88	0.99	**0.96**	0.96	0.96	0.96
	LR	0.99	0.94	0.94	0.94	0.94	0.99	**0.96**	0.96	0.96	0.96
	RF	0.96	0.89	0.88	0.89	0.89	0.99	**0.96**	0.95	0.95	0.96
Leukemia_4C	KNN	0.96	0.85	0.82	0.80	0.85	0.97	**0.89**	0.86	0.84	0.89
	LR	0.99	0.92	0.91	0.92	0.92	0.99	0.92	0.91	0.92	0.92
	RF	0.94	0.79	0.76	0.75	0.79	0.98	**0.89**	0.87	0.86	0.89
Harvard Lung_5	KNN	0.98	0.93	0.92	0.93	0.93	0.98	0.93	0.93	0.94	0.93
	LR	0.99	0.95	0.95	0.95	0.95	0.99	0.95	0.95	0.95	0.95
	RF	0.98	0.89	0.88	0.90	0.89	0.99	**0.92**	0.91	0.93	0.92
MLL_3C	KNN	0.97	0.90	0.90	0.90	0.90	0.99	**0.94**	0.94	0.95	0.94
	LR	1.00	0.97	0.97	0.97	0.97	1.00	0.97	0.97	0.97	0.97
	RF	0.98	**0.98**	0.93	0.93	0.93	0.98	0.86	0.86	0.86	0.86

Table 3 Leukemia 3 class confusion matrix before and after feature selection for KNN

Before feature selection						After feature selection					
Predicted class						Predicted class					
Actual class	Class	AML	B-cell	T-cell	∑	Actual class	Class	AML	B-cell	Tcell	∑
	AML	**6**	2	0	8		AML	**6**	1	0	7
	B-cell	0	**12**	0	12		B-cell	0	**14**	0	14
	T-cell	0	1	**3**	4		T-cell	0	0	**3**	3
	∑	6	15	3	24		∑	6	15	3	24

Table 4 Leukemia 3 class confusion matrix before and after feature selection for LR

Before feature selection						After feature selection					
Predicted class						Predicted class					
Actual class	Class	AML	B-cell	T-cell	∑	Actual class	Class	AML	B-cell	Tcell	∑
	AML	**7**	1	0	8		AML	**7**	1	0	8
	B-cell	1	**13**	0	14		B-cell	0	**14**	0	14
	T-cell	0	0	**2**	2		T-cell	0	0	**2**	2
	∑	8	14	2	24		∑	7	15	2	24

Table 5 Leukemia 3 class confusion matrix before and after feature selection for RF

Before feature selection						After feature selection					
Predicted class						Predicted class					
Actual class	Class	AML	B-cell	T-cell	∑	Actual class	Class	AML	B-cell	Tcell	∑
	AML	**8**	1	0	9		AML	**9**	0	0	9
	B-cell	1	**11**	0	12		B-cell	0	**11**	0	11
	T-cell	0	1	**2**	3		T-cell	1	0	**3**	4
	∑	9	13	2	24		∑	10	11	3	24

along the diagonal line. Confusion matrix of RF classifier is shown in Table 8 which also shows improvement in three of its classes. However, the problem of multi-class cancer data classification is visible here which neglects the class *PB* which is a minority class. The minority class neglection behavior in multi-class problem is also observed for the class label PB in the confusion Matrix of KNN as shown in Table 6 that all elements of the PB class are completely ignored and misclassified and allocated in the off diagonal spaces.

The performance of the proposed model registers improved result when it is evaluated using receiver operating characteristic (ROC) curve. ROC curve is used in addition to other performance metrics which has high degree of tolerance in classifying data with low class imbalance. ROC Curve for three of the classifiers

Table 6 Leukemia 4 class confusion matrix before and after feature selection for KNN

Before feature selection							After feature selection						
	Predicted class						Predicted class						
Actual class	Class	B-cell	BM	PB	T-cell	$\sum$	Class		B-cell	BM	PB	T-cell	$\sum$
	B-cell	11	0	0	0	11	Actual class	B-cell	**12**	0	0	0	**12**
	BM	2	**6**	0	0	8		BM	1	**7**	0	0	8
	PB	1	1	**0**	0	2		PB	0	2	**0**	0	2
	T-cell	0	0	0	**3**	3		T-cell	0	0	0	**2**	2
	$\sum$	14	7	0	3	24		$\sum$	13	9	0	2	24

Table 7 Leukemia 4 class confusion matrix before and after feature selection for LR

Before feature selection							After feature selection						
	Predicted class						Predicted class						
Actual class	Class	B-cell	BM	PB	T-cell	$\sum$	Class		B-cell	BM	PB	T-cell	$\sum$
	B-cell	**12**	0	0	0	**12**	Actual class	B-cell	**12**	0	0	0	**12**
	BM	0	**7**	0	0	**7**		BM	0	**7**	0	0	**7**
	PB	0	1	**1**	0	**2**		PB	0	1	**1**	0	**2**
	T-cell	0	0	0	**3**	**3**		T-cell	0	0	0	**3**	**3**
	$\sum$	**12**	**8**	**1**	**3**	**24**		$\sum$	**12**	**8**	**1**	**3**	**24**

Table 8 Leukemia 4 class confusion matrix before and after feature selection for RF

Before feature selection							After feature selection						
	Predicted class						Predicted class						
Actual class	Class	B-cell	BM	PB	T-cell	$\sum$	Class		B-cell	BM	PB	T-cell	$\sum$
	B-cell	**9**	0	1	0	10	Actual class	B-cell	**10**	0	0	0	10
	BM	0	**7**	0	1	8		BM	0	**8**	0	0	8
	PB	0	1	**1**	0	2		PB	0	1	**1**	0	2
	T-cell	2	0	0	**2**	4		T-cell	2	0	0	**2**	4
	$\sum$	11	8	2	3	24		$\sum$	12	9	1	2	24

KNN, Logistic Regression, and Random Forest on leukemia cancer dataset with three classes after feature selection is shown in Fig. 2. ROC curve for leukemia 4 class dataset is shown in Fig. 3. Moreover, ROC of lung cancer with 5 class and MLL leukemia with 3 class shown in Figs. 4 and 5 respectively. Percentage performance of all the classifiers is presented in Table 2.

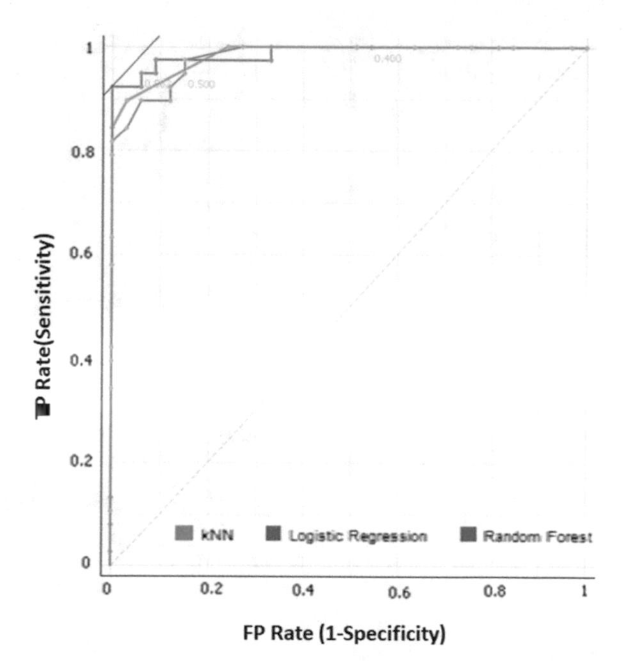

Fig. 2 ROC curve for leukemia 3 class dataset

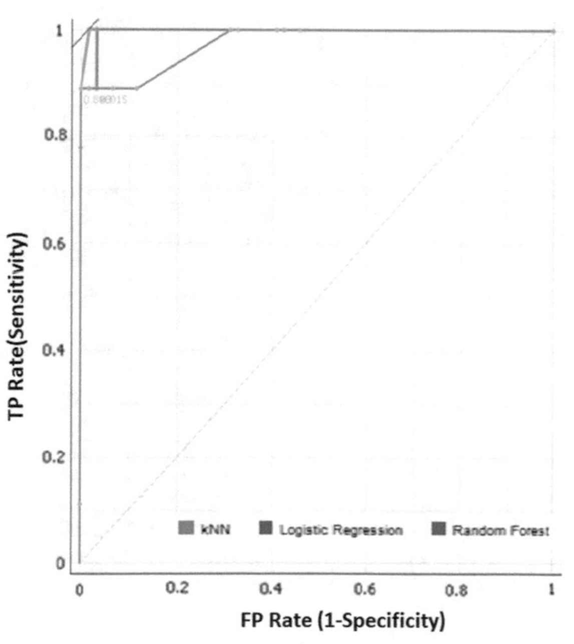

Fig. 3 ROC curve for leukemia 4 class dataset

Comparison of our work with some of the selected related works is presented in terms of classification accuracy. Our work is showing better accuracy on MLL and Lung cancer datasets. In the case of Leukemia with three-class dataset, one of

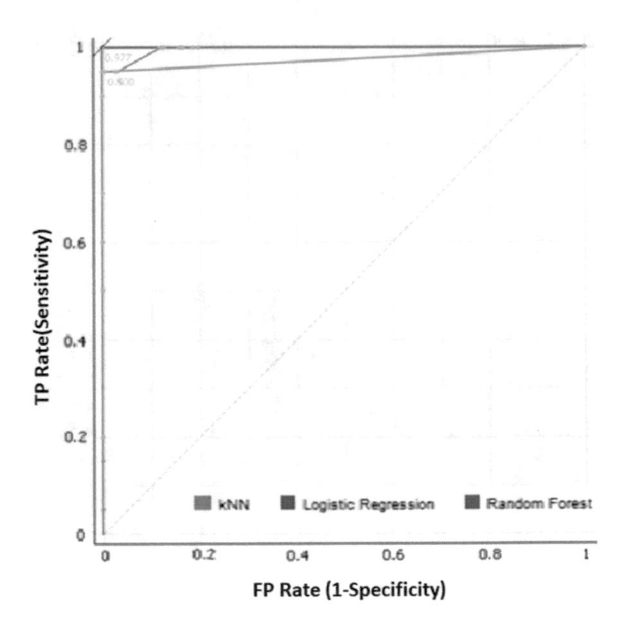

Fig. 4 ROC curve for harvard lung cancer

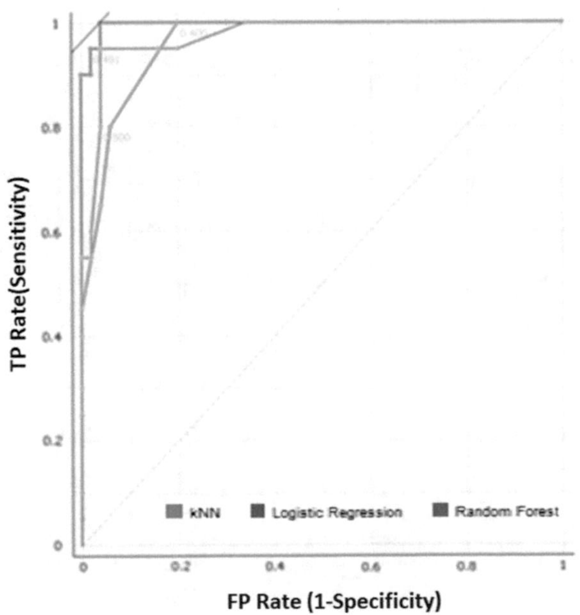

Fig. 5 ROC curve for MLL 3 class dataset

the latest work [20] shows better result with a unit difference. Results with better accuracy are indicated in boldface as shown in Table 9.

Table 9 Comparison of classification accuracy with related works

Authors	Method	MLL	Leukemia 3 class	Lung cancer
Our work	IG + KNN	0.944	0.96	0.931
	IG + LR	**0.972**	0.958	**0.951**
	IG + RF	0.861	0.958	0.921
Rabia et al. [2]	PCA + SVM	–	0.76	–
	PCA + FBFE + NB	–	0.914	–
	ICA + NB	–	0.86	–
Guo et al. [10]	LDA	0.949	–	0.94
Yang et al. [20]	PAM (for Leukemia 3 class) NB (for MLL), 1-SVM (Lung cancer)	0.958	**0.97**	0.833

6 Conclusion

In this paper, ensemble learning method based multi-class microarray cancer data classification is introduced. Ensemble learning method considers more than one classifier at a time on one or more dataset to enhance performance of a single classifier. To overcome the problem of model overfitting, feature selection is carried out using Information Gain (IG). tenfold cross validation method was used to train and test the model which trains the model 9 times and the test is carried out in the 10th iteration. Each sample is subjected to testing at least once. The model was trained and tested on four different microarray cancer datasets. From the results, it is noticed that both feature selection and cross-validation contributes a lot towards an improvement of performance measures of the specified classifiers. The proposed model is comparable to some related state of the art works. However, the behavior of misclassification of elements in minority classes is observed in some of the datasets such as leukemia 4 class, where the elements of the class PB are misclassified in the KNN classifier. From the observation we believe that the sophisticated and accurate model need to be devised to reduce the misclassification of minority classes. To evaluate the performance of the model, metrics such as Classification Accuracy (CA), recall, precision, F1-measure, receiver operating characteristic (ROC), and area under the curve (AUV) are used. The confusion matrix is also used to display correctly classified and misclassified samples in both before and after feature selection. The proposed model's performance result is comparable to recent works in the same domain.

References

1. Alberts, B., Johnson, A., Lewis, J., Roberts, K., Walter, P., Raff, M.: Cancer as a microevolutionary process (2002)
2. Aziz, R., Verma, C.K., Srivastava, N.: A fuzzy based feature selection from independent component subspace for machine learning classification of mi- croarray data. Genomics Data **8**, 4–15 (2016)
3. Bolón-Canedo, V., Sánchez-Marono, N., Alonso-Betanzos, A., Benítez, J.M., Herrera, F.: A review of microarray datasets and applied feature selection methods. Inf. Sci. **282**, 111–135 (2014)
4. Cerf, L., Gay, D., Selmaoui-Folcher, N., Crémilleux, B., Boulicaut, J.F.: Parameter-free classification in multi-class imbalanced data sets. Data Knowl. Eng. **87**, 109–129 (2013)
5. Chen, K.H., Wang, K.J., Wang, K.M., Angelia, M.A.: Applying particle swarm optimization-based decision tree classifier for cancer classification on gene expression data. Appl. Soft Comput. **24**, 773–780 (2014)
6. Das, A.K., Das, S., Ghosh, A.: Ensemble feature selection using biobjective genetic algorithm. Knowl.-Based Syst. **123**, 116–127 (2017)
7. Dashtban, M., Balafar, M.: Gene selection for microarray cancer classification using a new evolutionary method employing artificial intelligence concepts. Genomics **109**(2), 91–107 (2017)
8. Ebrahimpour, M.K., Eftekhari, M.: Ensemble of feature selection methods: a hesitant fuzzy sets approach. Appl. Soft Comput. **50**, 300–312 (2017)
9. Farid, D.M., Al-Mamun, M.A., Manderick, B., Nowe, A.: An adaptive rule-based classifier for mining big biological data. Expert Syst. Appl. **64**, 305–316 (2016)
10. Guo, S., Guo, D., Chen, L., Jiang, Q.: A l1-regularized feature selection method for local dimension reduction on microarray data. Comput. Biol. Chem. **67**, 92–101 (2017)
11. Kar, S., Sharma K.D., Maitra, M.: Gene selection from microarray gene expression data for classification of cancer subgroups employing PSO and adaptive K-nearest neighborhood technique. Expert Syst. Appl. **42**(1), 612–627 (2015)
12. Kumar, M., Rath S.K.: Classification of microarray using MapReduce based proximal support vector machine classifier. Knowl.-Based Syst. **89**, 584–602 (2015)
13. Lin, T.C., Liu, R.S., Chen, C.Y., Chao, Y.T, Chen, S.Y.: Pattern classification in DNA microarray data of multiple tumor types. Pattern Recognit. **39**(12), 2426–2438 (2006)
14. Moayedikia, A., Ong, K.L., Boo, Y.L., Yeoh, W.G., Jensen, R.: Feature selection for high dimensional imbalanced class data using harmony search. Eng. Appl. Artif. Intell. **57**, 38–49 (2017)
15. Nematzadeh, Z., Ibrahim, R., Selamat, A.: Comparative studies on breast cancer classifications with k-fold cross validations using machine learning techniques. In: 2015 10th Asian Control conference (ASCC), pp. 1–6 (2015)
16. Nguyen, T., Khosravi, A., Creighton, D., Nahavandi, S.: A novel aggregate gene selection method for microarray data classification. Pattern Recognit. Lett. **60**, 16–23 (2015)
17. Pedregosa, F., Varoquaux, G., Gramfort, A., Michel, V., Thirion, B., Grisel, O., Blondel, M., Prettenhofer, P., Weiss, R., Dubourg, V., Vanderplas, J., Passos, A., Cournapeau, D., Brucher, M., Perrot, M., Duchesnay, E.: Scikit-learn: machine learning in python. J. Mach. Learn. Res. **12**, 2825–2830 (2011)
18. Piao, Y., Piao, M., Ryu, K.H.: Multiclass cancer classification using a feature subset-based ensemble from microRNA expression profiles. Comput. Biol. Med. **80**, 39–44 (2017)
19. Salem, H., Attiya, G., El-Fishawy, N.: Classification of human cancer diseases by gene expression profiles. Appl. Soft Comput. **50**, 124–134 (2017)
20. Yang, S., Naiman, N.Q.: Multiclass cancer classification based on gene expression comparison. Stat. Appl. Genet. Mol. Biol. **13**(4), 477–496 (2014)
21. Zhu, Z., Ong, Y.-S., Dash, M.: Markov blanket-embedded genetic algorithm for gene selection. Pattern Recognit. **40**(11), 3236–3248 (2007)

A Study of Applying Different Term Weighting Schemes on Arabic Text Classification

D. S. Guru, Mostafa Ali, Mahamad Suhil and Maryam Hazman

Abstract In this paper, a study on the effect of different term weighting techniques on Arabic text complaints' categorization is made. Farmers' complaints written in unstructured and ungrammatical way are analyzed to be classified with respect to crop name. Initially, the complaints are preprocessed by removing stop words, correcting writing mistakes, and stemming. Some of the domain-specific special cases which may affect the classification performance are handled. Different term weighting schemes like TF, TF–IDF, and TF–ICF are used to form representative vectors for the complaints to train a classifier. Finally, the trained classifier is used to classify an unlabeled complaint. Moreover, a dataset contains more than 5300 Arabic complaints pertaining to 8 crops has been created. KNN classifier has been used for classification. The experiments show that there is stability difference between term weighting techniques. Further, a comparison analysis among the four feature selection techniques has been demonstrated.

Keywords Term weighting schemes · Arabic text complaints
Text classification approach · Feature selection techniques · TF–IDF

D. S. Guru · M. Ali (✉) · M. Suhil
Department of Studies in Computer Science, University of Mysore,
Manasagangotri, Mysuru 570006, Karnataka, India
e-mail: mam16@fayoum.edu.eg

D. S. Guru
e-mail: dsg@compsci.uni-mysore.ac.in

M. Suhil
e-mail: mahamad45@yahoo.co.in

M. Hazman
Central Laboratory for Agricultural Experts Systems,
Agricultural Research Center (ARC), Giza, Egypt
e-mail: maryam.hazman@gmail.com

P. Nagabhushan et al. (eds.), *Data Analytics and Learning*,
Lecture Notes in Networks and Systems 43,
https://doi.org/10.1007/978-981-13-2514-4_25

1 Introduction

With the huge amount of text being produced every day, there is an increasing demand to design new techniques to analyze text automatically and to extract the helpful knowledge. Machine learning is the technique that is used in automatic process for text analyzing. Text Classification (TC) is one of the data mining tasks which are used for routing an unlabeled document to a labeled class based on the learning process. There is many state-of-the-art classifiers like k-Nearest Neighbor (KNN) classifier [1]. A term weighting technique is used in computing the weight of a term with respect to a document. Term Frequency–Inverse Document Frequency (TF–IDF) is one of the widely used term weighting schemes for text classification [2]. Arabic language is one of the major international languages spoken by more than 300 million people and it has many features which do not exist in any other language [3]. There is a lack of research on Arabic language text because of the complexity in context and also in morphology of Arabic language words that makes it difficult to extract features to apply text mining tasks. Besides, analysis of Arabic farmers' complaints is a very important task to infer useful knowledge to recommend suitable solutions to the farmers. Classification of Arabic complaints into different crops is difficult task as they are written in unstructured way.

A farmer can describe his problem in Arabic language and he should wait for an expert to response with the answer. Agriculture portals allow farmers to solve their problems and it is noticed that most of the farmers describe their complaints in slang Arabic text. The major objectives of this work are to analyze Arabic slang complaints by applying suitable preprocessing steps and to make a comparison among term weighting schemes that can be used in estimating the weight of the terms of each complaint to come out with a stable and better performance of classification process. Three-term weighting techniques are considered in our work, namely Term Frequency (TF), Term Frequency–Inverse Document Frequency (TF–IDF), and Term Frequency–Inverse Class Frequency (TF–ICF). For selecting the most discriminating features for each class, four feature selection techniques are used, namely Chi-Square (Chi) [4], Information Gain (IG) [5], Bi-Normal Separation (BNS) [4], and Weighted Log-Likelihood Ratio (WLLR) [6].

The rest of the paper is organized as follows; Sect. 2 presents a literature survey on the term weighting schemes used in text classification. Section 3 presents a brief description about the term weighting techniques used for this study. Section 4 gives a review of the four feature selection techniques which are considered in our work. Section 5 refers to the Arabic complaints dataset. Section 6 explains the classification approach, and finally Sect. 7 shows the experiments and results.

2 Literature Survey

Due to the vital role played by the terms in text classification, we can trace many works on text classification by using different weighting schemes. The authors in [7] showed that the result of text classification depends on the choice of term weighting schemes by proposing a new single term weighting technique. In [8], the authors made a comparison study for TF * IDF, LSI, and multi-word for text representation and they applied on Chinese and an English document collection to evaluate the three methods. The authors in [2] introduced the new term Inverse Category Frequency (ICF) and then they proposed two novel approaches which are; tf.icf and ICF-based supervised term weighting schemes. In [9], the authors proposed an enhancement on TF–IDF by modifying the weight of each word based on its length. The authors in [10] proposed new three supervised weighting schemes namely, qf * icf, iqf * qf * icf, and vrf for question categorization with a comparison with other supervised and unsupervised techniques. A comparison among term weighting schemes is made [11, 12]. The authors in [11] also present a new weighting scheme Term Frequency–relevance Frequency (TF–RF) that achieved better performance than other schemes. The authors in [13], have proposed a new scheme called Synonyms-Based Term (SBT) weighting scheme. This scheme computes the Inverse Document Frequency (IDF) of the term according to the synonyms based. In [14], the authors used TF–IDF for reducing features dimensionality and used ANN as a classification technique to classify Arabic text documents. In [15], the authors exploited two basic factors of Importance of Terms in a Document (ITD) and Importance of Terms for expressing Sentiment (ITS). In [16], the authors proposed a framework for selecting a most relevant subset of features by ranking the features in groups instead of ranking individual features for text categorization.

3 Term Weighting Schemes

3.1 *Term Frequency–Inverse Document Frequency (TF–IDF)*

TF–IDF is used to estimate the importance of a term for a given document of interest with respect to all other documents in the corpus. As a weighting term, *TF–IDF* is considered as unsupervised technique [2]. The *TF–IDF* computes the weight for a term t_i based on its frequency within the document d and its weight within the whole corpus D. Hence, the ability of the term t_i in discriminating a class from the other is not captured during the computation of its weight. The value of *TF–IDF* will be highest, if the term t_i has many occurrences within a small number of documents and this gives high discriminating power to those documents. It will become lower if the term t_i has a few occurrences in a document or if occurs in many documents. The term *TF–IDF* can be shown in Eq. (1).

$$TF\text{-}IDF(t_i, d, D) = \left\{\frac{f(t_i, d)}{|N_d|}\right\} \times \log\left\{\frac{N}{|d \in D : t_i \in d|}\right\} \tag{1}$$

where N_d is total number of term frequencies in the document and N is the total number of documents in the corpus $N=|D|$.

3.2 Term Frequency–Inverse Class Frequency (TF–ICF)

As *IDF* is the inverse relative document frequency of the term with respect to other documents D in the corpus, *ICF* means the inverse relative class frequency of the term with respect to all classes C. The part *ICF* can be defined as shown in Eq. (2).

$$ICF(t_i, C) = \log\left\{\frac{K}{k = |c \in C : t_i \in c|}\right\} \tag{2}$$

where $K=|C|$ is the total number of classes in the corpus and k is the number of classes contains the term t_i. The term *ICF* is previously suggested as a supervised weighting term by [2] to classify standard news dataset, but here it is applied in an Arabic complaints dataset. So, the term *TF–ICF* can be shown in Eq. (3).

$$TF\text{-}ICF(t_i, d, C) = \left\{\frac{f(t_i, d)}{|N_d|}\right\} \times \log\left\{\frac{K}{|k = c \in C : t_i \in c|}\right\} \tag{3}$$

4 Feature Selection Techniques

In this section, a brief description about four feature selection methods considered for our study is presented which are *CHI, IG, BNS, and WLLR*. These techniques are widely used as feature selection techniques [4, 16]. If there is N number of documents, N_j is the number of documents belonging to class C_j, a feature t_i, a_{ij} is the number of documents in C_j and containing t_i, b_{ij} is the number of documents in C_j and not containing t_i,f_{ij} is the number of documents that do not belong to C_j and containing t_i and d_{ij} is the number of documents that do not belong to C_j and not containing t_i, the feature selection technique can be described as follows.

4.1 Chi-Square (Chi)

The feature t_i can be selected based on its correlation with a category C_j and it can be computed as in [4]. It is defined in Eq. (4).

$$x^2(t_i, C_j) = \frac{N \times (a_{ij}d_{ij} - b_{ij}f_{ij})^2}{(a_{ij} + b_{ij}) \times (a_{ij} + f_{ij}) \times (b_{ij} + d_{ij}) \times (f_{ij} + d_{ij})} \tag{4}$$

4.2 *Information Gain (IG)*

It is used in machine learning as a term goodness criterion. By knowing the presence or the absence of a term in the document, the number of bits required is measured for predicting the category [5]. IG is shown in Eq. (5).

$$\begin{aligned} IG(t_i) = &-\sum \Pr(C_j) \log \Pr(C_j) + \Pr(t_i) \sum \Pr(C_j|t_i) \log \Pr(C_j|t_i) \\ &+ \Pr(\bar{t}_i) \sum_i \Pr(C_j|\bar{t}_i) \log \Pr(C_j|\bar{t}_i) \end{aligned} \tag{5}$$

4.3 *Bi-Normal Separation (BNS)*

BNS is defined as difference between the inverse cumulative probability function of true positive ratio (TPR) and false positive ratio (FPR). This measures the separation between these two ratios [4]. It is shown in Eq. (6).

$$BNS(t_i, C_j) = F1(tpr) - F1(fpr) \tag{6}$$

where F1 is the normal inverse cumulative distribution function. It is considered as a feature selection metric.

4.4 *Weighted Log-Likelihood Ratio (WLLR)*

WLLR is proportional to the frequency measurement and the logarithm of the ratio measurement [6]. WLLR can be shown in Eq. (7).

$$WLLR(t_i, C_j) = \frac{a_{ij}}{N_j} \log \frac{a_{ij} * (N - N_j)}{b_{ij} * N_j} \tag{7}$$

5 Arabic Farmers' Complaints

As a service from the agricultural governmental centers in Egypt, agricultural portals have been developed to publish the recommendation guides to the farmers. There is also ability for a farmer to describe his complaint freely and waiting an agriculture specialist expert to provide him by how to solve his problem. These complaints are written in unstructured way of Arabic text which is not well formatted. All the complaints present different causes that affect the plant. A dataset has been created by collecting more than 5300 Arabic text complaints related to eight different crops from VERCON website [17]. An example of the complaint is "نوفمبر فظهر حشائش عريضة ورفيعة فما هى المكافحة المناسبة مع العلم بان الارض رملية ؟ارض قمح تم زراعتها فى منتصف" which means "**A farming land has been planted with wheat crop. The plants has been cultivated in mid-November, then broad and high grasses appeared, what is the appropriate control mechanism to handle these weeds in a sandy land?**". After solving these agricultural complaints, the Egyptian agricultural extension officers used it to solve other farmers' complaints. Agricultural extension officers take time to find the nearest farmer's complaint to use in solving new one. So, in our work, we classify these complaints to help them to recommend the suitable solutions for a new farmer's complaint.

6 Classification Approach

The architecture of the classification approach is shown in Fig. 1.

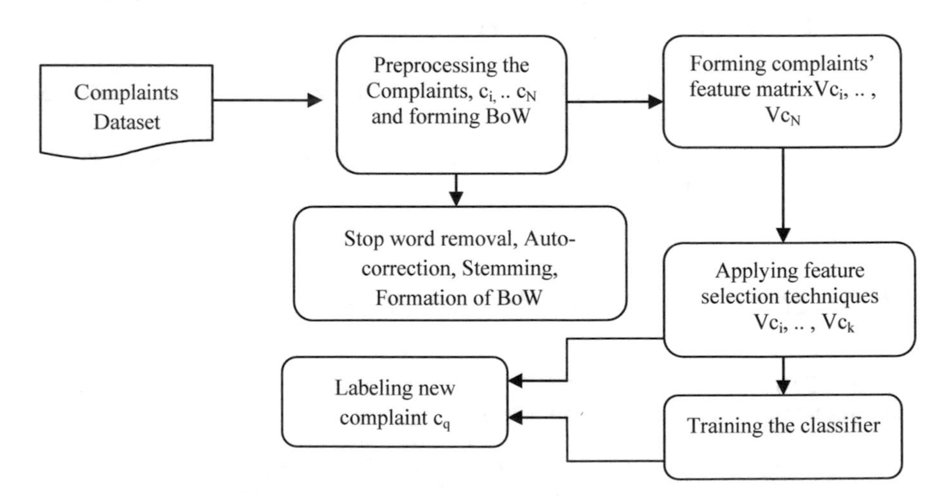

Fig. 1 Classification approach for classifying Arabic complaints

6.1 Preprocessing Phase

In this section, the complaints are analyzed automatically to be ready for classification phase. Four processes are applied on the complaints viz., stop words removal, complaints auto-correction, stemming, and formation Bag of Words (BoW).

Stop Words Removal: The most frequent unwanted words must be removed from the complaints because they affect weighting the discriminating features. The Arabic stop words like; من, الى, عن, على, فوق, تحت, etc., are eliminated by using the technique proposed in [18].

Auto-correction: The complaints described by the farmers may have writing mistakes that affect the classification results. One of those mistakes is writing the crop name or any other keywords in a slang way that make it difficult to be extracted. For example, the corn crop name "ذره" may be incorrectly written as "زره" and the rice crop name "أرز" may be incorrectly written in the slang way as "رز". So, to handle this problem, we suggest a regular expression to replace the incorrect term with the correct one. The second issue that there is a disease name in some crops that is same as the name of a crop. For example, in the complaint " والسيقان ملمسها قطني وجود بقع بيضاء على الورقة والسنابل" which describe one problem related to the "Wheat" crop, the crop is affected by the disease "بقع بملمس قطني" which means "spots with cotton feel". This disease mostly affects many crops. The complaints describing this disease will generally contain the word "cotton" which is actually the name of a crop. Hence it adversely affects the classification performance. Therefore, we removed such terms from all other crop classes using a list of crop names.

Stemming: An Arabic stemmer [19] which has high accuracy by the experimentation done by the authors in [20] is used to stem the terms of each complaint.

Formation of BoW: After analyzing the terms of the complaints, a BoW is formed by combining all the terms from all the complaints. The combined BoW is considered as a common bunch of words that can represent all the complaints in the dataset.

6.2 Classification Phase

In this section, we present a classification approach used for training and classification of the complaints. After completing preprocessing phase, the complaints are represented in the form of a feature matrix and different feature selection techniques are applied to reduce the dimensionality of the feature matrix.

Forming the Feature Matrix of the Complaints: Based on the combined BoW, the weight of every term in each complaint is evaluated by using the TF, TF–IDF, and TF–ICF to form the vector of the complaint $Vc_i = \{w_1, w_2, w_3, \ldots, w_n\}$, where Vc_i is the weight vector for the complaint c_i, w_i is the weight for the term t_i with respect to the complaint c_i, and n is the number of terms in the BoW. Hence, complaints are represented by three different feature matrices corresponding to the three-term weighting schemes.

Feature Selection Process: After representing the complaints in the form of a feature matrix, four feature selection techniques namely, *CHI, IG, BNS* and *WLLR* are used to select the most discriminating features with respect to each class. The resultant feature matrix after this step contains only m columns corresponding to m features selected.

Training and Classification: The complaints of different classes represented in the form of m-dimensional feature vectors are used to train a classifier. In our model, k-Nearest-Neighbor (KNN) classifier has been employed. Given a test complaint cq, it is represented in the form of a similar m-dimensional vector and it is given as input to the trained KNN classifier for classification.

7 Experimentation, Results and Discussion

In this section, we present the results of the experimentation conducted on our dataset of 5300 complaints to evaluate the proposed model. The three-term weighting schemes (TF, TF–IDF, and TF–ICF) presented in Sect. 2 are used along with the four feature selection techniques (CHI, IG, BNS, and WLLR). The experiments are conducted by varying the number of features from 20 to 400 in steps of 20. For experimentation, 50% of the dataset is used as training set to train the KNN classifier with K = 3 and the remaining 50% is used for testing.

7.1 Results with TF as Term Weighting

This experiment shows the performance of classifying the complaints by using the TF term weighting scheme in evaluating the terms of each complaint. Table 1 shows the results of applying the classifier with four different feature selection techniques on the complaints in TF matrix form.

It can be noticed from the Table 1 that the accuracy of classification starts between 81 and 82% for all the four feature selection techniques at number of features equal to 20. Then, the accuracy of all the techniques appeared to be increased to reach the maximum (given in bold) as the number of features increased and then starts decreasing. It is seen that the performance does not increase stably with the increase in the number of features. Figure 2 shows the graphical representation of the same.

Table 1 Accuracy of classification with TF as term weighting scheme

# of selected features	Classification accuracy			
	CHI	IG	BNS	WLLR
20	81.68	81.79	81.12	82.01
40	83.64	83.51	83.14	83.58
60	84.59	**85.13**	83.95	84.53
80	85.18	85.05	**84.07**	**84.77**
100	85.1	84.23	83.91	84.33
120	**85.46**	83.89	83.69	83.88
140	85.4	83.19	83.41	83.43
160	85.14	[illegible]	83.08	82.3
180	84.69	81.3	82.04	81.84
200	84.3	80.6	81.54	81.68
220	83.59	80.27	80.85	81.33
240	82.94	80.05	80.42	80.84
260	82.71	79.76	80.33	80.25
280	82.32	79.44	80.06	79.79
300	81.26	79.33	79.79	79.59
320	80.94	79.15	79.7	79.22
340	80.66	78.91	79.66	78.99
360	80.13	78.74	79.52	78.72
380	79.95	78.65	79.13	78.61
400	79.9	78.52	78.99	78.63

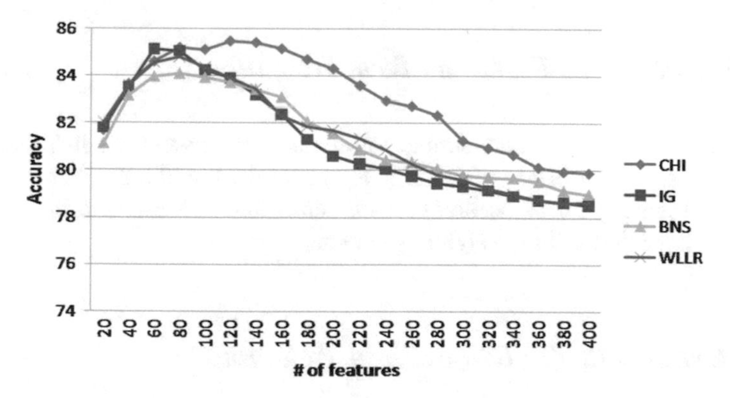

Fig. 2 The performance of our proposed classification using TF as term weighting scheme for different feature selection techniques

Table 2 Accuracy of classification with TF–IDF as term weighting scheme

# of selected features	Classification accuracy			
	CHI	IG	BNS	WLLR
20	81.63	81.81	81.16	81.69
40	83.35	83.85	82.82	83.78
60	84.47	**85.27**	**82.96**	**84.84**
80	84.77	84.34	82.81	84.28
100	84.53	83.67	82.03	83.64
120	**84.82**	83.49	81.27	83.2
140	84.61	83.1	80.8	83.05
160	84.26	82.65	80.33	82.7
180	83.83	82.44	80.1	82.31
200	83.62	81.91	79.85	82
220	83.31	81.4	79.64	81.58
240	82.99	81.14	79.25	81.08
260	82.72	80.9	79.27	80.62
280	82.31	80.5	79.12	80.37
300	81.84	80.31	79.17	80
320	81.52	79.94	78.99	79.37
340	81.33	79.71	78.82	79.01
360	80.8	79.35	78.48	78.71
380	80.14	78.95	78.3	78.54
400	79.9	78.81	78.08	78.28

7.2 Results with TF–IDF as Term Weighting

Here, the performance is more similar to previous experiment. As in Table 2, it is observed that the performance stability is not resulted as the number of features increased. Figure 3 shows accuracy of classification with increase in number of features using TF–IDF term weighting scheme.

7.3 Results with TF–ICF as Term Weighting

The result of classification with TF–ICF being the term weighting scheme is presented in Table 3. It is clear from the Table 3 that the performance of classification in this experiment is the best among all the three experiments. Although the accuracy at beginning is almost same as previous results, the accuracy with the increase number of features continued to increase till 87.56, 87.35, 86.25, and 86.3 for the selection

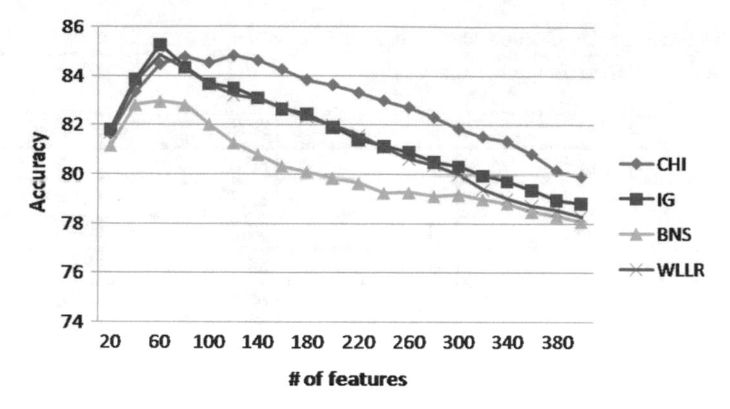

Fig. 3 Performance of our proposed classification using TF–IDF as term weighting scheme for different feature selection techniques

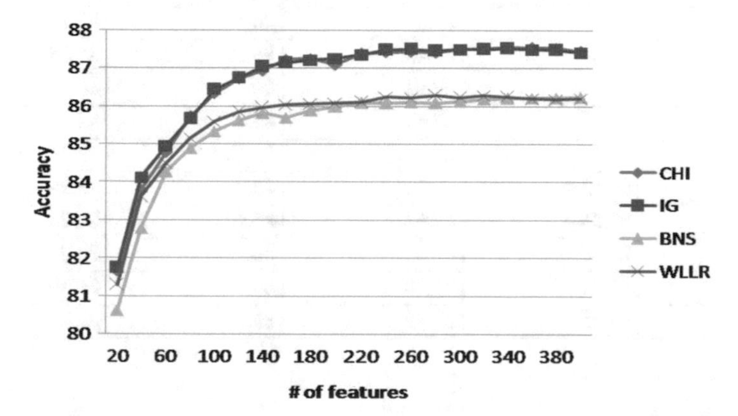

Fig. 4 Performance of our proposed classification using TF–ICF as term weighting scheme for different feature selection techniques

techniques Chi, IG, BNS, and WLLR, respectively. Moreover, the performance of TF–ICF-based classification is the best in terms of stability as the number of selected features is increased.

The accuracy of classification using TF–ICF is graphically shown in Fig. 4. It can be observed from Fig. 4 that not only the performance of the weighting scheme TF–ICF is better than other two weighting schemes used, but also the performance is more stable.

Table 3 Accuracy of classification with TF–ICF term weighting scheme

# of selected features	Classification accuracy			
	CHI	IG	BNS	WLLR
20	81.66	81.75	80.62	81.3
40	83.8	84.12	82.79	83.62
60	84.78	84.95	84.27	84.5
80	85.73	85.67	84.89	85.16
100	86.34	86.44	85.34	85.6
120	86.72	86.74	85.63	85.85
140	86.94	87.05	85.82	85.97
160	87.24	87.15	85.7	86.05
180	87.26	87.22	85.89	86.07
200	87.1	87.26	86	86.09
220	87.39	87.36	86.1	86.12
240	87.43	87.49	86.09	86.25
260	87.44	87.51	86.12	86.23
280	87.42	87.47	86.1	**86.3**
300	87.5	87.5	86.14	86.24
320	87.53	87.51	86.2	86.29
340	**87.56**	**87.53**	86.22	86.26
360	87.55	87.5	86.23	86.21
380	87.54	87.49	86.23	86.19
400	87.45	87.42	**86.25**	86.21

8 Conclusion

In this work, an approach for farmers' complaints classification into different crops has been proposed. Three different schemes are used to compute the weight of the terms with respect to complaints namely; TF, TF–IDF and TF–ICF. Moreover, four different state-of-the-art feature selection techniques namely; Chi, IG, BNS and WLLR are used in the study. Finally, a KNN classifier is applied to classify the complaints based on the crop name. Experiments are conducted on our own dataset of 5300 complaints for each term weighting scheme. The results show that the accuracies of using TF and TF–IDF as a term weighting schemes are nearly similar and the performance is not stable as the number of features is increased. On the other hand, it is observed that the classification accuracy with TF–ICF as term weighting is better and more stable than other two weighting schemes used.

References

1. Cover, T.M., Hart, P.E.: Nearest neighbor pattern classification. IEEE Trans. Inf. Theory **13**(1), 21–27 (1967)
2. Wang, D., Zhang, H., Wu, W., Lin, M.: Inverse-category-frequency based supervised term weighting schemes for text categorization. J. Inf. Sci. Eng. **29**, 209–225 (2013)
3. Hammo, B., Abu-Salem, H., Lytinen, S.: QARAB: a question answering system to support the Arabic language. In: Proceedings of the ACL-02 Workshop on Computational Approaches to Semitic Languages, pp. 1–11 (2002)
4. Forman, G.: An extensive empirical study of feature selection for text classification. J. Mach. Learn. Approach **3**, 1289–1305 (2003)
5. Yang, Y., Pedersen, J.O.: A comparative study on feature selection in text categorization. In: Proceedings of the Fourteenth International Conference on Machine Learning (ICML-97), pp. 412–420 (1997)
6. Nigam, K., McCallum, A., Thrun, S., Mitchell, T.: Text classification from labeled and unlabeled documents using EM. Mach. Learn. **39**(2/3), 103–134 (2000)
7. Salton, G., Buckley, C.: Term-weighting approaches in automatic text retrieval. Inf. Process. Manage. **24**(5), 513–523 (1988)
8. Zhang, W., Yoshida, T.: A comparative study of TF-IDF, LSI and multi-words for text classification. Expert Syst. Appl. **38**, 2758–2765 (2011)
9. Fawaz, S., AbuZeina, D.: A new enhanced variation of TF-IDF scheme for arabic text classification. In: 3rd International Conference on Innovative Engineering Technologies (ICIET '2016), pp. 24–28 (2016)
10. Quan, X., Liu, W., Qiu, B.: Term weighting schemes for question categorization. IEEE transactions on pattern analysis and machineintelligence **33**(5), 1009–1021 (2011)
11. Lan, M., Sung, S.-Y., Low, H.-B., Tan, C.-L.: Acomparative study on term weighting schemes for textcategorization. In: Neural Networks, 2005. IJCNN '05. Proceedings. 2005 IEEE International Joint Conference, vol. 1, pp. 546–551. IEEE (2005)
12. Lan, M., Tan, C.L., Su, J., Lu, Y.: Supervised and traditional term weighting methods for automatic text categorization. IEEE Trans. Pattern Anal. Mach. Intell. **31**(4), 721–735 (2009)
13. Kumari, M., Jain, A., Bhatia, A.: Synonyms based term weighting scheme: an extension to TF.IDF. Proc. Comput. Sci. **89**, 555–561 (2016)
14. Zaghoul, F.A., Al-Dhaheri, S.: Arabic text classification based on features reduction using artificial neural networks. In: Computer Modelling and Simulation (UKSim), UKSim 15th International Conference on IEEE (2013)
15. Deng, Z.H., Luo, K.H., Yu, H.L.: A study of supervised term weighting scheme for sentiment analysis. Expert Syst. Appl. **41**(7), 3506–3513 (2014)
16. Guru, D.S., Suhil, M., Raju, L.N., Vinay Kumar, N.: An alternative framework for univariate filter based feature selection for text categorization. Pattern Recognit. Lett. **103**, 23–31 (2018)
17. http://www.vercon.sci.eg/. Accessed Nov 2017
18. Ibrahim, A.: Effects of stop words elimination for arabic information retrieval: a comparative study. Int. J. Comput. Inf. Sci. **4**(3), 119–133 (2006)
19. Khoja, S., Garside, R.: Stemming Arabic Text. Computing Department, Lancaster University, Lancaster, UK (1999)
20. Sawalha, M., Atwell: Comparative evaluation of Arabic language morphological analysers and stemmers. In: 22nd International Conference on Computational Linguistics, pp. 107–110 (2008)

A Survey on Different Visual Speech Recognition Techniques

Shabina Bhaskar, T. M. Thasleema and R. Rajesh

Abstract In automatic speech recognition (ASR) visual speech information plays a pivotal role especially in the presence of acoustic noise. This paper provides a short review of the different methods for visual speech recognition systems (VSR). Here, we discuss the different stages of VSR including the face and lip localization techniques and different visual feature extraction techniques. We also provide the details of audio-visual database related to this study.

Keywords Visual speech recognition VSR
Audio-visual speech recognition AVSR · Discrete cosine transform DCT
Automatic lip reading (ALR) · Discrete wavelet transform DWT
Hidden markov model HMM

1 Introduction

Speech recognition provides an easy communication between man and machine, but the performancc of speech recognition system degrades when working under noisy condition. Adding visual information obtained from mouth movements and lip shapes can reduce the effect of acoustic noisy background condition and helps to improve the speech recognition system performance [1]. This visual information is known as lip reading or speech reading [2] and this technique is down pat especially by people with hearing problems. This lip-reading ability helps them to reduce the communication gap between other people. Visual speech recognition (VSR) or Automatic Lip reading (ALR) is a technique in which speech recognition is done by the using visual cues

S. Bhaskar (✉) · T. M. Thasleema · R. Rajesh
Central University of Kerala, Kasaragod, Kerala, India
e-mail: shabinabhaskar@hotmail.com

T. M. Thasleema
e-mail: thasnitm1@hotmail.com

R. Rajesh
e-mail: rajeshr@cukerala.ac.in

P. Nagabhushan et al. (eds.), *Data Analytics and Learning*,
Lecture Notes in Networks and Systems 43,
https://doi.org/10.1007/978-981-13-2514-4_26

obtained from the images of visible articulators like mouth, teeth, tongue, lips, chin, and any other related part of the face. Visual speech recognition has been used in wide variety of applications ranging from audio-visual speech recognition (AVSR), speaker recognition, human computer interaction (HCI), talking heads, sign language recognition to video surveillance. This study focuses on lip-reading techniques for speech recognition and most of the related works have been accomplished through audio-visual speech recognition researches. In VSR system recognition of spoken word is done by processing the visual signals that formed during speech. Visual signal processing involves the area such as artificial intelligence, image processing, pattern recognition, object detection, statistical modeling, etc.

The relevant visual speech information for speech recognition is obtained from the mouth and lip region of face therefore, it is important for any lip-reading system to concentrate on the lip area. In lip-reading systems, for extracting lip features some approaches directly locate the lips of the speakers [3], but in other work first perform face localization based on prior knowledge and only after that the lip localization is performed [4]. However, perfect lip and face localization are difficult problem due to variations in sensor quality, light conditions, background, lips dynamic, pose, shadowing, facial expressions, scale, rotation and occlusion. An accurate lip localization combined with the robustness of the extracted features is vital in determining the accuracy of the VSR system. So, face localization and lip localization are important steps in VSR system to extract the correct features for speech recognition.

The existing visual feature extraction techniques can be categorized as geometric feature based, image based, motion based, and model based [4, 5]. In image based, either the raw pixel values used straightly or the image undergoes some transformation. Whereas in motion based, facial movements during uttering consist of significant speech information. Geometric feature based uses the geometrical attributes such a height, width of mouth as features. In model-based technique, models of visible articulators are constructed and its configuration is described by small set of parameters used as visual features [1].

This paper is arranged as follows: Sect. 2 deals with the architecture of VSR and different methods associated with each stage. Section 3 discuss what are the observations made and Sect. 4 is conclusion (Fig. 1).

2 Architecture of VSR

Most VSR approaches are based on recognizing the visual unit called visemes. A viseme is the smallest visual unit and it represents mouth shape or mouth appearance or a group of mouth dynamics that are required to generate a phoneme in the visual domain [6]. A viseme is the visual equivalent of phoneme speech, and a phoneme is the smallest (shortest) audible component of speech. A detailed study of viseme classification methods is described in [6, 7]. Stages of a typical VSR system include image/video acquisition, face and lip localization, feature extraction, then visemes recognition and word recognition is done by using appropriate classifier and lan-

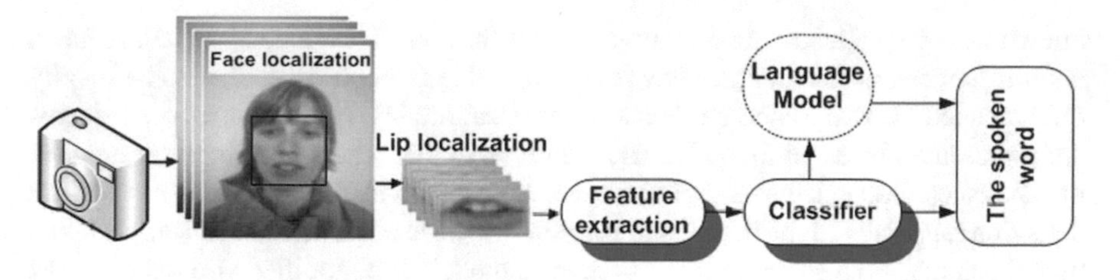

Fig. 1 Architecture of VSR (from Hassanat)

guage model. The first stage is a preprocessing stage and in second step, the system perform the face localization from an image or video frame. Face localization is an essential step because it decreases the search area for the lip localization process and increases the accuracy of lip localization. The accuracy of the final stages, visual feature extraction and recognition depends on the lip localization stage hence lip localization plays an important role in VSR systems.

2.1 Face Localization

Face detection plays a vital role in many face related applications. In face localization an assumption is made that it is guaranteed to find a face in an image, but the face detection algorithm needs to find out whether there is actually a face or not, and then locates the face. The face detection accuracy depends upon the parameters such as pose, orientation and lighting or recording condition. Face detection is not a problem related to this study because most audio-visual databases are based on the assumption that there is only one face, upright, looking at the camera. There are many methods that are available for face detection and localization that include template based [8], knowledge based and appearance-based methods [1]. Most commonly used method is the Viola–Jones algorithm it is based on the Adaboost classifier which is used for any object detection including human face [9].

2.2 Lip Localization

The location of the mouth and the lip region or simply the region of interest (ROI) is important for the extraction of vital visual speech information. It is therefore paramount to locate certain landmarks, e.g., lip corners, nose, eye in order to be able to locate the mouth accurately and then crop off the ROI and normalize its size for further extraction of visual features. Other methods include active shape model (ASM), active appearance model (AAM), active contour model (ACM). ACM's are general shape detection methods such as snake detectors and have a variation known as the localized active contour model that was introduced in [10]. It has a natural framework that enables any region-based segmentation energy to be locally reformulated. Active Shape models are statistical models of the shapes, i.e., shapes are represented by a number of labeled landmarks that adjust iteratively in order to fit the detected object into a digital image. It has successfully been used in feature detection especially in the mouth and lip region. On the other hand, active appearance model contains both shape model and texture model. The modified version of AAM in [5, 7, 11] uses the gray scale and shape information of whole image.

The difference in color of the lips and the face region around the lips is the determinant factor used in the color-based approach in lip localization and is described in [3]. An algorithm is proposed for mouth region determination from face ROI [4] which does not assume any geometric model and complex procedures. Another method to increase the lip tracking accuracy is by marking fiducials on the speaker's lips [5]. By using this method, the lip visual feature will be independent of the distance between camera and speaker. In [8] Lip localization is done by converting gray scale image into binary image and counting the black pixel in binary image.

2.3 Visual Feature Extraction

As discussed earlier in visual speech recognition, visual features can be mainly classified as shape-based features and appearance-based. In appearance-based feature extraction, the entire mouth region is assumed to be informative about speech. To make computationally feasible appearance-based model we have to perform the transformation like PCA, DCT, DWT, etc., to convert into fewer dimensions. By applying any of the statistical or parametric lip tracking algorithms, the shape-based features can be extracted from the inner and outer lip contours.

Visual feature extraction using both lip shape and intensity information was done in [1]. The visual modules, identify and track the speaker lip movements and extract relevant speech features. The lip localization is done by using appearance-based lip-model. Visual speech features are represented by the lip contour and mouth area gray level information. The noise-robust acoustic features are extracted from the acoustic signal in acoustic module. The joint temporal modeling of the acoustic-visual feature is done in the sensor fusion module and is realized using multi stream Hidden Markov Model (HMM).

Geometrical visual features and their best combinations are compared and analyzed for both visual only and audio-visual speech recognition [5]. In this, the effect of modeling parameters of HMM on the recognition result of each single visual feature and combinations of geometrical visual features are experimented for audio-visual speech recognition. Finally, the lip vertical aperture is identified as the best visual feature in their study. For audio-visual speech recognition Timothy [2] introduced a segment-based modeling methodology. Segment consisting of variable length segment hypothesis which each corresponds to a potential phonetic unit. Here they used appearance-based visual feature extraction method and by using AVCSR Toolkit the features are extracted from the mouth region. Using histogram equalization each image is first normalized then PCA is applied, and the top 32 coefficients are concatenated to create one 96-dimensional vector per frame. Audio and visual feature streams are integrated using segment constrained HMM.

Comparison and study of different image transform-based features for VSR is done in [3] and compared the performance of DCT, DWT, PCA, FDCT (fast discrete curvelet transform) and LDA under clean and noisy condition. FDCT is based on anisotropic scaling principle and it gives important and helpful features from lip contour for speaker recognition study. Other methods capture the lip information that are suitable for speech recognition purpose. Also, the dynamic visual features included in this work which provides information about the change or rate of change of the static features over time. The dynamic features used in this experiment were calculated as first and second derivatives of cubic splines constructed for each feature in utterance.

Puvisan and Palanivel [4] introduced a lip-reading method and the participants selected for this study were people with hearing difficulty. In this work for feature extraction they performed DCT and DWT transform on mouth images and these features are applied to inputs of HMM for recognizing the visual speech. A review of various viseme recognition methods and algorithms are presented in [6]. The paper proposed both geometrical and texture parameters calculation algorithm. The textual parameters are based on the determination of histograms for ROI and of the DCT transform. For the extraction of geometrical features, the points describing the contour of lips were used and the geometrical parameters divided into three types: the distance, the angle and the surface. The two classifiers used are HTK package and WEGA package (SMO-SVM extended by Sequential Minimal Optimization). Bimodal Hindi speech recognition work [8] made a comparative study of visual feature such as 2D-DCT, 2D-DWT followed by DCT, PCA and 2D-DWT followed by PCA. Comparison of different classifiers for lip reading with CUAVE and TULIPS database is another work [9] that used DCT, DWT as feature extraction methods.

Morade and Patnaik [10] proposed localized active contour model for contour extraction and different geometrical parameters such as area, width and height of lip during lip-reading process can be calculated by using this model. Also, the dynamic information such as changes in area, height, and width of lips are considered here and it tested using HMM with three states and five states. Astik et al. [11] presented a multiple camera AVSR experiment by using active appearance model. Here the jaw and lip region are important for extracting the shape and appearance information. Modeling is done by using four stream VSR combined with single stream acoustic HMM to build five-stream AVSR.

TCD-TIMIT [12] consists of high-quality audio and video footage of 62 speakers. A set of extracted ROIs is provided with TCD-TIMIT database and is extracted using nostril tracking algorithm. And they suggest that 44-length DCT coefficient vectors are adequate for representing ROIs. This database provides useful baselines for other researchers exploring audio-visual speech data.

Matthews et al. [13] in their work on extraction of visual features of lip-reading, the major problem in this system that is general to all computer vision systems to generate visual features which are in enormous quantity of data in video sequences. The feature extraction as done using low level statistical model to extract 60 features from 80 × 60 mouth images. The PCA have been used to identify orthogonal directions by their relative variance contribution. Multiple visual features [14] such as AAM-shape and AAM-texture parameters, DCT coefficients calculated from the inner and outer lip contours, histogram inner and outer parameters etc. are provided in MODALITY corpus. For the process of speech recognition, triphone based left–right HMM with five hidden states were used.

Paper [15] describes a new method by using a skin color filter followed by a border following method for geometric lip feature extraction from mouth images. The convex hull technique is also applied for the lip outline shape extraction. The extraction of outline of lip shape is important for getting good quality geometric features like width, height, perimeter, etc.

2.4 Databases

The major obstacle in visual speech processing study is the unavailability of suitable database. Only a very less number databases are publicly available with most of the database consists of restricted number of participants with limited vocabulary [16]. Next section deals with different databases associated with this study.

CUAVE (Clemson University Audio-Visual Experiments)
CUAVE is a standard database with video frame rate 30 frames/sec and consists of two significant sections. In first section, 36 participants (17 male and 19 female) participated in database recording in two different manner. For that each participant was requested utter 50 isolated digits while they were in standing position and no head movements were allowed. One more 30 isolated digits recorded in conditions like moving side-to-side, forward and backward, or moving the head. From that point forward, the participant was encircled from both profile views while uttering 20 isolated digits and the individual then uttered 60 connected digits while facing the camera again. The second section, the participants selected as pairs and 20 pair of participants engaged in recording. The recording is done in this manner such that from each pair one participant uttered a connect digit sequence followed by the other. For the second time, the participants order reversed and finally both speakers simultaneously utter their own digit sequences.

TULIPS
A Tulips is a low-scale audio-visual database developed by Movellan's laboratory at the Department of Cognitive Science, UCSD. It consists of total 96 utterance from 12 speakers uttering one to four digits in English. Each utterance was hand segmented so that the video and audio channels extended to one frame either side of an interval containing the significant audio energy.

XM2VTS (Extended Multi Modal Verification for Teleservices and Security Applications)

The XM2VTS database was developed by the University of Surrey for personal identification. Total 295 subjects participated in recording and the language materials include three utterances recorded in four sessions. Each speaker was requested to utter two continuous digit strings as well as one phonetically balanced sentence and for all sessions the utterance remained same.

AV-TIMIT (Audio-Visual Texas Instruments Massachusetts Institute of Technology)
The corpus composed of audio and video recordings of 233 speakers in which 117 speakers were male and 106 speakers were female. They selected the language materials as phonetically balanced sentences from TIMIT corpus. This database is helpful for the research areas such as person identification, automatic lip reading and multimodal speech recognition. First 20 sentences read by each speaker after that nine different speakers read the same sentence. These different speakers select any of the sentence uttered by all the speakers.

AVCAR (Audio-Visual Speech in Car)
This project selected a car environment for their recording. They arranged four cameras on car dashboard for getting four synchronized video streams with different views. As the space in the car is very limited, the angles between the views relative to the speaker were modest and the actual degrees unknown. Total 50 female and 50 male speakers were involved in the recording and but for downloading data of 86 of them are available. During recording they arranged five noise conditions. Each participant was asked to first speak isolated digits and letters twice under each condition. After that each speaker asked to read 20 sentences which were randomly selected from TIMIT corpus and 10 digits twenty phone numbers.

MODALITY (Multimodal Database of Speech for English Language)
The language materials consist of numbers, days and months name and a set of computers controlling verbs and nouns. The corpus consists of 35 speakers, 26 male and 9 female speakers. The corpus contains both native and non-native English speakers. A noise set up is created during recording by placing four loudspeakers in the room corners.

3 Discussions

The aim of feature extraction techniques for VSR systems is to represent the relevant speech information for the recognition speech by suppressing the other irrelevant information. In geometric features, geometry of mouth is captured but the other significant features such as tongue, teeth, etc., (cavity) information is lost and also geometric features depends on the lip contour extraction accuracy. The image transformation-based features capture both cavity information with geometrical parameters.

As compared to image transformation-based features, the geometrical features are less complex and easy to extract. But the combination of geometrical features such as the vertical distance between the points in chin and nose, their ratio, their first order derivatives, area, angle of lips gives a better recognition rate than the single geometric feature [5].

In the case of image transformation-based feature the recognition rates also depend on the classifier combination and the database [4, 9]. HMM with DWT based features performs better than HMM with DCT based features [4]. If we consider the CUAVE database the SVM performance is better as compared to BPNN, KNN, RFT, and Naïve Bayes. While in case of TULIPS database, BPNN is identified as the best classifier for the recognition of digits. There is small additional benefit was achieved by associating dynamic features in clean condition but the main benefits of dynamic features can be achieved through while testing in noisy condition. Comparison of different visual features are given in Table 1.

The database availability is the main problem related to the study of visual based application. Only a few databases are publicly available for visual only and audio-visual ASR study. Most of the currently available database contain limited number of speakers but as the number of speakers increases the feature extraction accuracy also increases [14, 16]. Furthermore, only very few databases considered acoustic and visual noise condition. If we include head pose variations and facial expressions by including the natural conversation then it will lead to more comprehensive AV speech database. The comparison of different database available is done in Table 2.

Table 1 Comparison of visual features

System	Visual features	Method	Classifier
Dupont 2002	Appearance-based features	Point distribution models are used to extract speech features	HMM
Mustafa 2004	Geometric features	Outer lip horizontal and vertical width, outer lip area, angle of outer lip corner	HMM
Seymour 2008	Image transform based	DCT, DWT of mouth region, dynamic features	HMM
Puvisan 2011	Image transform based	DCT, DWT of mouth region	HMM
Sunil 2015	Image transform based	DWT of lip portion	SVM, KNN, RFT, Naïve Bayes
Astik 2016	Shape-based	Active appearance model	HMM
Andrzej 2017	Shape-based	Active appearance model	HMM

Table 2 Comparison of different databases

Database	Resolution	Frame rate (fps)	Language material	Additional features	No. of speakers
TULIPS	100 × 75	30	Numerals 1–4	No	12
CUAVE	720 × 480	29.97	Connected numerals	Simultaneous speech	30
XM2VTS	720 × 576	25	Sentences–numerals and words	Head rotations	295
AV-TIMIT	512 × 384	25	10 TIMIT sentences	Noise	43
AVCAR	360 × 240	29.97	Isolated numerals, letters, phone numbers, TIMIT sentences	Automotive noise	86
MODALITY	1920 × 1080	100	168 commands	Varying noise	35

4 Conclusion

We have made a short review on the face and lip detection methods, visual feature extraction techniques and databases related to the visual speech recognition (VSR). The quality of VSR mainly depends upon the database and visual features selected. The review reveals that only the two databases AVCAR and XM2VTS considered about the speaker number above 70 and most of them considered the acoustic noise only. For a better database design for VSR, we have to include large number of speakers, natural conversations and facial expressions, both acoustic and visual noise. Finally, the comparative analysis of visual feature extraction techniques shows that the transform-based approaches include more visual information than the geometric feature-based approaches. But the combination of both transform and geometric-based technique will lead to a better visual feature extraction approach.

References

1. Dupont, S., Luettin, J.: Audio-visual speech modelling for continuous speech recognition. IEEE Trans. Multimed. **2**(3), 141–151 (2000)
2. Hazen, T.J.: Visual modal structures and asynchrony constraints for audio-visual speech recognition. IEEE Trans. Audio Speech Lang. Process. **14**(3) (2006)
3. Seymour, R., Stewart, D., Ming, J.: Comparison of image transform based features for visual speech recognition in clean and corrupted videos. EURASIP J. Image Video Process. **2008**(14) (2008)
4. Puvisan, N., Palanivel, S.: Lip reading of hearing impaired persons using HMM. Int. J. Expert Syst. Appl. **38**(4) (2011)
5. Kaynak, M.N., Cheok, A.D., Sengupta, K., Jian, Z., Chung, K.C.: Lip geometric features for human-computer interaction using bimodal speech recognition: comparison and analysis. Speech Commun. **43**(1–2), 1–16 (2004)

6. Jachimski, D., Czyzewski, A., Ciszewski, T.A.: Comparative study of English viseme recognition methods and algorithms. Multimed. Tools Appl. (2017)
7. Hassanat, A.B.: Visual words for automatic lip reading. Ph.D. thesis, Buckingham, UK, University of Buckingham (2009)
8. Upadhyaya, P., Farooq, O.: Comparative study of visual feature for bimodal Hindi speech recognition. Arch. Acoust. 609–619 (2015)
9. Morade, S.S., Patnaik, S.: Comparison of classifiers for lip reading with CUAVE and TULIPS database. Int. J. Light Electr. Opt. **126**(24) (2015). Elsevier
10. Morade, S.S., Patnaik, S.: A novel lip-reading algorithm by using localized ACM and HMM: tested for digit recognition. Int. J. Light Electr. Opt. **125**(18) (2014). Elsevier
11. Astik, B., Sahu, P.K., Chandra, M.: Multiple camera audio visual speech recognition using active appearance model in car environment. Int. J. Speech Technol. **19**(1) (2016). Springer
12. Harte, N.: TCD-TIMIT: an audio-visual corpus of continuous speech. IEEE Trans. Multimed. (2015)
13. Matthews, I., Cootes, T.F., Banbham, J.A., Cox, S., Harvey, R.: Extraction of visual features of lip reading. IEEE Trans. Pattern Anal. Mach. Intell. **24**(2) (2002)
14. Czyzewski, A., Kostek, B., Bratoszewski, P., Kotus, J., Szykulski, M.: An audio-visual corpus for multimodal automatic speech recognition. J. Intell. Inf. Syst. **49**, 167 (2017)
15. Ibrahim, M.Z., Mulvaney, D.J.: Geometric based lip-reading using template probabilistic multi-dimension dynamic time warping. J. Vis. Commun. Image Represent. **30** (2015)
16. Zhu, Z., Zhao, G., Hong, X., Pietikainen, M.: A review of recent advances in visual speech decoding. Int. J. Image Vis. Comput. **32**(9) (2014)

Activity Recognition from Accelerometer Data Using Symbolic Data Approach

P. G. Lavanya and Suresha Mallappa

Abstract With the advent of smart phones, data collection from low cost sensors like accelerometer has become easy and inexpensive. Accelerometer data plays a very important role in activity recognition of individuals. Recognition of human activity is useful in a variety of applications like monitoring physical activity, providing context-aware services and providing assistance to the elderly to name a few. This provides us with a challenging task of analysing the data thus collected which has an inherent uncertainty. In this study, we explore a symbolic data representation method for generation of features from the raw time series data collected by accelerometer sensors present in Android based phones. The experiments are carried out on a benchmark dataset for activity recognition. The data generated is validated for the activity recognition task using different individual classifiers and also ensemble classifiers. The proposed data representation which is space efficient gives promising results and the experiments conducted also lay emphasis on the use of ensemble classifiers.

Keywords Activity recognition · Accelerometer data · Classification · Sensor data Time series data

1 Introduction

Human Activity Recognition (HAR) is evolving as an important research area which plays a vital role in elderly care, security and assisted living which are being considered as the next era of personal care in the field of healthcare [1]. The present era of ubiquitous and pervasive computing has become possible due to efficient low cost sensors. These sensors can be light sensors, audio sensors, GPS sensors, image sensors, etc., which are becoming part of our daily life as they come with our smart

P. G. Lavanya (✉) · S. Mallappa
DoS in Computer Science, University of Mysore, Mysuru, India
e-mail: lavanyarsh@compsci.uni-mysore.ac.in

S. Mallappa
e-mail: sureshasuvi@gmail.com

P. Nagabhushan et al. (eds.), *Data Analytics and Learning*,
Lecture Notes in Networks and Systems 43,
https://doi.org/10.1007/978-981-13-2514-4_27

phones, cameras, cars, and other smart devices [2]. Smart phones are becoming very common and hence can be considered as the best low-cost sensor data providers. The data transfer is no longer cumbersome and can be done automatically, thanks to Internet of Things. One such common sensor present in almost all smart phones is the accelerometer, which measures acceleration in all three orthogonal directions. There are a number of applications using accelerometers which range from pedometers to vibration detectors. A pedometer measures the number of steps taken by a person and calculates the calories burned which is used by many people to monitor their daily exercise routine. There are applications which send this daily data to websites, which further analyse the data and prepare charts and graphs regarding the activity of the person over a period of time.

Human Activity recognition is an active research area where recognition can be treated as a classification problem with certain set of known activities. Human Activity recognition can also be viewed as a simple form of context awareness which is the primary requirement to deliver assistance for elderly or disabled people during emergency, preventing further complications. The number of older people, i.e., in age group above 60 is continuously increasing and is expected to double by 2050 [3]. This, in turn, leads to problems associated with ageing and hence research in the area of assisted living is very much essential.

In this work, a tri-axial accelerometer based dataset is used for activity recognition. This work focuses on applying Symbolic data representation technique, which will allow us to transform the raw time series data and store it in a compact manner without compromising on quality of recognition. As a huge amount of data is involved in such activity recognition problems, data compaction is an essential technique which helps us to reduce the storage required. Symbolic data representation is suitable for sensor data as it can handle the inherent uncertainty in the data collected under unsupervised environment. The experimental evaluation of the techniques is carried out on a publicly available dataset, WISDM [4] which is well documented and has been evaluated by different researchers. The intention of the study is to check the feasibility of reducing the number of features compared to the WISDM dataset. The contribution of this work includes a simple yet effective data representation and validation of the same using different classifiers including ensemble classifiers.

The paper is structured as follows. Section 2 discusses about the related work in the area. Section 3 describes the proposed method of data generation. Section 4 gives the details of the experiments and conclusion is given in Sect. 5.

2 Related Work

Human Activity Recognition is an area of research which is attracting the research community as it caters to the present day requirement of personal recommendations based on context awareness. A simple example can explain this better. If a mobile wearable device can track the activity of the person, the data collected can be analysed and suitable suggestions be given to the user. If the user is sitting or walking for a longer time than is usual, the device can remind him to take a break. A more beneficial application would be to track the activity of elders who need assistance where a caretaker can be alerted if there is a deviation in the elderly person's routine activity which can save a life. In this direction, lot of work is being carried out in this field. A dataset which comprises of features generated using accelerometer data collected has been publicly made available in [4]. Classification techniques have also been performed on the dataset given for activity recognition thus making it a benchmark dataset. Similar work is done in [5] by extracting 12 features but the data is limited to only two subjects. An activity recognition system data generated from accelerometers positioned at the ankle is discussed in [6] which also adds a reclassification step to improve the accuracy of the recognition. A comprehensive survey of the various techniques used in extracting data from raw time series is given in [7]. An approach to recognise user's activity based on acceleration, orientation and magnetic fields using Mixture of Experts model is given in [8]. Both labelled and unlabeled data are handled using Global Local Co-Training algorithm. An integration of supervised and unsupervised learning is provided in [9] where a cluster-based classification system is proposed to deal with activity recognition. Data transformation using PCA to reduce dimensionality is proposed in [10] on the Wisdm_actitracker dataset. In [11], the problem of activity recognition is addressed using data from three motion sensors and adaptive windowing. An ensemble of classifiers approach is proposed in [12] which uses the WISDM dataset and shows that the results improve compared to single classification technique. Meta Adaboost M1 classifier with different decision tree classification techniques is used in [13] which also uses the same WISDM dataset. Activity decision algorithm using ANN based on the features of IoT in a smart home environment is proposed in [14]. We can observe that various approaches can be implemented to achieve the task of activity recognition. Nevertheless, the need is for a simple yet efficient model which is not computationally intensive and performs activity recognition accurately.

3 Proposed Method

As the raw time series data cannot be directly used in classification algorithms, it has to be transformed into features which are also referred to as examples. In this work, we propose a model which uses Symbolic Data representation approach to generate features from raw time series data collected from accelerometer present in mobile phones. The proposed method explores the use of Symbolic data representation of the commonly calculated statistical measures which are simple computations. The number of features generated in our method is considerably less compared to the existing data representation which gives promising results. Different Classifiers are applied on the features thus generated to accomplish activity recognition. Examples are generated with the same duration as in the WISDM dataset, i.e. 10 s segment which is approximately 200 raw data samples as one example. Extensive experiments

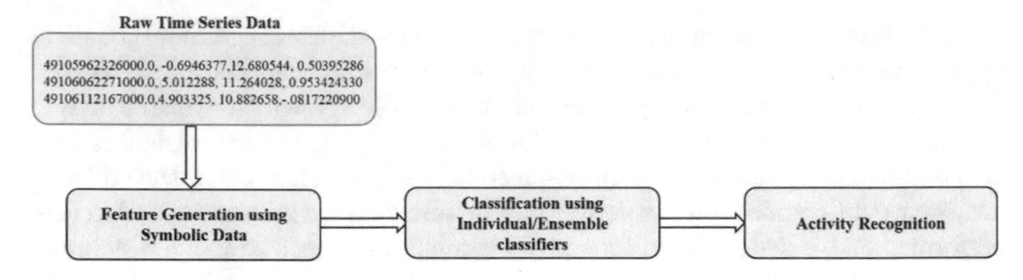

Fig. 1 Overall flow of the proposed method

are carried out with all classifiers and the results are evaluated. The flow of the method is given in Fig. 1.

3.1 WISDM Dataset

The raw accelerometer data collected has 10,98,207 samples with six features which consists of values for the three axes/dimensions and the other three being Userid, timestamp and activity label. The data is collected from 36 users for six activities. The activities considered are labelled as "Walking, Jogging, Upstairs, Downstairs, Sitting and Standing".

In [4], the variants of basic six features, i.e. mean and standard deviation are used and a total of 43 features are generated. The raw time series data is transformed into examples with a duration fixed to a 10 s segment after comparing the 10 s segment with 20 s segment as the former gave slightly better results.

In the WISDM transformed dataset, examples contain 5424 instances with 46 features. Out of 46 features, 43 features are generated from the raw data. The 43 features generated are Average (3), Standard Deviation (3), Average Absolute Difference (3), Average Resultant Acceleration (1), Time Between Peaks (3) and Binned Distribution (30). The numbers in brackets indicate the number of features generated with respect to each type of feature.

3.2 Feature Generation Using Symbolic Data

In this work, we analyse the possibility of representing the features in interval data form. While dealing with large datasets, it is necessary to summarise the underlying data by extracting knowledge and representing the same in a more complex form than the standard ones as they contain internal variation [15]. Symbolic data is one such popular approach which captures the essential information present in the data. Symbolic data can be of different types like categorical, multi-valued, modal and interval. In this work, the features are generated using interval data form which reduces the number of features and also improves the accuracy of activity recognition.

An interval variable is represented using a set of bounded intervals [16]:

$$Y_i = \left[y_i^-, y_i^+\right], i = 1\ldots n \tag{1}$$

In this method, $\mu - \sigma$ represents y_i^- and $\mu + \sigma$ represents y_i^+ where μ is mean of the data sample and σ is the standard deviation.

A total of twelve features are generated in the proposed approach as explained below:

(1) The symbolic interval data for each of the three axes—six features
(2) Range of the data sample—three features
(3) Root Mean Square (RMS)—three features

Mean and Standard Deviation

The mean of the data sample considered is an important summary statistic and can be calculated easily with minimum computation [7]. The mean of a sample $a_1, a_2 \ldots a_n$ is given by

$$\mu = \frac{1}{n}\left(\sum_{1}^{n} a_i\right) \tag{2}$$

Standard Deviation is another simple yet important metric which gives the measure of the spread. These two commonly used measures are used in different combinations in our proposed method.

$$\sigma = \sqrt{\frac{1}{N}\sum_{i=1}^{N}(x_i - \mu)^2} \tag{3}$$

Range

The range of the data is considered as the difference between the highest and lowest value of the given data samples. Range suggests the spread of the values and has been used to discriminate between the activities like walking and running. It has also been used as an input to detect steps [7]. Range of a set $A = \{a_1, a_2, \ldots a_n\}$ is given by:

$$R = \max(A) - \min(A) \tag{4}$$

RMS

The root mean square (RMS) of a signal x_i that represents a sequence of n discrete values $\{x_1, x_2, \ldots, x_n\}$ is given by:

$$X_{RMS} = \sqrt{\frac{x_1^2 + x_2^2 + \cdots + x_n^2}{n}} \tag{5}$$

This measure is also a simple yet significant measure and is used in activity recognition and especially in gait recognition as it changes with the walking style and speed. It is also used in measuring gait abnormality [17].

The memory occupied by the data generated from the proposed method is reduced by almost 60% compared to the existing dataset. This compactness is a required quality for working with handheld devices having memory constraints.

3.3 Classification Methods

The classification techniques used with the WISDM dataset are J48, Logistic Regression, and Multi-layer Perceptron. Along with this we have experimented our proposed data representation with Vote and LibD3C techniques which are ensemble classifiers. The Vote technique has been proposed in [12] and hence used for comparison.

J48: J48 algorithm is the well-known decision tree algorithm C4.5 implemented in Java and is available in Weka. It is the extension of ID3 algorithm and is based on the Information gain. It is known as a statistical classifier and features in the top ten classifiers. It can be applied on continuous and categorical attributes. Misclassification errors are reduced in this method as it has a better method for tree pruning compared to its predecessors.

Logistic Regression: A well-known statistical method which is used to analyse a dataset comprising of two or more independent variables based on which the outcome is predicted. Logistic regression is more robust and does not require the data to be normally distributed. It is less prone to over-fitting as it has low variance.

Multi-layer Perceptron (MLP): MLP is a popular ANN classifier based on back propagation to classify instances. It can be used to distinguish data that is not linearly separable. It consists of minimum three layers—input layer, hidden layer and the output layer. The hidden layer is responsible for the transformation of the input data using a nonlinear transformation. This makes it a universal approximator and hence a good classifier.

Vote: This is an ensemble of classifiers. Different classifiers are combined and a decision is taken based on the combination rules. The different rules applied are minimum probability, maximum probability, majority voting, etc. This ensemble classifier with J48, Logistic Regression and MLP is explored well with different rules on the same dataset in [12]. They have concluded that average of probabilities is the best suitable and hence it has been used in our experiments also.

LibD3C: This is also an ensemble classifier which is developed for multi-class classification problems. It uses the hybrid approach for selective ensemble technique based on ensemble pruning. A combination of k-means clustering and dynamic selection and circulating combination is used. K-means is used to eliminate redundant classifiers whereas dynamic selection is used to improve the ensemble performance [18].

4 Experiments and Results

The experiments are carried out on a publicly available dataset which has a large number of samples compared to other accelerometer datasets which are available or experimented upon. The experiments were carried out on an Intel Core i5 4210@ 1.70 GHz machine with 8 GB RAM. The proposed feature generation method is implemented in MATLAB. The classification techniques are performed using Weka software as the results can be compared with the existing results. The dataset mentioned in [4] has 29 users and the results are based on that data. The latest WISDM dataset downloaded is modified and has 36 users with more samples. Hence the methods used earlier are also experimented with the present data to compare the recognition accuracy. In all the experiments, default settings for the classifiers are used. Tenfold cross-validation technique is used in all the experiments. The performance of the classifier is evaluated using overall accuracy. As accuracy is not a sufficient measure in the case of imbalanced datasets, we have also validated the performance using Precision, Recall and F-measure. The performance of Symbolic data approach compared to the WISDM dataset with respect to various classifiers is given in Tables 1, 2, 3, 4 and 5. The Vote ensemble technique and the Adaboost classifier with Decision Stump, Hoeffding Tree, Random Tree, REP Tree, J48, and Random Forest are also experimented upon with the proposed symbolic data representation.

It is observed from the results, that the proposed symbolic WISDM dataset performs equally well when compared with the WISDM dataset. The number of features in the proposed feature generation method is 12 and the number of features extracted in the WISDM dataset is 43. This is almost one-third of the WISDM feature set and still keeps up with the performance. The time taken for different classifiers with the WISDM dataset and the proposed symbolic dataset is shown in Table 6.

It is also observed that the choice of the classifier plays an important role and the LibD3C classifier which is an ensemble classifier gives a very good accuracy compared to others. Results show that activities Upstairs and Downstairs are difficult to recognise and hence has a lower F-measure compared to other activities. But from the results, it is observed that the LibD3C classifier performs better in these cases also.

The ROC curves for the performance of LibD3C classifier with the WISDM dataset and the proposed symbolic dataset is shown in Figs. 2 and 3.

The ROC curves for the activity ‘Walking’ for different classifiers with WISDM dataset and the proposed symbolic dataset are shown in Figs. 4 and 5.

Table 1 Performance comparison—overall accuracy for different classifiers

Dataset	No. of features	J48	Log Reg	MLP	Vote	LibD3C	Adaboost classifier					
							Decision stump	Hoeffding tree	Random tree	REP tree	J48	Random forest
WISDM dataset	**46**	89.46	84.94	92.65	93.47	92.93	63.7	75.67	89	94.56	94.04	94.44
Symbolic WISDM dataset (Proposed)	**13**	87.36	81.84	86.13	88.36	91.4	62.74	71.94	87.85	90.84	92.29	92.27

Table 2 Performance of LibD3C classifier

Activity	WISDM dataset			Symbolic WISDM dataset (proposed)		
	Precision	Recall	F-Measure	Precision	Recall	F-Measure
Walking	0.94	0.99	0.96	0.90	0.97	0.94
Jogging	0.98	0.98	0.98	0.96	0.99	0.97
Upstairs	0.81	0.76	0.78	0.83	0.73	0.78
Downstairs	0.79	0.71	0.75	0.85	0.63	0.72
Sitting	0.99	0.98	0.99	0.98	0.98	0.98
Standing	0.99	0.95	0.97	0.96	0.97	0.97
Weighted Avg	0.927	0.929	0.928	0.911	0.914	0.91

Table 3 Performance of J48 classifier

Activity	WISDM dataset			Symbolic WISDM dataset (proposed)		
	Precision	Recall	F-Measure	Precision	Recall	F-Measure
Walking	0.94	0.96	0.95	0.90	0.92	0.91
Jogging	0.96	0.96	0.96	0.96	0.96	0.96
Upstairs	0.68	0.68	0.68	0.67	0.68	0.67
Downstairs	0.68	0.63	0.66	0.65	0.57	0.61
Sitting	0.99	0.96	0.98	0.97	0.97	0.97
Standing	0.97	0.98	0.97	0.96	0.97	0.97
Weighted Avg.	0.89	0.90	0.89	0.87	0.87	0.87

Table 4 Performance of logistic regression

Activity	WISDM dataset			Symbolic WISDM dataset (proposed)		
	Precision	Recall	F-Measure	Precision	Recall	F-Measure
Walking	0.86	0.95	0.90	0.79	0.94	0.86
Jogging	0.99	0.99	0.99	0.95	0.96	0.95
Upstairs	0.54	0.50	0.52	0.58	0.55	0.57
Downstairs	0.53	0.36	0.43	0.46	0.12	0.19
Sitting	0.94	0.94	0.94	0.97	0.95	0.96
Standing	0.95	0.91	0.93	0.94	0.96	0.95
Weighted Avg.	0.84	0.85	0.84	0.79	0.82	0.79

Table 5 Performance of Multi-layer Perceptron

Activity	WISDM dataset			Symbolic WISDM dataset (proposed)		
	Precision	Recall	F-Measure	Precision	Recall	F-Measure
Walking	0.98	0.97	0.98	0.86	0.94	0.90
Jogging	1.00	0.99	0.99	0.94	0.98	0.96
Upstairs	0.72	0.82	0.77	0.70	0.63	0.66
Downstairs	0.73	0.64	0.69	0.64	0.39	0.48
Sitting	0.97	0.95	0.96	0.96	0.94	0.95
Standing	0.94	0.94	0.94	0.89	1.00	0.94
Weighted Avg.	0.98	0.97	0.98	0.86	0.94	0.90

Table 6 Time taken for different classifiers

Classifier	WISDM dataset	Symbolic WISDM dataset (proposed)
J48	2.95	0.78
Log Reg	196.65	5.78
MLP	266.04	39.16
Vote	654.01	46.75
LibD3C	44.36	39.21

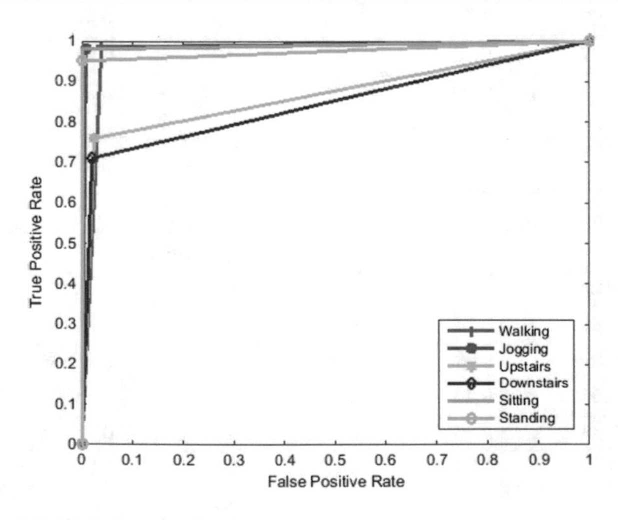

Fig. 2 ROC for LIBD3C classifier for WISDM dataset

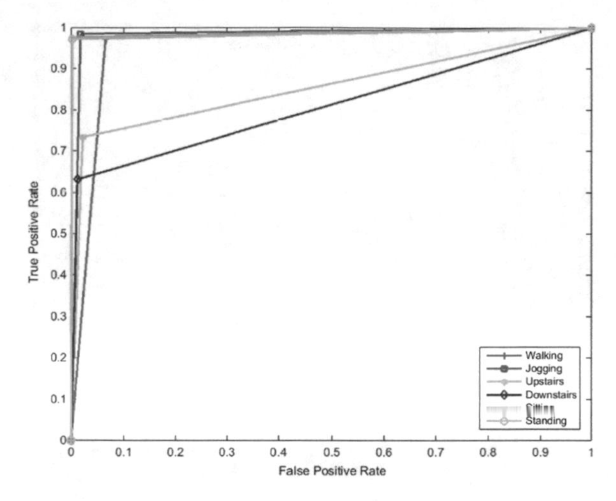

Fig. 3 ROC for LIBD3C classifier for proposed symbolic dataset

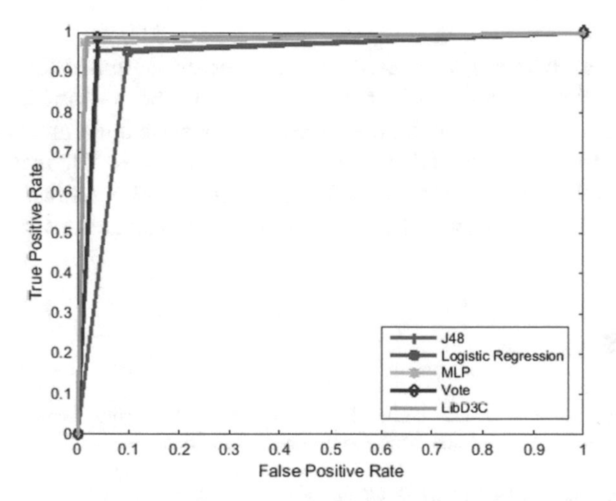

Fig. 4 Performance of different classifiers for 'Walking' activity—WISDM dataset

5 Conclusion

Human Activity Recognition is being considered as an important area of research as it helps in context aware recommendations and assisted living for the elderly or patients. In this work, the raw sensor data collected is represented using symbolic data representation approach along with simple statistical measures. Experiments are carried out to validate the proposed symbolic data representation using classifiers and the results are compared with the existing dataset. The number of features generated by the proposed method is significantly less compared to the number in the existing

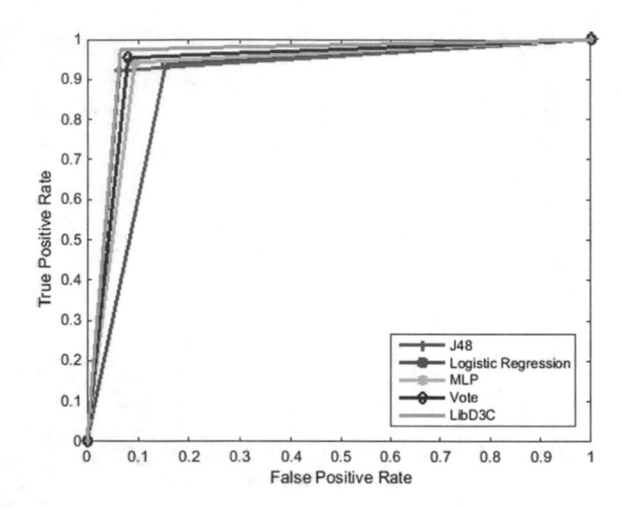

Fig. 5 Performance of different classifiers for 'Walking' activity-symbolic dataset (proposed)

WISDM dataset, whereas the classification accuracy is almost the same. As the number of samples continuously increase in number and becomes a big data problem, data representation which is computationally inexpensive and compact is useful as these computations have to be done using hand held devices. The results encourage us to explore different data representation techniques and combination of existing individual classifiers to come out with much more efficient activity recognition.

References

1. Brezmes, T., Gorricho, J.-L., Cotrina, J.: Activity recognition from accelerometer data on a mobile phone. In: International Work-Conference on Artificial Neural Networks, pp. 796–799 (2009)
2. Chen, K.-H., Chen, P.-C., Liu, K.-C., Chan, C.-T.: Wearable sensor-based rehabilitation exercise assessment for knee osteoarthritis. Sensors. **15**, 4193–4211 (2015)
3. http://www.un.org/en/development/desa/population/publications/pdf/ageing/WPA2015_Report.pdf
4. Kwapisz, J.R., Weiss, G.M., Moore, S.A.: Activity recognition using cell phone accelerometers. ACM SIGKDD Explor. Newsl. **12**, 74–82 (2011)
5. Ravi, N., Dandekar, N., Mysore, P., Littman, M.L.: Activity Recognition from Accelerometer Data, pp. 1541–1546. Aaai (2005)
6. Bernecker, T., Graf, F., Kriegel, H.-P., Moennig, C., Dill, D., Tuermer, C.: Activity recognition on 3d accelerometer data. Technical Report (2012)
7. Figo, D., Diniz, P.C., Ferreira, D.R., Cardoso, J.M.: Preprocessing techniques for context recognition from accelerometer data. Pers. Ubiquit. Comput. **14**, 645–662 (2010)
8. Lee, Y.-S., Cho, S.-B.: Activity recognition with android phone using mixture-of-experts co-trained with labeled and unlabeled data. Neurocomputing **126**, 106–115 (2014)
9. Abdallah, Z.S., Gaber, M.M., Srinivasan, B., Krishnaswamy, S.: CBARS: cluster based classification for activity recognition systems. In: International Conference on Advanced Machine Learning Technologies and Applications, pp. 82–91 (2012)

10. Kuspa, K., Pratkanis, T.: Classification of mobile device accelerometer data for unique activity identification (2013)
11. Shoaib, M., Bosch, S., Incel, O.D., Scholten, H., Havinga, P.J.: Complex human activity recognition using smartphone and wrist-worn motion sensors. Sensors **16**, 426 (2016)
12. Catal, C., Tufekci, S., Pirmit, E., Kocabag, G.: On the use of ensemble of classifiers for accelerometer-based activity recognition. Appl. Soft Comput. **37**, 1018–1022 (2015)
13. Walse, K.H., Dharaskar, R.V., Thakare, V.M.: A study of human activity recognition using AdaBoost classifiers on WISDM dataset. Inst. Integr. Omics and Appl. Biotechnol. J. **7**, 68–76 (2016)
14. Bourobou, S.T.M., Yoo, Y.: User activity recognition in smart homes using pattern clustering applied to temporal ANN algorithm. Sensors **15**, 11953–11971 (2015)
15. Diday, E., Esposito, F.: An introduction to symbolic data analysis and the SODAS software. Intell. Data Anal. **7**, 583–601 (2003)
16. Signoriello, S.: Contributions on Symbolic Data Analysis: A Model Data Approach (2008)
17. Sekine, M., Tamura, T., Yoshida, M., Suda, Y., Kimura, Y., Miyoshi, H., Kijima, Y., Higashi, Y., Fujimoto, T.: A gait abnormality measure based on root mean square of trunk acceleration. J. Neuroeng. Rehabil. **10**, 118 (2013)
18. Lin, C., Chen, W., Qiu, C., Wu, Y., Krishnan, S., Zou, Q.: LibD3C: ensemble classifiers with a clustering and dynamic selection strategy. Neurocomputing **123**, 424–435 (2014)

Automated IT Service Desk Systems Using Machine Learning Techniques

S. P. Paramesh and K. S. Shreedhara

Abstract Managing problem tickets is a key issue in any IT service industry. The routing of a problem ticket to the proper maintenance team is very critical step in any service desk (Helpdesk) system environment. Incorrect routing of tickets results in reassignment of tickets, unnecessary resource utilization, user satisfaction deterioration, and have adverse financial implications for both customers and the service provider. To overcome this problem, this paper proposes a service desk ticket classifier system which automatically classifies the ticket using ticket description provided by user. By mining historical ticket descriptions and label, we have built a classifier model to classify the new tickets. A benefit of building such an automated service desk system includes improved productivity, end user experience and reduced resolution time. In this paper, different classification algorithms like Multinomial Naive Bayes, Logistic regression, K-Nearest neighbor and Support vector machines are used to build such a ticket classifier system and performances of classification models are evaluated using various performance metrics. A real-world IT infrastructure service desk ticket data is used for this research purpose. Key task in developing such a ticket classifier system is that the classification has to happen on the unstructured noisy data set. Out of the different models developed, classifier based on Support Vector Machines (SVM) performed well on all data samples.

Keywords Machine learning · Natural language processing (NLP)
Ticket classification · Service desk (Helpdesk) · SVM classification
Term frequency inverse document frequency (TF-IDF)

S. P. Paramesh (✉) · K. S. Shreedhara
Department of Studies in CS & E, U.B.D.T College of Engineering,
Davanagere 577004, Karnataka, India
e-mail: sp.paramesh@gmail.com

K. S. Shreedhara
e-mail: ks_shreedhara@yahoo.com

P. Nagabhushan et al. (eds.), *Data Analytics and Learning*,
Lecture Notes in Networks and Systems 43,
https://doi.org/10.1007/978-981-13-2514-4_28

1 Introduction

Service desk systems are generally the place where customers, end users, employees, or any other resources receive business support relating to their organization services. Most of the current helpdesk systems are web-based and will have predefined user interface for interaction and have fields like ticket category, priority, description, submitter, etc. Using this interface, user log the problem tickets and gets resolution for the same within a specified Service Level Agreements (SLA's) associated with each ticket. The structure of the sample tickets raised by end user from one of the real word enterprise service desk is given in Fig. 1.

Most of the large IT service industries will have a dedicated service desk web portal in order to manage the problems faced by the customers, employees, or any other end users. User's faces a lot of issues regarding the IT infrastructure like network issues, email delivery issues, hardware problems, OS problems, etc., and they get the resolution for the problem by logging a ticket in the helpdesk system. When end user needs to create the problem ticket using the service desk web portal, user manually selects the ticket category from the list of predefined categories. Manual selection may results in wrong selection of category due to lack of knowledge about the problem ticket and since it depends on the user understanding of the problem. Due to wrong selection of problem category, ticket will be assigned to wrong resolver group and hence delays the resolution time due to ticket reassignments. So there is a need to develop a system which auto categorizes the helpdesk tickets.

In this paper, we have developed a ticket classifier system that automatically identifies the category of ticket by mining the ticket description entered by end user and thereby the ticket is assigned to the correct resolution group to handle the ticket. So the end users are required to provide only the problem description in natural language and the proposed system automatically categorizes the ticket. Different supervised machine learning techniques are used to build classifier models and performances of these models are evaluated using various accuracy measures. To train the classifiers, historical data dump of IT infrastructure service desk tickets containing ticket description and associated category is used. Historical ticket data contains many other fields like submitter, priority, resolution description, etc., but only the ticket description and the category fields are used to train the classifier model. Key focus areas of this paper are

Ticket ID	Ticket Category	Priority	Submitter	Ticket Description	Status	Resolved by	Resolution description
100	Network Issue	High	235123	Unable to login into the LAN	open	none	none
101	Hardware Problem	Medium	6532	Hard disk crashed	In Progress	none	In progress
102	Software Installation	High	17234	Need to install Eclipse	open	none	none

Fig. 1 Sample service desk ticket data

- Service desk ticket data preprocessing.
- Construction of feature vector representation of the ticket data.
- Building the classifier models using various classification algorithms.
- Evaluation and comparison of classifiers performance.

Text classifiers works well for well-cleaned data. The service desk dataset considered for our research purpose had lot of noisy and unclean data and thus making the development of ticket classifier system a challenging task.

2 Prior Research

Some of the previously performed studies on classification of service desk tickets through supervised and unsupervised machine learning techniques are given below.

Lucca et al. [1] used different classification approaches like vector space model, probabilistic model, SVM, Decision trees (CART), and K-nearest neighbor to perform the classification of software maintenance request tickets based on the ticket description. K-cross validation technique is used to compare the performance of each of the classifier models. SVM classifier performed well on the considered service desk ticket data. The training data considered for this research purpose had almost 6000 tickets collected over a period of two years and were divided into eight predefined categories.

Dasgupta et al. [2], proposed an approach for classification of noisy and unstructured ticket into predefined ticket category. Proposed methodology defines the logical structure of ticket description and then uses the contextual information present in the logical structure to classify the tickets. The ticket analysis tool developed using this approach is called Bluefin. For performance analysis, the tool is compared with another ticket analysis tool (Smart Dispatch) which is based on SVM text classification.

The work by Kadar et al. [3] proposes a methodology to develop a tool which classifies the user submitted change requests (CR) into one of the activities in a catalogue. It uses information retrieval techniques and multinomial logistic regression classifier for classification of change requests. Active learning approaches were used to reduce the cost of building such a classifier system.

Agarwal et al. [4] developed automated ticket dispatch tool called Smart dispatch. The tool analyzes the ticket description and identifies the appropriate resolution group by using a combination of SVM classification and a discriminative term weighting schemes.

Daio et al. [5] suggested a rule-based crowd-sourcing approach to classify issue tickets. A Rule based classifier system uses a set of rules authored by experts and each rule contains IF-THEN clause. IF clause usually takes the Boolean expression and when it evaluates to true THEN clause returns the ticket category. The proposed method is then compared with supervised learning techniques such as Naive Bayes and the results shows that the rule based classifier performed well when insufficient labeled data are available.

Wei et al. [6] proposed an approach to automatically structure issue tickets consisting of unstructured data. Proposed method uses the Conditional Random Fields (CRF's) to analyze and structure the information in the raw ticket data. The problem ticket structuring system based on CRF involves segmentation and annotation phases. Segmentation converts the ticket description into a format that can be labeled and Annotation assigns a label to each unit.

Text classification has also been used for troubleshooting from past history of tickets. One of the most recent works that is very close in spirit to this is a system called NetSieve [7] that analyses user ticket description text in network tickets.

Shimpi et al. [8] used unsupervised machine learning techniques for classification of service tickets. Whenever there exist unlabeled ticket data and it is difficulty to classify the unlabeled tickets manually, a clustering approach is followed to classify the tickets.

Sakolnakorn et al. [9] proposed a framework based on ID3 decision tree classification to identify the resolver group for the incident ticket in banking business. Various decision tree algorithms like ID3, random tree, J48, random forest, NBTree, and REP tree were used and k-cross validation is used to test the accuracies of these models. Results showed that ID3 performed well on the considered data set.

Xu et al. [10] propose SVM ensemble classification model called STI-E to identify the situation tickets from the collection of tickets. Identification of situation tickets helps to upgrade monitoring system configurations to reduce the system faults. The proposed model uses the domain word discovery algorithm to identify the domain words from the historical tickets. Using the identified domain words, a selective labeling policy is used to select a sample from the collection. Imbalanced data present in the manual tickets is handled by using the random under sampling and oversampling techniques and finally STI-E is used to identify the situation tickets.

Imbalanced datasets greatly affects the accuracy performances of the traditional classifiers. Sotiris [11] brings out the various approaches to handle such class imbalance problems. Both data and algorithmic levels solutions are discussed. Data level solutions includes random under sampling and oversampling techniques. Algorithm level solutions include Threshold method, one-class learning, etc.

Roy et al. [12] proposed a framework to cluster incident tickets based on ticket description. Clustering of service tickets helps in manual categorization of tickets and saves the time and improves resource utilization. Clustering is then followed by labeling of each cluster where each cluster label gives concise and proper information about each cluster. The proposed method uses the partition-based clustering method to identify the cluster and label of the new incoming ticket.

Zeng et al. [13] uses hierarchical multi-label classification along with the domain knowledge to identify the ticket category and to provide auto resolution for the problem ticket in IT environments.

Son et al. [14] discuss the automation of helpdesk system called XSEDE ticket system using the Multinomial Naïve Bayes (MNB) and Softmax Regression Neural Network (SNN) text mining algorithms.

Frenay et al. [15] discusses the techniques for identification and removal of label noise.

In [16] the authors propose a probabilistic model for the analysis of IT service desk tickets. Analysis involves identifying the root cause for the problem tickets so as to reduce the amount of future tickets. The framework uses probabilistic topic modeling and clustering techniques for the analysis of tickets.

Our research problem is basically a text document classification problem. Sebastiani [17] discusses the review of text classification using various machine learning approaches. In the literature, many supervised and unsupervised techniques were proposed to solve text and document classification problems [18–22].

3 Proposed Methodology

Ticket classification is a document classification problem wherein each ticket description represents a document and ticket category represents the label of the document. Historical ticket data dump containing tickets description and the associated category of the tickets is used for training purpose. To build the ticket classification model, the training data must be preprocessed and converted into feature vector representation. We then employ several well established classification techniques to build the ticket classifier model. Figure 2 represents the key blocks of the proposed system for the classification of service desk tickets.

The key blocks of the proposed system are briefly explained as follows.

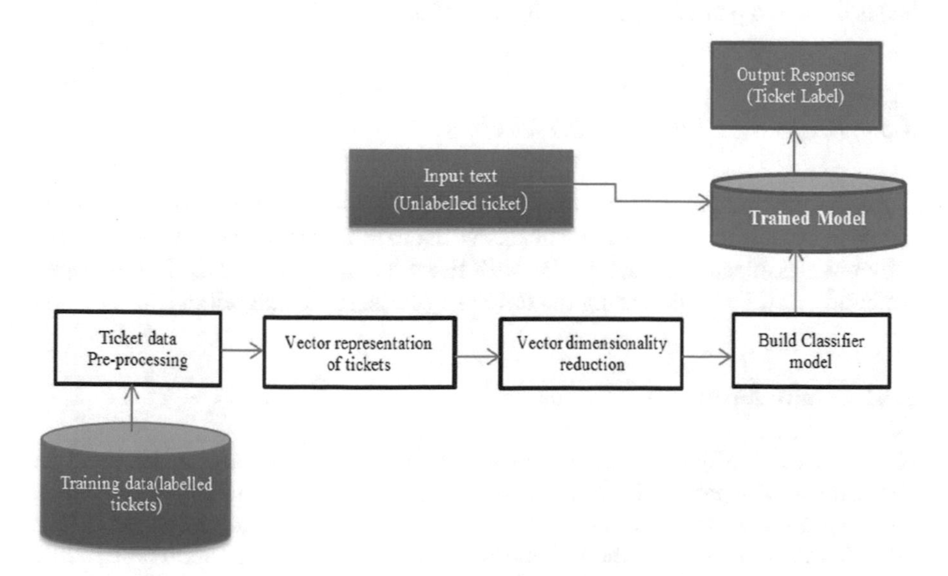

Fig. 2 Solution diagram for classification of unlabeled tickets

3.1 Data Preprocessing

The preprocessing of training data is necessary to handle all noisy and unwanted data and to make sure that the data is suitable for data mining. The given training data is analyzed to understand its characteristics. The service desk dataset considered for this research had lot of unclean data such as stop words, functional words, phone numbers, email address, date, time, numbers, special characters, etc. The preprocessing block removes all such unwanted data since they do not contribute in identifying the category of the ticket and only the relevant features are retained. The dataset also had imbalanced and wrongly labeled data. Proper machine learning techniques are used to handle such noise from the dataset.

3.2 Vector Representation of Ticket Data and Dimensionality Reduction

After the preprocessing of the ticket data, important features are extracted. To build the ticket classification model, the preprocessed ticket data must be represented in numerical form. In this research, a vector space model is used for the representation of ticket data where each ticket description is represented by a vector and each independent term in the training data are considered as features. A feature vector construction is done on the tickets using TF-IDF term weighting scheme. After the TF-IDF matrix is constructed using the ticket data, feature selection by Chi-Squared test is done as a part of dimensionality reduction.

3.3 Building Classification Models

Once the data preprocessing and vectorization is done, next step is to build the classifier models to categorize the tickets with unknown labels. The performance of different classification models varies with the different kinds of data. In this paper, we build classifier models using the following classification algorithms.

3.3.1 Naive Bayes

Naive Bayes classifier is a simplest probabilistic classification algorithm based on Bayes theorem of probability. This classifier is used in text or document classification problems due to its independence rule, i.e., words in a document are unrelated to each other. In this paper, we used the Multinomial Naive Bayes [14] algorithm for building the classifier model which considers the frequency of the words in a document.

We used the framework in [20] for building classifier model. In MNB model, a document d_i is a collection of words from the same vocabulary V. Let N_{it} be the count or frequency of the word w_t occurs in a document d_i. The probability of a text document d_i given its class c_j is simply the multinomial distribution:

$$P(d_i|c_j;\theta) = P(|d_i|)|d_i|!\prod_{t=1}^{V}\frac{P(w_t|c_j;\theta)^{N_{it}}}{N_{it}!} \tag{1}$$

The estimate of the probability of word w_t in class c_j is given by:

$$\theta_{w_t|c_j} = \mathrm{P}(w_t|c_j;\theta) = \frac{1+\sum_{i=1}^{|D|} N_{it}P(c_j|d_i)}{|V| + \sum_{s=1}^{|V|}\sum_{i=1}^{|D|} N_{is}P(c_j|d_i)} \tag{2}$$

The category of the test document can be obtained by determining the posterior probabilities of each class c_j and then selecting the class having the highest posterior probability. Posterior probability of each class c_j given the test document d_i can be obtained by using the Eq. 3.

$$P(c_j|d_i;\theta) = \frac{P(c_j|\theta)P(d_i|c_j;\theta_j)}{P(d_i|\theta)} \tag{3}$$

3.3.2 Logistic Regression

Logistic Regression (LR) is one of the discriminative classifier used for text classification problems. It is basically a binary classifier but can be generalized for multiclass problems. Given a test document X, the estimate of the conditional probability of mapping a class label y to the test example X using the logistic regression is given by:

$$P(y|X) = \frac{1}{1 + exp(w_0 + \sum_{i=1}^{n} w_i X_i)} \tag{4}$$

3.3.3 K-Nearest-Neighbor (KNN)

In order to classify the new test document, KNN method finds the k most similar documents based on the distance metric and then assigns the majority class label to the new document. It uses the Euclidian distance between two points as a distance metric to find the K similar documents. The classification process determines similar documents using the distance between the documents which is given by Eq. 5.

$$d(x, y) = \sqrt{\sum_{r=1}^{N} (a_{rx} - a_{ry})^2}, \tag{5}$$

where $d(x, y)$ represents the distance between documents x and y, N is the number of unique terms in the corpus, a_{rx} is the term weight in document x and a_{ry} is the term weight in document y.

3.3.4 Support Vector Machines (SVM)

Support vector machines classify new test instances by determining a hyperplane that separates two groups of classes [21]. SVMs are supervised machine learning classifiers suitable for text classification problems since they can handle large sets of features. Linear SVM classifier is used for our research purpose. We construct multiclass SVM classifier model by combining the results of several binary classifiers using one against the rest method.

3.4 Trained Model

Accuracy performance of each of the above classifier models is then evaluated using various performance metrics used for classification. At the end of this module, the classifier model which performs better is selected and stored in the trained model store. Whenever a new unlabeled ticket arrives, the category of the ticket is automatically predicted using the trained model.

4 Implementation Approach and Experimental Results

Python 2.7 version and its supported libraries like Numpy, Pandas, Matplotlib, and Scikit-learn are used to implement this research topic.

4.1 Data Set Analysis

A real-time enterprise IT infrastructure service desk ticket dataset (around 11,000 tickets) collected over a period of one month is used for our research purposes. Each ticket contains various fields but only ticket description and its category (label) were used for training purpose and rest of the fields are ignored. Initial data analysis revealed the below details.

- No of tickets: 10,742 records
- No of Classes (Category): 18.

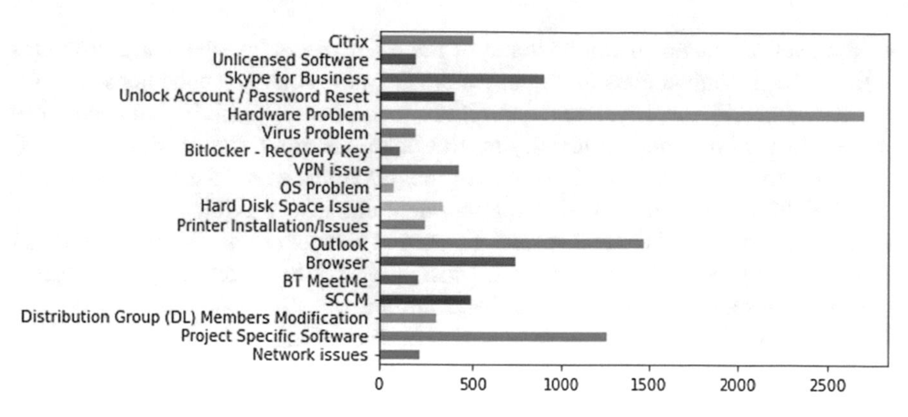

Fig. 3 Distribution of tickets across different classes

Distribution of the data across different classes is given in Fig. 3. Initial data analysis also revealed that, the historical data contains lot of unwanted text, wrongly labeled data (noisy label) and imbalanced data. Unwanted text such as stop words, functional words, phone numbers, email address, date, time, numbers, special characters, etc., were present in large number.

Training data considered for this research work also had imbalanced classes with some classes having more instances and rest having too less tickets. With the help of the domain knowledge, careful observation of the dataset also revealed that there are some wrongly labeled instances. Appropriate machine learning algorithms are used to handle both imbalanced and wrongly labeled data.

4.2 *Data Preprocessing*

The different techniques implemented to handle the noise in the unstructured ticket data are as follows.

4.2.1 Handling the Unwanted Ticket Data

Each ticket description contains lot of unwanted text data. To handle these noisy descriptions of the data, first the text will be tokenized into separate words and then the following techniques are used to remove various unwanted entities.

- Stop word elimination: Commonly used English stop word list is used to filter out the stop words from the training data.
- Removal of email ids, phone numbers, date, time: All these entities do not contribute in identifying the problem category. Regular expressions or Pattern recognizers are developed using Python to remove such entities.

- Removal of functional words: Some of the determiners (ex: the, that), pronouns (she, they), conjunctions (and, but), modals (may, could), prepositions (in, of), auxiliary verbs (be, have), and quantifiers (some, both) could be removed since they do not contribute in identifying the ticket category. Part of Speech (POS) tagging to each word is used to filter out the functional words from the ticket data. NLTK library of Python is used to implement this POS tagging.
- White spaces, numbers and special characters are also removed from the ticket data as they do not contribute to identification of the ticket category. Regular expressions are written to exclude these entities.
- Stemming: It is one of the most common preprocessing steps in any text mining problems. Stemming is the technique of reducing each word to its base form. For example, the words "computer", "computing" and "computerized" can be replaced by "computer". In this research, we used Standard porter stemming algorithm to achieve stemming.

4.2.2 Handling the Imbalanced Dataset

The dataset considered for our research had lot of imbalanced classes, i.e., variations in the number of tickets available for each classes. Imbalanced dataset affects the performance of the classifier models as they tend to classify new incoming ticket into the class containing more number of instances. So there is need to ensure that classes with less number of tickets must also suitably represented. Techniques like Random under sampling and oversampling are used to handle imbalanced data problems [11].

4.2.3 Handling the Wrongly Labeled Data

With the help of domain knowledge, initial data analysis revealed that some instances of our dataset had some noisy labels for ex: some instances of "Hard Disk Space issue" category were wrongly categorized as "Hardware Problem". Wrongly labeled instances or noisy labels could severely degrade the performance of the classifier model. So these outliers need to be handled before building the actual classification models. Concept of classification filter [15] is used to handle wrongly labeled data. Using SVM classifier on the initial training data, mislabeled instances are identified and then manually relabeled using IT infrastructure domain knowledge.

4.3 Vector Representation of Ticket Data and Feature Selection

After the preprocessing of the training data, important features are extracted and feature vector representation is done on the preprocessed data. Vector space model is used to represent the ticket description in the form of feature vector representation. In the vector space model, each document (ticket description) $d \in D$ of the training dataset D is represented as a feature vector $x(d) = \left(x_{d,1}, \ldots, x_{d,m}\right)$. Each vector element specifies the unique term in the document corpus and is represented using tf-idf weighting scheme. Tf-idf specifies the importance of particular word in the document. Formally, tf-idf is given by $tf - idf(t, d, D) = tf(t, d) \times \log \frac{|D|}{|\{d \in D : t \in d\}|}$. Here $tf(t, d)$ is the word count of the term t in the document d, |D| is the total count of documents in the dataset D, and $|\{d \in D : t \in d\}|$ is the number of documents containing the term t. Feature vector representation is then followed by compression to lower the dimensionality by using chi-square test.

4.4 Building and Evaluation of the Classifier Models

After the feature vector representation and proper feature selection, our next task is to build the accurate classifier. Our dataset would be split using 80 : 20 percentage split ratio with 80% of tickets used for training the classifier and the remaining instances used for testing the classification accuracy.

One cannot possibly know which algorithm will perform best on the given dataset beforehand. So different classifier models are generated using different classification algorithms on the training dataset and accuracy performances of these algorithms are analyzed. As discussed above, below classification algorithms are considered for our research purposes:

- Multinomial Naive Bayes (MNB)
- Logistic Regression (LR)
- K-nearest neighbor (KNN) (with K = 5)
- Support vector machines (SVM).

K-cross validation (with k = 10) is used to measure the average performances of each classifier models on the training set (8593 tickets). In k-cross validation, the entire collection is randomly splitted into k-partitions of same size. Out of the k-partitions, one partition is used as test set to test the classifier and the remaining k − 1 subsamples are used to build the classifier. This procedure is repeated k times, with each of the k-partitions used as test data exactly once. The average of the k folds is then used to find the accuracy of the classifier. Average accuracy performances of classifiers using k-cross validation on the training set are presented in Fig. 4.

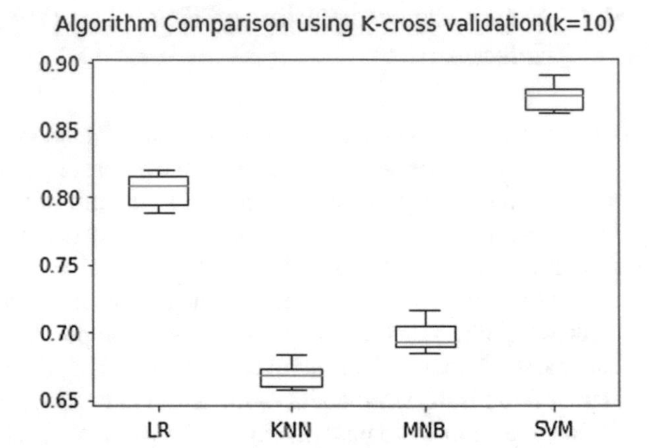

Fig. 4 Accuracy performances of classification methods using K-cross validation

SVM with average accuracy of 87% performed well on all random data samples of the training set when compared to other classification algorithms. Logistic regression (81%) performed well when compared to Multinomial Naïve Bayes (69%) and KNN (67%). Next, we analyze the performances of all the models on the test data set (2149 tickets).

Classifier accuracy is the correct number of predictions made out of the total number of predictions. Sometimes accuracy measure can be misleading so additional metrics are also required to evaluate a classifier. For our research purpose, along with accuracy measure, other classification performance metrics like precision, recall, and F1-score are also evaluated. Precision refers to the number of documents retrieved that are relevant and Recall specifies the number of relevant documents that are retrieved. F1-score is the weighted mean of precision and recall. Mathematically, these metrics are defined as follows:

$$precision = \frac{true\,positive\,(TP)}{true\,positive\,(TP) + false\,positive\,(FP)} \tag{6}$$

$$\mathrm{rec}all = \frac{true\,positive\,(TP)}{true\,positive\,(TP) + false\,negative\,(FN)} \tag{7}$$

$$F1\text{ - }score = 2 \times \frac{precision \times recall}{precision + recall} \tag{8}$$

True positives (TP) are the positive tuples that are correctly classified by the classifier and the correctly classified negative tuples are called True negatives (TN). When the classifier model incorrectly predicts the positive class then its outcome is False positives (FP) and when the classifier model incorrectly predicts the negative class then it results False negatives (FN). The performance results of running different classifiers on the test set are shown in Fig. 5.

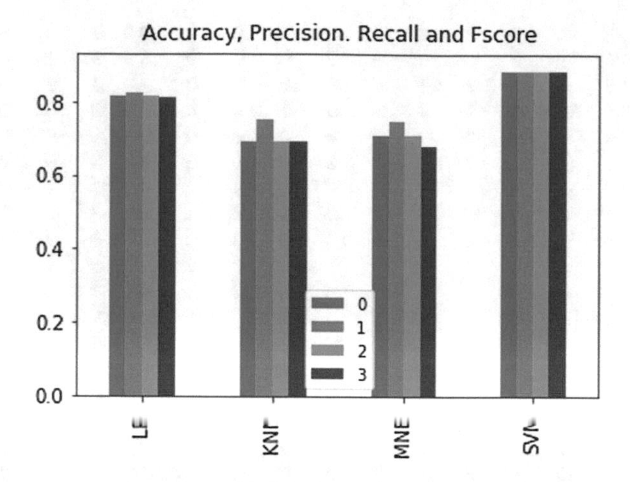

Fig. 5 Comparison of accuracy, precision, recall and F1-score for all classifiers on test set

The classification report

	precision	recall	f1-score	support
BT MeetMe	0.90	0.88	0.89	42
Bitlocker - Recovery Key	0.87	0.87	0.87	23
Browser	0.82	0.67	0.74	147
Citrix	0.89	0.83	0.86	99
Distribution Group (DL) Members Modification	0.88	1.00	0.94	60
Hard Disk Space Issue	0.79	0.75	0.77	67
Hardware Problem	0.86	0.89	0.88	538
Network issues	0.79	0.95	0.87	44
OS Problem	1.00	0.88	0.93	16
Outlook	0.90	0.92	0.91	295
Printer Installation/Issues	0.93	0.86	0.90	50
Project Specific Software	0.91	0.92	0.91	252
SCCM	0.90	0.98	0.94	97
Skype for Business	0.96	0.89	0.92	178
Unlicensed Software	1.00	1.00	1.00	40
Unlock Account / Password Reset	0.89	1.00	0.94	79
VPN issue	0.87	0.88	0.88	84
Virus Problem	0.84	0.84	0.84	38
avg / total	0.89	0.89	0.89	2149

Fig. 6 Classification report of SVM classifier on test set

It is found that Support Vector Machines (SVM) classifier (with 89% accuracy) performed well on the test data when compared with other classifiers. So in our research, SVM model is used as trained model and stored in the model store to predict the category of the new incoming unlabeled ticket. The individual class level performance of the SVM classifier over the test data is given in the Fig. 6 and Fig. 7 in the form of classification report and confusion matrix respectively.

```
[[ 37   0   2   1   0   0   0   1   0   0   0   1   0   0   0   0   0   0]
 [  0  20   0   0   0   0   3   0   0   0   0   0   0   0   0   0   0   0]
 [  1   0  98   1   2   2  20   0   0  10   0   3   1   0   0   6   1   2]
 [  0   0   3  82   0   1   5   0   0   1   0   4   2   0   0   0   1   0]
 [  0   0   0   0  60   0   0   0   0   0   0   0   0   0   0   0   0   0]
 [  0   0   3   1   0  50  13   0   0   0   0   0   0   0   0   0   0   0]
 [  0   1   8   2   0  10 479   3   0  10   3   7   4   5   0   1   4   1]
 [  0   0   0   0   0   0   0  42   0   0   0   2   0   0   0   0   0   0]
 [  0   2   0   0   0   0   0   0  14   0   0   0   0   0   0   0   0   0]
 [  1   0   1   0   4   0   8   4   0 271   0   0   1   0   0   3   1   1]
 [  0   0   1   0   0   0   5   0   0   0  43   1   0   0   0   0   0   0]
 [  2   0   0   3   0   0   5   3   0   2   0 231   1   1   0   0   4   0]
 [  0   0   0   0   0   0   2   0   0   0   0   0  95   0   0   0   0   0]
 [  0   0   1   0   2   0  11   0   0   4   0   2   0 158   0   0   0   0]
 [  0   0   0   0   0   0   0   0   0   0   0   0   0   0  40   0   0   0]
 [  0   0   0   0   0   0   0   0   0   0   0   0   0   0   0  79   0   0]
 [  0   0   0   1   0   0   2   0   0   2   0   3   0   0   0   0  74   2]
 [  0   0   2   1   0   0   1   0   0   0   0   0   1   1   0   0   0  32]]
```

Fig. 7 Confusion matrix of SVM classifier

5 Conclusion

Ticket routing plays an important role in any kind of service desk systems. Misrouting of the ticket to wrong maintenance group results in significant delays in getting the resolution for the problem. We proposed a ticket classifier system for an IT infrastructure service desk, which automatically routes the ticket to correct resolution group. We used four different well-known text classification algorithms and evaluated the classifier performances on the ticket data. A real word historical dataset consisting of ticket description and label is used to train classifiers. The classifier performances vary directly related to the classification algorithm and the dataset. A reasonably good accuracy has been achieved using Support vector machines (SVM) classifier model since it worked well for all samples of our service desk ticket data. The key challenge in ticket classification is handling the initial unstructured noisy data. We used various data cleaning techniques to remove the noise in the unstructured ticket data. The proposed automated ticket classifier system results in improved end user experience and customer satisfaction, effective support resource utilization, quicker ticket resolution time, and growth in business.

6 Future Work

A domain specific keyword list and stop word list could be extracted so that only relevant features can be used for mining. We plan to explore alternate classification techniques like ensemble of classifiers to further improve the performance of the model. Bi-grams, tri-grams, etc., can be used as features to further improve the performance of the system. We would need to explore more machine learning techniques to handle data related challenges. A further focus would be required to handle even larger training data.

References

1. Lucca, G.A.D., Penta, M.D., Gradara, S.: An approach to classify software maintenance requests. In: ICSM (2002)
2. Dasgupta, G., Nayak, T.K., Akula, A.R., Agarwal, S., Nadgowda, S.J.: Towards auto-remediation in services delivery: context-based classification of noisy and unstructured tickets. In: Service-Oriented Computing—12th International Conference, ICSOC 2014, November 3–6, 2014, Proceedings, pp. 478–485. Paris, France (2014)
3. Kadar, C., Wiesmann, D., Iria, J., Husemann, D., Lucic, M.: Automatic classification of change requests for improved it service quality. In: Proceedings of the 2011 Annual SRII Global Conference, SRII '11, pp. 430–439. Washington, DC, USA, IEEE Computer Society (2011)
4. Agarwal, S., Sindhgatta, R., Sengupta, B.: Smartdispatch: enabling efficient ticket dispatch in an it service environment. In: Proceedings of the 18th ACM SIGKDD International Conference on Knowledge Discovery and Data Mining, KDD '12, pp. 1393–1401, New York, NY, USA, ACM (2012)
5. Diao, Y., Jamjoom, H., Loewenstern, D.: Rule-based problem classification in IT service management. In: IEEE International Conference on Cloud Computing (2009)
6. Wei, X., Sailer, A., Mahindru, R., Kar, G.: Automatic structuring of IT problem ticket data for enhanced problem resolution. In: Integrated Network Management, IM 2007. 10th IFIP/IEEE International Symposium on Integrated Network Management, 21–25 May 2007, pp. 852–855. Munich, Germany (2007)
7. Potharaju, R., Jain, N., Nita-Rotaru, C.: Juggling the jigsaw: towards automated problem inference from network trouble tickets. In: 10th USENIX Symposium on Networked Systems Design and Implementation (NSDI 13), pp. 127–141. Lombard, IL, USENIX (2013)
8. Shimpi, V., Natu, M., Sadaphal, V. et al.: Problem identification by mining trouble tickets. In: Proceedings of the 20th International Conference on Management of Data, Computer Society of India, pp. 76–86 (2014)
9. Sakolnakorn, P.P.N., Meesad, P., Clayton, G.: Automatic resolver group assignment of IT service desk outsourcing in banking business (2007)
10. Xu, J., Tang, L., Li, T.: System situation ticket identification using SVMs ensemble. Expert Syst. Appl. **60**, 130–140 (2016)
11. Handling imbalanced data—a review by Sotiris Kotsiantis et al. GESTS Int. Trans. Comput. Sci. Eng. **30** (2006)
12. Roy, S., Muni, D.P., Yan, J.J.Y.T. et al.: Clustering and labeling IT maintenance tickets. In: International Conference on Service-Oriented Computing, pp. 829–845. Springer International Publishing (2016)
13. Zeng, C., Zhou, W., Li, T., Shwartz, L., Grabarnik, G.Y.: Knowledge guided hierarchical multi-label classification over ticket data. IEEE Trans. Netw. Serv. Manag. **14**(2), 246–260 (2017)
14. Son, G., Hazlewood, V., Peterson, G.D.: On automating XSEDE user ticket classification. In: Proceedings of the 2014 Annual Conference on Extreme Science and Engineering Discovery Environment, p. 41. ACM (2014)
15. Frenay, B., Verleysen, M.: Classification in the presence of label noise: a survey. TNNLS **25**(5), 845–869 (2014)
16. Jan, E.E., Chen, K.Y., Idé, T.: Probabilistic text analytics framework for information technology service desk tickets. In: 2015 IFIP/IEEE International Symposium on Integrated Network Management, (IM), pp. 870–873. IEEE (2015)

17. Sebastiani, F.: Machine learning in automated text categorization. ACM Comput. Surv. **34**(1), 1–47 (2002)
18. Ikonomakis, M., Kotsiantis, S., Tampakas, V.: Text classification using machine learning techniques. WSEAS Trans. Comput. **8**(4), 966–974 (2005)
19. Khan, A., Baharudin, B., Lee, L.H., Khan, Kh: A review of machine learning algorithms for text-documents classification. J. Adv. Inf. Technol. **1**(1), 4–20 (2010)
20. McCallum, A., Nigam, K. et al.: A comparison of event models for naive bayes text classification. In: AAAI-98 Workshop on Learning for Text Categorization, vol. 752, pp. 41–48. Citeseer (1998)
21. Joachims, T.: text categorization with support vector machines: learning with many relevant features, technical report 23. Universitat Dortmund, LS VIII (1997)
22. Allahyari, M., Pouriyeh, S., Assefi, M., Safaei, S., Trippe, E.D., Gutierrez, J.B., Kochut, K.: A brief survey of text mining: classification, clustering and extraction techniques. arXiv:1707.02919 (2017)

Automatic Segmentation and Breast Density Estimation for Cancer Detection Using an Efficient Watershed Algorithm

Tejas Nayak, Nikitha Bhat, Vikram Bhat, Sannidhi Shetty, Mohammed Javed and P. Nagabhushan

Abstract In the current world, breast cancer is a fatal disease that is causing women mortality at a larger rate. The central idea in preventing this fatal disease is by early diagnosis through the use of computer aided mammography, which can be accomplished by developing intelligent algorithms and software techniques to process these mammography images automatically, and eventually extract information assisting the physicians in detecting and diagnosing the abnormal growth of the cancer cells. The current research work is an effort in the aforesaid direction, where automatic breast cancer detection is made feasible at early stages, with the image processing technique of segmentation and breast density estimation through mammographic images using the well-known Watershed Algorithm. The watershed transformation is a well-known technique for contour based feature extraction and region segmentation, which provided us the motivation to explore it further for assisting the detection of breast cancer. The proposed technique is tested with publicly available MIAS dataset, and accuracy thus obtained is comparable with the state-of-the-art techniques available in the literature with improved computational efficiency.

Keywords Breast cancer · Mammography · Tumor detection
Watershed transformation

1 Introduction

Breast cancer is a well-known fatal disease that is increasing the women mortality at a rapid rate worldwide. The National Cancer Institute [1] has estimated that 1 out of 8 women in the US will face this fatal disease in her lifetime. In 2017 itself, 252,710

T. Nayak (✉) · N. Bhat · V. Bhat · S. Shetty
Department of CSE, NMAM Institute of Technology, Nitte, India
e-mail: veernayak14@gmail.com

M. Javed · P. Nagabhushan
Department of IT, Indian Institute of Information Technology, Allahabad, Allahabad, India
e-mail: pnagabhushan@hotmail.com

P. Nagabhushan et al. (eds.), *Data Analytics and Learning*,
Lecture Notes in Networks and Systems 43,
https://doi.org/10.1007/978-981-13-2514-4_29

new cases of invasive breast cancer are going to be detected in women as per the survey made in USA [1]. Every year, roughly 144,937 fresh cases of breast cancer are detected, and 70,218 women face casualty in India alone [2]. However, early diagnosis of breast cancer has proved helpful for the success of the treatment and reducing the mortality rate caused by this disease [2]. The process of diagnosis of breast cancer involves detecting the areas of high intensities in the mammographic images, which confirm the growth of benign or malignant tumors. X-ray and Magnetic Resonance Imaging (MRI) based imaging are the advanced medical imaging techniques that are generally preferred to generate good quality images of human body parts while treating tumors in the breast, brain, ankle, and foot. X-rays and MRI of the breast are commonly used tool for screening breasts with the technique of mammography or ultrasound imaging. The X-ray photograph of the human breast is called as mammogram. This technology is largely used for screening women who have the high risk of breast cancer, and subsequently help the doctor in evaluating the growth rate of cancer and accordingly take preventive measures in overcoming the abnormalities.

During the past few decades, different set of methodologies have been proposed to segment breast images based on breast density. Oliver et al. [3] and Petrodi and Brady [4] have used texture-based models to extract mammographic appearances within the breast region by statistical distribution of clusters and parenchymal density patterns. Boyd et al. [5] and Sivaramakrishna et al. [6] used optimal thresholding technique as key criteria to bifurcate fatty and the dense regions in the breast. Oliver et al. [7] in his review paper gave a detailed report on the advantages and disadvantages associated with different kinds of breast mass density detection techniques using Receiver Operating Characteristic (ROC) and Free-Response Receiver Operating Characteristic (FROC) Curves. Liu et al. [8] have shown that combining watershed algorithm and level set method increases the performance of the segmentation. Also, if over-segmentation is caused by watershed, a noise reduction technology has been introduced. Berber et al. [9] have come up with a new segmentation algorithm, called Breast Mass Contour Segmentation (BMCS) method, which is tested against various segmentation methods like level set segmentation, seed based region growing technique and watershed segmentation. Ireaneus Anna Rejani and Thamarai Selvi [10] used the method of local thresholding and nearest neighborhood clustering for the detection of cancer in the breast. Ireaneus Anna Rejani and Thamarai Selvi [10] have used wavelet enhancement methods to detect breast cancer. The approach was to enhance the mammogram and feature extract the tumor area by making use of discrete wavelength transform (DWT) using the support vector machine classifier. A sensitivity of 88.75% was obtained using this approach for the mini-MIAS dataset which was tested over 75 images. Kavanagh et al. [11] worked on screening mammography and symptomatic status and showed the variation in sensitivity among different age groups of women. Oliver et al. [12] used pixel-based segmentation by using spatial information and Breast Imaging-Reporting and Data System (BIRADS) [13] classification for analyzing dense, glandular and fatty regions of the breast. In the BIRAD-IV stage [13] the mass formation is denser and indicates the malignant stage or suspicious abnormality, which can be used for early diagnosis of cancerous tissues [12].

Increase in breast density can be a strong indicator of diagnosing cancer in a women [13]. So, this paper is focused on development of image processing technique for automatic segmentation and extraction of breast density using an efficient Watershed Algorithm. The Watershed Transformation is a well-known technique for both contour analysis and region growing based segmentation [14, 15]. Therefore, an attempt is made to detect cancerous tissues in the breast using morphological watershed algorithm. Region based breast cancer detection basically involves detecting contoured region of malignant and benign tumors. This research paper is organized as follows: Sect. 2 gives a brief description of the dataset used and further introduces the proposed model of the breast density segmentation and estimation. Discussions and Experimental results are provided in Sect. 3, and Sect. 4 concludes the paper.

2 Proposed Technique

The automatic segmentation and breast density estimation technique proposed in the paper uses Mammographic Image Analysis Society (MIAS) dataset [16] which contains 322 images extracted from UK National Breast Screening Program. The publicly available MIAS Database has been modified to 200-μ pixel edge and appended, so that all the images are of same size as 1024×1024 pixels

The American College of Radiology uses BIRAD Classification which is widely accepted in risk assessment and quality assurance in Mammography, ultrasound, and MRI. There are different stages into which BIRAD can be classified [13]: (i) BIRAD I and BIRAD II are starting of benign stages like breast lipomas and breast cysts, (ii) BIRAD III indicates the stage is probably benign, (iii) BIRAD IV indicates suspicious abnormality and leading to malignant stages. The BIRAD classification shows the pattern of increase in breast density in each stage in the MIAS dataset. The database has three classes of breast density: fatty, fatty-glandular and dense-glandular and is based on BIRAD Classification (see Fig. 2 for more details). In the benign stage, usually a lump is formed near the breast region which shows high density in comparison to its neighborhood and the shape of the breast is like its surrounding. In the malignant stage, the density of the region where the lump was formed in benign stage significantly increases and there is distortion in the shape of the breast.

It is also well-established fact in the medical community that as the breast density increases the chance of cancer also increases [13], and most of the breast cancers will lead to malignant stage in a quick span of time. However, it is noted that the efficacy of the sophisticated systems developed in assisting breast cancer decrease due to increase in breast density [13]. Therefore, automatic segmentation of breast densities and subsequent quantification of the dense regions will surely assist doctors in early detection and diagnosis of cancers.

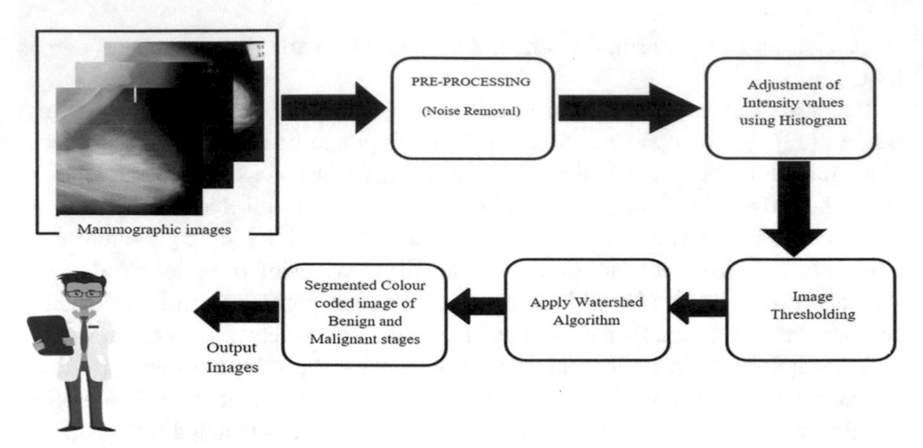

Fig. 1 Working model of proposed methodology

This paper aims at developing a computer tool using watershed segmentation that helps the radiologist in having second opinion on the diagnosed result. The working diagram is already depicted in Fig. 1. A set of input images, which are taken from the MIAS dataset are shown in Fig. 2 with the increasing density indicated by white regions.

In the proposed model, the X-ray image is read using MATLAB software for the detection of breast cancer. Let $g(x, y)$ denote the input image. Further Preprocessing the X-ray images is necessary as they contain noise. During X-ray imaging noise may appear because of scattering of radiation and also due to any leakage. An important morphology based transformation is to carry out filtering using top-hat that removes the background from the source image. Top-hat is the subtraction of an opened image from the source image. Top-hat transformation for a greyscale image $g(x, y)$ is defined as,

$$T_{hat}(g) = g - (g \, o \, b),$$

where b is the structuring element. The very significance of using top-hat filtering is that it automatically corrects the effects that occur because of illumination in non-uniform manner. Morphological opening operation in gray scale images has the power to remove the elements that are smaller than structuring element (SE) [17].

Pectoral muscle present in the breast is also one form of noise which is removed during the preprocessing stage by estimation using the straight line method similar to the work of [12]. Note that, if pectoral muscles are not separated then proper segmentation results cannot be obtained. So using a preprocessing stage pectoral muscles from the original image are removed. The next task is to vary the intensity values of the image using histogram stretching. The technique performs mapping of the intensity values of the input image to the output image using new values. This will produce good contrast in the final image [18].

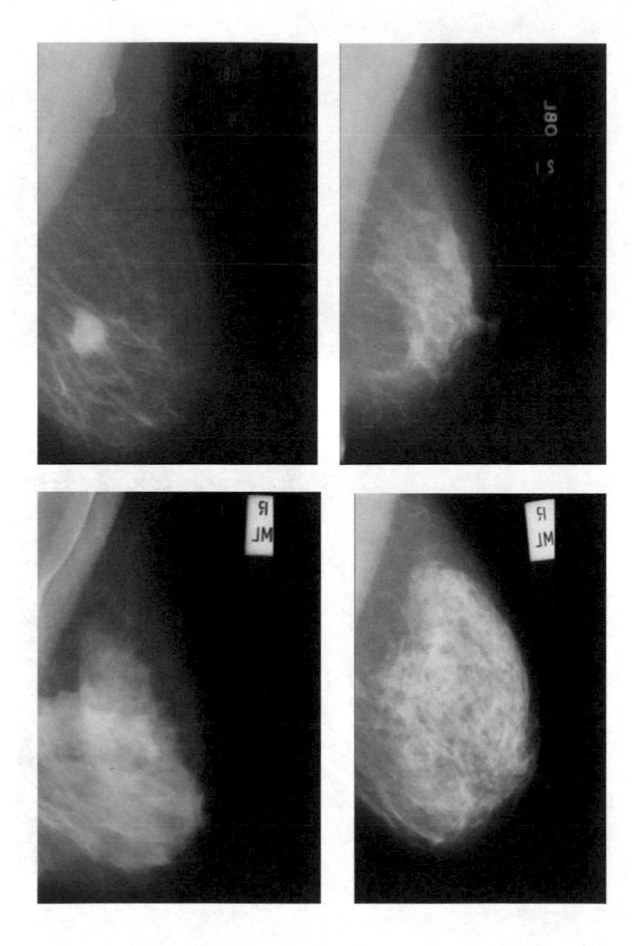

Fig. 2 Sample mammographic images from the MIAS dataset that shows increased breast density (white region)

Image segmentation is a technique that separates or breaks the input image into different meaningful parts depending on the type of the object. We usually segment regions by identifying common properties. In the same manner, we need to detect contours using the differences across different regions. The basic clue in identifying region separators is to use pixel intensity values. Therefore, thresholding technique can be used to separate dark and light regions [18]. Thresholding-based segmentation basically works with gray scale intensity values and it uses a threshold to separate the required classes [15, 17]. Thresholding value at a point (x, y) on the input image $g(x, y)$ is considered as object if $g(x, y) > D$; else it is considered as a background pixel at a selected Threshold D. Threshold-based segmented image $S(x, y)$ is given as,

$$S(x, y) = \begin{cases} 1 \; if \;\; g(x, y) > D \\ 0 \; if \;\; g(x, y) \leq D \end{cases}$$

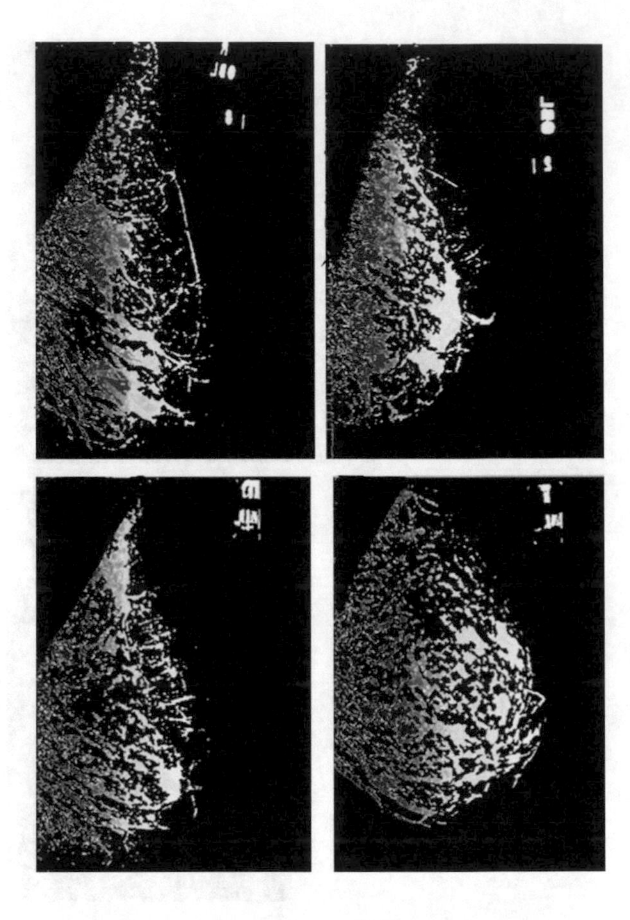

Fig. 3 Watershed segmented result of examples sown in Fig. 2

When watershed is applied, basically three different regions are generated. The first type of points are got from the various minima present in the image region. The second aspect here is the development of catchment basin. The final aspect is formation of points by separation of catchment basins, and these are coined as watershed detected lines.

Segmentation is accomplished by classifying the pixels into two classes based on the threshold value. The threshold is value always pre-computed before the actual segmentation is carried out. Thresholding-based segmentation is quite useful technique that does the job quickly separating the background from the foreground regions. The final image is a binary image, and this type of technique is very useful in images that have high contrast. Further, watershed algorithm is used to get better segmentation results. Watershed segmentation is a region dependent method that uses morphological operations and provides a complementary approach to the segmentation of objects [4]. The watershed segmentation results on the running sample images are shown in Fig. 3.

A sample image shown in Fig. 4 is taken from google.com for the purpose of demonstrating the working of our proposed method with any random mammographic image. Figure 4 depicts all the intermediate stages of the proposed method. Watershed works on the points belonging to the regional minimum [14]. The main motive of applying watershed transform in doing segmentation by transforming the image into other image domain whose catchment basins are the main entity that we want to detect. Let m_1, $m_2, \ldots, m_q$ denote the coordinate points in the minima region of the threshold segmented image S(x, y). Let $E(m_i)$ denote the coordinate points of the detected catchment basins forming objects of interest. Let T[n] be set of coordinates (u, v) for the watershed image such that: $T[n] = \{(u, v)|g(u, v) < n\}$ [14]. Applying watershed transformation directly may lead to over-segmentation. To avoid this, it is advised to use markers, where marker represents a connected object that is the part of the image. The objects are indicated by the presence of internal markers and external markers show strong association with the pixels in the background. The colored portion illustrated in Fig. 4, shows the presence of tumor region in the breast part by means of watershed algorithm. The region marked in red and different shades of red in Fig. 4 shows the region of highest intensity indicating dense region formation in the breast. Yellow region indicates the lower intensity values and can be considered as harmless or least possibility of tumor formation in that region.

3 Experimental Results and Discussion

This section is dedicated to showcase the experimental results and performance-related parameters associated with the watershed technique proposed in this paper. The final goal here is to extract breast density related parameters from the segmented regions of the mammographic images. A sample result for a set of images using the proposed method is depicted in Fig. 4. The segmented image which is a part of the processing stage is compared with the ground truth image for finding accuracy of the algorithm developed. The following three state-of-the-art performance metrics are proposed to evaluate the obtained results [12].

3.1 Right Classification (P1)

This performance measure gives the right classification of the pixels in the image by classifying them in the fatty and dense categories as Right Positives (RP), Right Negatives (RN), Wrong Positives (WP), and Wrong Negatives (WN). Here, RP and WN are the number of correct classifications, RN and WP are the number of false classifications. The P1 measure is computed as percentage of correct classification as given below,

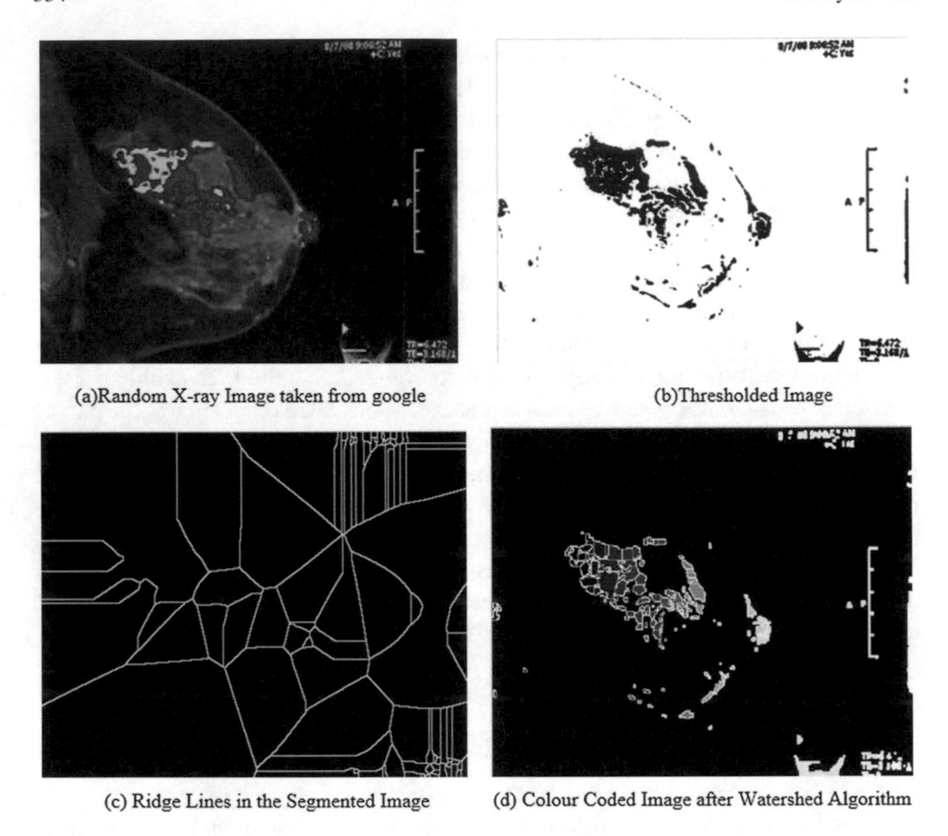

(a)Random X-ray Image taken from google

(b)Thresholded Image

(c) Ridge Lines in the Segmented Image

(d) Colour Coded Image after Watershed Algorithm

Fig. 4 Random image taken from www.google.com representing water ridge lines and color coded image after watershed algorithm

$$P1 = \frac{RP + RN}{RP + RN + WP + WN}$$

3.2 Overlapped Area (P2)

The area overlap performance measure considers the dense pixels for classification. Let R1 and R2 be the two segmentation results obtained, then P2 is defined with the total pixels intersecting R1 and R2 to the total pixels representing the union of R1 and R2. The mathematical formulation is shown as follows:

$$P2 = \frac{|R1 \cap R2|}{|R1 \cup R2|} = \frac{RP}{RP + WP + WN}$$

3.3 Method of Dice Coefficient (P3)

In P3 measure, correctly classified dense pixels are given more preference in both segmentations.

$$\mathrm{P3} = \frac{2|\mathrm{R1} \cap \mathrm{R2}|}{|\mathrm{R1}| + |\mathrm{R2}|} = \frac{2 \times \mathrm{RP}}{2 \times \mathrm{RP} + \mathrm{WP} + \mathrm{WN}}$$

The Dice Coefficient is not only a measure of how many positives are found, but it also penalizes for the false positives that the method finds. Dice Coefficient takes values between 0 and 1, in which 1 is identical segmentation and 0 is non-identical segmentation [12] (Table 1).

Sensitivity is another measure to estimate the correctness of proposed methodology. Sensitivity is defined as

$$\mathrm{Sensitivity} = \frac{\mathrm{RP}}{\mathrm{RP} + \mathrm{WN}}$$

The proposed approach in our work with image processing techniques proved its performance via performance metrics with an average sensitivity of 94.24%. The comparison of sensitivity among different researcher works as quoted in Table 2 along the features used, shows that we have obtained greater sensitivity value.

The performance reported by the proposed technique is between the two models demonstrated by [12] using LDA and PCA strategies. But, it should be noted that

Table 1 Experimental results for the proposed method using the performance metrics of P1, P2 and P3

Performance measure	Proposed method values	PCA strategy [12]	LDA strategy [12]
P1	0.886	0.916 ± 0.038	0.890 ± 0.031
P2	0.876	0.900 ± 0.122	0.842 ± 0.196
P3	0.925	0.943 ± 0.077	0.876 ± 0.139

Table 2 Sensitivity measure compared with different research works

Papers	Features	Sensitivity of mammograms
Proposed method	Efficient watershed algorithm	94.24% (88–97%)
Rejani et al. [10]	SVM classifier	88.75%
Baines et al. [19]	Unstated	69% (69–78%)
Bone et al. [20]	Histopathological correlation	93%
Kavanagh et al. [11]	Screening mammography and symptomatic status	60–68%
Vikhee and Thool [17]	Wavelet processing and adaptive thresholding	85%

Table 3 Time complexity of the proposed algorithm for selected images from the MIAS

Image from the dataset [16]	Computation time (in ms)	Size (in MBs)
mdb001	0.2424	1
mdb005	0.3435	1
mdb010	0.2618	1
mdb012	0.2787	1
mdb013	0.2714	1
mdb015	0.2467	1
mdb069	0.2868	1
mdb072	0.2219	1
mdb075	0.2461	1
Figure 3. (Random Image taken from www.google.com)	0.0401	0.011

implementation wise, the LDA and PCA strategies are computationally expensive operations. Whereas, the time complexity of Watershed Algorithm is O(n) which has linear time computation in comparison with LDA and PCA strategies where the time complexity is $O(\min(p^3, n^3))$, where there are "n" data points consisting of "p" features each. Table 3 shows the time complexity of selected images from the dataset as well the sample image selected from google.com. The average computational time obtained on the 20 images of MIAS dataset [16] is 0.2643 s, and for the image taken from google.com, the computational time was found to be 0.011 s. The computation time is calculated for images containing both benign and malignant tissues.

4 Conclusion

In this paper, an efficient morphological watershed algorithm is developed to help early detection of tumors in the breast through image processing techniques of segmentation and breast density estimation. In the experimental results, MIAS ground truth details are compared with the output image to obtain the accuracy of the proposed technique. The experimental results reported are competent with the existing literature methods with an increased computational efficiency. Based on the current research work, future scope would be to improve the accuracy of the algorithm developed and carry out detailed classification of dense and fatty regions based on the BIRAD classification rules.

References

1. U.S. Breast Cancer Statistics. http://www.breastcancer.org/symptoms/understand_bc/statistics. Accessed 20 July 2017
2. Trends of breast cancer in India. http://www.breastcancerindia.net/bc/statistics/stati.hm. Accessed 20 July 2017
3. Oliver, A., Freixenet, J., Zwiggelaar, R.: Automatic classification of breast density. In: Proceedings IEEE International Conference on Image Processing, 2005, vol. 2, pp. 1258–1261
4. Petroudi, S., Brady, M.: Breast density segmentation using texture. Lect. Notes Comput. Sci. **4046**, 609–615 (2006)
5. Boyd, N.F., Byng, J.W., Jong, R.A., Fishell, E.K., Little, L.E., Miller, A.B., Lockwood, G.A., Tritchler, D.L., Yaffe, M.J.: Quantitative classification of mammographic densities and breast cancer risk: results from the Canadian national breast screening study. Nat. Cancer Inst. **87**, 670–675 (1995)
6. Sivaramakrishna, R., Obuchowski, N.A., Chilcote, W.A., Powell, K.A.: Automatic segmentation of mammographic density. Acad. Radiol. **8**(3), 250–256 (2001)
7. Oliver, A., Lladó, X., Pérez, E., Pont, J., Denton, E., Freixenet, J., Martí, J.: A review of automatic mass detection and segmentation in mammographic images. Med. Image Anal. **2010**(14), 87–110 (2010)
8. Liu, J., Chen, J., Liu, X., Chun, L., Tang, J., Deng, Y.: Mass segmentation using a combined method for cancer detection. In: BMC System Biology, vol. 5, Supplementary 3, p. S6 (2011). https://doi.org/10.1186/1752-0509-5-S3-S6
9. Berber, T., Alpkocak, A., Balci, P., Dicle, O.: Breast mass contour segmentation algorithm in digital mammograms. Comput. Methods Programs Biomed. **110**(2), 150–159 (2013). https://doi.org/10.1016/j.cmpb.2012.11.003
10. Ireaneus Anna Rejani, Y., Thamarai Selvi, S.: Early detection of breast cancer using SVM classifier technique. Int. J. Comput. Sci. Eng. **1**(3), 127–130 (2009)
11. Kavanagh, A., Giles, G., Mitchell, H., Cawson, J.N.: The sensitivity, specificity, and positive predictive value of screening mammography and symptomatic status. J. Med. Screen. **7**, 105–110 (2000)
12. Oliver, A., Lladó, X., Pérez, E., Pont, J., Denton, E., Freixenet, J., Martí, J.: A statistical approach for breast density segmentation. J. Digit. Imaging **23**(5), 527–537 (2010). https://doi.org/10.1007/s10278-009-9217-5
13. BIRAD Classification. https://radiopaedia.org/articles/breast-imaging-reportingand-data-system-birads. Accessed 20 July 2017
14. Gonzalez, R., Woods, R.: Digital Image Processing, 3rd edn, pp. 627–675. Prentice Hall, Upper Saddle River, N.J. (2008)
15. Alhadidi, B., Mohammad, H.: Mammogram breast cancer edge detection using image processing function. Inf. Technol. J. **6**(2), 217–221 (2007). ISSN-1812-5638
16. Mammographic Image Analysis Society, Dataset. http://peipa.essex.ac.uk/ipa/pix/mias. Accessed 1 June 2017
17. Vikhe, P.S., Thool, V.R.: Mass detection in mammographic images using wavelet processing and adaptive threshold technique. J. Med. Syst. **40**(4), 82–97 (2016)
18. Hypermedia Image Processing Reference. https://homepages.inf.ed.ac.uk/rbf/HIPR2/copyrght.htm. Accessed 15 June 2017
19. Baines, C., Miller, A.B., Wall, C., McFarlane, D.V., Simor, I.S., Jong, R., Shapiro, B.J., Audet, L., Petitclerc, M., Ouimet-Oliva, D.: Sensitivity and specificity of first screen mammography in the Canadian national breast screening study: a preliminary report from five centres. Radiology **160**(2), 295–298 (1986)

20. Bone, B., Aspalin, P., Bronge, L., Isberg, B., Perbek, L., Veress, B.: Sensitivity and specificity of MR Mammography with histopathological correlation in 250 breasts. Acad. Radiol. **37**, 208–213 (1996)

Offline Signature Verification Based on Partial Sum of Second-Order Taylor Series Expansion

B. H. Shekar, Bharathi Pilar and D. S. Sunil Kumar

Abstract This paper presents a novel feature extraction technique which is based on partial sum of second-order Taylor Series Expansion (TSE) for offline signature verification. Partial sum of TSE is calculated with finite number of terms within a small neighborhood of a point, yields approximation for the regular function. This essentially an effective mechanism to extract the localized structural features from signature. We propose kernel structures by incorporating the Sobel operators to compute the higher order derivatives of TSE. Support Vector Machine (SVM) classifier is employed for the signature verification. The outcome of experiments on standard signature datasets demonstrates the accuracy of the proposed approach. We performed a comparative analysis for our approach with some of the popular other approaches to exhibit the classification accuracy.

Keywords Signature verification · Taylor series expansion · Higher order derivatives · Support vector machine

1 Introduction

Signature has been considered as one of the most trusted biometrics for authentication as well as identification of individuals due to its legal acceptance in commercial and banking sectors. Depending on the signature acquisition method used, automatic signature verification systems can be classified into two categories: Offline

B. H. Shekar · D. S. Sunil Kumar
Department of Computer Science, Mangalore University, Mangalore, Karnataka, India
e-mail: bhshekar@gmail.com

D. S. Sunil Kumar
e-mail: dssunil6@gmail.com

B. Pilar (✉)
Department of Computer Science, University College, Mangalore, Karnataka, India
e-mail: bharathi.pilar@gmail.com

P. Nagabhushan et al. (eds.), *Data Analytics and Learning*,
Lecture Notes in Networks and Systems 43,
https://doi.org/10.1007/978-981-13-2514-4_30

signature verification and online signature verification. In offline signature verification, signature samples are collected from the signer on paper and is fed for verification. In case of online signature verification, signature samples are captured from the signer using digital gadgets like scanner, pen tablet, etc. The offline signature verification becomes more challenging when it comes to classify skilled forgeries. This is because, as opposed to the online case, offline signatures lack dynamic information. Hence, designing a powerful feature extraction technique is essential in offline signature verification approaches.

We have seen many feature extraction techniques, which extract significant features from signatures. A feature extraction technique can be applied on either ways, one is a global where features are extracted from the entire input image and the second is local where the input image is partitioned into different segments and from each segment, features are extracted. Finally, the entire signature is represented by combination of these extracted features. We made an attempt in this approach to extract dominant features from the signature image based on partial sum of second-order Taylor series expansion. Partial sum of TSE with finite terms is obtained on a small neighborhood of a point that can estimate the function efficiently and it gives a robust mechanism for the localized features of signature. In order to verify the proposed method, we employed an efficient classifier, namely, Support Vector Machine (SVM).

Next section of this paper is followed by Sect. 2 that contains literature review of some of the well-known methods. Section 3 contains detailed explanation of the proposed approach. Section 4 contains the details related to experiments followed by the experimental findings and lastly Sect. 5 is about conclusion.

2 Review of the Literature

Research in the area of signature verification has brought a lot of innovations from the researchers over the last two decades. The local histogram feature approach [18] divides the signature into number of zones with the help of both the Cartesian and polar coordinate systems. For every zone, Histogram of Oriented Gradients (HOG) and histogram of Local Binary Patterns (LBP) histogram features are computed. A variant of LBP, namely, Blockwise Binary Pattern (BBP) [15], divides signature into 3×3 blocks to extract local features. Writer-independent [11] method emphasizes on shape and texture features of a signature. This method extracts the black pixels concentrated along with candidate pixel. The proposed method employs MLP and SVM classifiers. The inter-point envelope based distance moments [9] capture structural and temporal information by computing inter-point distances, which is the distance between reference point and other points in an extracted envelope. Shikha and Shailja [16] proposed an offline signature recognition system, which is based on neural network architecture. The proposed method uses Self-Organizing Map (SOM) as a learning algorithm and Multilayer Perceptron (MLP) for classification of patterns.

Shekar et al. [13] proposed a grid-structured morphological pattern spectrum. The proposed method divides the signature into equally sized eight grids. Then pattern spectrum is obtained for every grid. Bhattacharya et al. [3] have proposed a Pixel Matching Technique (PMT). The proposed approach verifies by comparing each pixel of sample signature against test signature. Kruthi and Shet [8] proposed support vector machine based offline Signature Verification System (SVM). SVM is employed for classification of signatures as objects. Guerbai et al. [5] proposed an approach for offline signature verification system, which is based on one-class support vector machine. This technique considers only genuine signature pattern of a signer. Soleimani et al. [17] proposed an approach of Deep Multi-task Metric Learning (DMML), and Shekar et. al. proposed Taylor series for iris recognition [14]. The proposed approach considers the similarities as well as dissimilarities of a signature.

Although we have seen a plethora of algorithms for offline signature verification, devising an accurate offline signature verification method is still a challenging issue. Therefore, we were inspired to devise a more accurate technique for offline signature verification system.

3 Proposed Approach

The proposed method uses partial sum of first finite number of terms of second-order Taylor series expansion technique for offline signature verification. This process involves three important stages: preprocessing, feature extraction, and classification. During preprocessing stage, the Otsu's binarization method is applied on signature samples. This binarization process adds some noise which was later removed by conducting morphological filter operations. Later, the thicknesses of strokes of signatures are normalized by performing morphological operations like thinning and dilating. The following section presents the process of feature extraction and classification.

3.1 Taylor Series Expansion

Let function $f(x)$ be the function that is continuously infinitely differentiable in a neighborhood. The Taylor series expansion for $f(x)$ on a point $x = a$ indicates the function in a small neighborhood of a point a. An expression for infinite Taylor series expansion for a function $f(x)$ at $x = a$ is

$$f(x) \equiv \sum_{n=0}^{\infty} \frac{f^n(a)}{n!}(x-a)^n \tag{1}$$

When the functional value and the values of its derivatives are known at $x = a$, then the function can be estimated at all the points of the neighborhood of a. Then above expression can also be expressed as

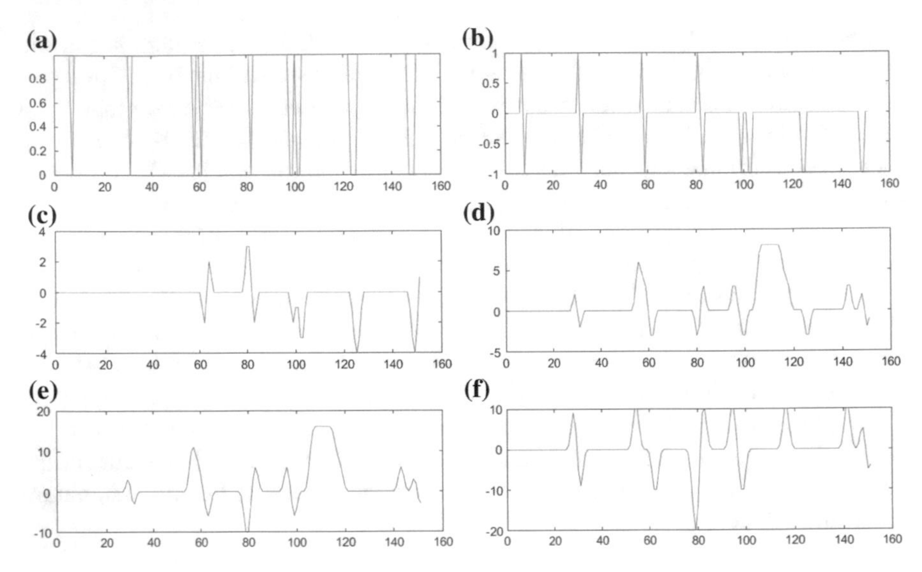

Fig. 1 **a** 1D signal of the signature sample from CEDAR database; **b**–**f** order of first–fifth derivatives of the signal. The Y-axis shows the range from which variation in derivatives. The range becomes wider as the order becomes higher

$$f(x) \equiv \sum_{n=0}^{N} \frac{f^n(a)}{n!}(x-a)^n + R_n \tag{2}$$

here R_n represents Taylor's remainder. Assuming R_n tends to zero as $n \to \infty$ for a small neighborhood, then the partial sum of TSE is calculated for N finite number of terms yields approximation of function $f(x)$. The advantages of TSE are explored here to extract the signature features in a local region. In [1] and [2], authors explored the second and third coefficients of TSE. These are based on the first- and second-order derivatives obtained separately at eight different scales and are then encoded on the basis of zero-crossings of the convolution outputs. First derivative reveals the minute local features, and the second derivative reveals features of concavity and convexity of the edges. It is a well-known fact that the higher order derivatives can well extract the global information within a neighborhood; hence, we have taken third-order derivatives also. Figure 1 shows the output derivatives of a 1D signal obtained for a signature sample of CEDAR database. The range of values over the vertical axis tells the range of variations in derivatives; this becomes wider for higher orders. It is verified empirically that derivatives above fifth order fail to contribute much. It is shown in Fig. 1 that there is small variation among fourth and fifth derivatives.

In order to calculate horizontal as well as vertical higher order derivatives, we have extended Sobel's kernel which uses coefficients of nth-order binomial expansion, i.e., the elements of the nth-order kernel are also elements of the nth row of Pascal triangle and are obtained by the coefficients of the binomial expansion $(a+b)^n$, a and

b becomes unity. For example, the kernels to calculate third-order derivatives along horizontal direction and vertical direction are given below.

$$\begin{bmatrix} -1 & -3 & -3 & -1 \\ 0 & 0 & 0 & 0 \\ 0 & 0 & 0 & 0 \\ 1 & 3 & 3 & 1 \end{bmatrix} \quad \begin{bmatrix} -1\,0\,0\,1 \\ -3\,0\,0\,3 \\ -3\,0\,0\,3 \\ -1\,0\,0\,1 \end{bmatrix}$$

The summation of horizontal and vertical partial sum of TSE forms the features for that point $x = a$. Histogram of 10 bins is used to represent the features.

3.2 Classification Based on SVM

Classification is performed by employing a well-known classifier, namely, Support Vector Machines (SVM). SVM works on the basis of decision planes that defines decision boundaries. A decision boundary is one that separates a set of objects which belongs to different classes. Here, SVM is to classify the samples by training the model. The trained model based on the features which essential estimates the values of test data [6, 10]. Let n be a training set $\{(x_i; y)\}_{i=1}^{n}$ where $x_i \in R^L$ is chosen from X domain and label y_i is a integer of $Y = \{0, 1\}$. Finally, the aim of SVM classifier is to build a model that essentially labels the unlabelled test sample. It is the process of learning a function f : $X \rightarrow$ Y which derives an instance x to an instance y of Y. For a given instance of pairs of labels (x_i, y_i) , $i = 1, \ldots, l$ where $x_i \in R^n$ and $y \in \{0, 1\}^l$, for the following optimization problem, the SVM classifier requires the solution:

$$min_{(W,b,\xi)} \frac{1}{2} W^T W + C \sum_{i=1}^{l} \xi_i \tag{3}$$

Subjected to

$$y_i \left(W^T \phi\,(x_i) + b \right) \geq 1 - \xi_i \tag{4}$$

and

$$\xi_i \geq 0. \tag{5}$$

Let the training vectors x_i are drawn into a hyperplane space by the function ϕ. The SVM draws a linear separating hyperplane which essentially widens the maximum margin. $C > 0$ is the penalty parameter of the error term, and kernel trick is used transform objects from lower dimensional space to higher dimensional space. Since SVM is a bilinear classifier, N SVM classifiers require to classify N classes. So, in this proposed method, we used one against all strategy which employs N number of SVM classifiers for N number of writers.

Table 1 Number of samples in the datasets

Dataset	No. of signers	Genuine signatures	Skilled forgery	Total signatures
CEDAR	55	24	24	2640
MUKOS	30	30	15	1350

4 Experimentation Results and Discussions

Here, we brought down experimental results carried out on the signature databases, namely, The Centre of Excellence for Document Analysis and Recognition (CEDAR) and Mangalore University Kannada Off-line Signature corpus (MUKOS) a regional language offline signature corpus. Both the databases contain different number of signers with various genuine and forge signatures. The experimental setup of both datasets is presented in Table 1. The experimentations are carried out in MATLAB R2017a software environment on DELL Laptop with Intel(R)Core(TM) i5-7200U CPU with 8 GB RAM on Windows-10.

The knowledge repository contains the TSE features extracted from every signature sample of the dataset. It includes both genuine signatures and skilled forge signatures. From datasets, signatures are considered into two sets: one is training set and second is testing set. The samples are varied in number. Experiments are conducted in four sets. For set-1, we considered first ten genuine signatures and first ten skilled forges signatures for training and are tested against the remaining signatures of same dataset. Set-2 considers first 15 genuine signatures along with first 15 skilled forge signatures for training and other samples of same dataset are tested. For set-3, we considered ten genuine signatures and ten forge signatures which are randomly selected to train and other signatures are to test. Set-4 considers 15 randomly chosen samples from the same datasets for training and other samples for testing. Experiments are repeated for 5 times for set-3 and set-4 to avoid randomness, and results are tabulated.

4.1 Experimental Details on CEDAR Dataset

We conducted experiments on publicly available standard offline signature dataset, namely, The Centre of Excellence for Document Analysis and Recognition(CEDAR), which is developed by SUNY Buffalo. CEDAR database consists of 2640 signature samples from 55 signers. Each signer contributed 24 genuine signature samples. Each of the 20 arbitrary chosen signers contributed 24 skilled forge signatures.

Experiments were started over set-1 along with set-3 setup, and we considered features of ten genuine signatures and ten skilled forge signatures for training. There are 14 genuine signatures and 14 skilled forge signatures to testing. For set-2 and set-4, we considered 15 genuine signatures and 15 skilled forge signature samples

Table 2 Experimentation results obtained for CEDAR dataset

Experimental setup	Accuracy	FRR	FAR
Set-1	94.28	5.97	5.45
Set-2	94.09	7.27	4.54
Set-3	95.25	5.6	3.83
Set-4	93.63	10.3	2.42

Table 3 Experimentation results obtained for CEDAR dataset—a comparison

Proposed by	Classifier	Accuracy	FAR	FRR
Kalera et al. [7]	PDF	78.50	19.50	22.45
Chen and Shrihari [4]	DTW	83.60	16.30	16.60
Kumar et al. [10]	SVM	88.41	11.59	11.59
Pattern spectrum [13]	EMD	91.06	10.63	9.4
Surroundedness [11]	MLP	91.67	8.33	8.33
Inter-point envelope [9]	SVM	92.73	6.36	8.18
Proposed approach	SVM	95.25	5.6	3.83

to train and 9 genuine and 9 forge signatures for testing. To avoid randomness, we repeated experiments on set-3 and set-4 datasets for over five intervals. The obtained results are validated using the metrics FAR and FRR and are tabulated in Table 2.

We have made a comparative study with some of the well-known offline signature verification methods, namely, Kalera et al. [7], Chen and Shrihari [4], and Kumar et al. [10], and the results are shown in Table 3.

4.2 *Experimentation on MUKOS Dataset*

We also conducted experiments on regional language dataset, namely, Mangalore University Kannada Off-line Signature [MUKOS] that is a Kannada regional language corpus. The database consists of total number of 1350 signatures from 30 contributors. From each signer contributed 30 genuine signatures and 15 skilled forgeries. All genuine signatures were collected using black ink pen on A4 size white sheet, and each sheet is fragmented into 14 boxes. Then, the contributors are allowed to practice forging. After a time gap, forgeries were obtained by imitating genuine signatures. Standard scanner was used to collect forgeries.

Then, the experiments are continued with set-1 and set-3. This setup consists of ten genuine and ten skilled forgery signatures, and from this feature vector was obtained. This forms the training model. Testing model is constructed with 15 genuine and 5 skilled forgery signatures. Similarly, experiments are also conducted to train from set-2 and set-4; here, 15 genuine and 15 skilled forge signatures are considered to obtain

Table 4 Experiment results for MUKOS dataset

Experimental setup	Accuracy	FRR	FAR
Set-1	96.8	0.8	5.6
Set-2	96.8	2.8	3.6
Set-3	98.26	1.3	2.13
Set-4	98.93	0.5	1.6

Table 5 Experiment results for MUKOS dataset—a comparative analysis

Method	Classifier	Accuracy	FAR	FRR
Shape-based eigen signature [12]	Euclidean distance	93.00	11.07	6.40
Pattern spectrum [13]	EMD	97.39	5.6	8.2
Proposed approach	SVM	98.93	0.5	1.6

the feature vector. Then, 15 genuine and 15 skilled forge signatures are considered to test. The accuracy of the proposed approach obtained from set-1 to set-4 is tabulated in Table 4. The experimentation for set-2 and set-4 is the average of five instances of experimentations with randomly chosen samples. A comparative observation on the MUKOS dataset with some of the past works is tabulated in Table 5.

5 Conclusion

This work presents a novel signature verification technique which uses partial sum of second-order Taylor series expansion (TSE) for offline signature verification. Finite sum of TSE computed on an arbitrary small neighborhood can approximate the function extremely well. This is a strong and feasible mechanism to extract the local features of signature. We present kernel structures by incorporating the Sobel operators to compute the higher order derivatives of TSE. The experimental results obtained on offline signature datasets, namely, CEDAR and MUKOS, demonstrate the improvements in the classification accuracy compared to some of the well-known offline signature verification methods.

Acknowledgements We would like to acknowledge Bharathi R. K. for the support rendered in terms of providing a regional language dataset, namely, Mangalore University Kannada Off-line Signature (MUKOS) dataset.

References

1. Bastys, A., Kranauskas, J., Krüger, V.: Iris recognition by fusing different representations of multi-scale Taylor expansion. Comput. Vis. Image Underst. **115**(6), 804–816 (2011)
2. Bastys, A., Kranauskas, J., Masiulis, R.: Iris recognition by local extremum points of multiscale Taylor expansion. Pattern Recognit. **42**(9), 1869–1877 (2009)
3. Bhattacharya, I., Ghosh, P., Biswas, S.: Offline signature verification using pixel matching technique. Proc. Technol. **10**, 970–977 (2013)
4. Chen, S., Srihari, S.: Use of exterior contours and shape features in off-line signature verification. In: ICDAR, pp. 1280–1284 (2005)
5. Guerbai, Y., Chibani, Y., Hadjadji, B.: The effective use of the one-class svm classifier for handwritten signature verification based on writer-independent parameters. Pattern Recognit. **48**(1), 103–113 (2015)
6. Hsu, C.W., Chang, C.C., Lin, C.J.: A Practical Guide to Support Vector Classification (2003)
7. Kalera, M.K., Srihari, S., Xu, A.: Off-line signature verification and identification using distance statistics. Int. J. Pattern Recognit. Artif. Intell. **18**, 228–232 (2004)
8. Kruthi, C., Shet, D.C.: Offline signature verification using support vector machine. In: 2014 Fifth International Conference on Signal and Image Processing (ICSIP), pp. 3–8. IEEE (2014)
9. Kumar, M.M., Puhan, N.: Inter-point envelope based distance moments for offline signature verification. In: 2014 International Conference on Signal Processing and Communications (SPCOM), pp. 1–6. IEEE (2014)
10. Kumar, R., Kundu, L., Chanda, B., Sharma, J.D.: A writer-independent off-line signature verification system based on signature morphology. In: Proceedings of the First International Conference on Intelligent Interactive Technologies and Multimedia IITM '10, pp. 261–265. ACM, New York, NY, USA (2010)
11. Kumar, R., Sharma, J., Chanda, B.: Writer-independent off-line signature verification using surroundedness feature. Pattern Recognit. Lett. **33**(3), 301–308 (2012)
12. Shekar, B.H., Bharathi, R.K.: Eigen-signature: a robust and an efficient offline signature verification algorithm. In: 2011 International Conference on Recent Trends in Information Technology (ICRTIT), pp. 134 –138, June 2011
13. Shekar, B.H., Bharathi, R.K., Pilar, B.: Local morphological pattern spectrum based approach for off-line signature verification. In: International Conference on Pattern Recognition and Machine Intelligence, pp. 335–342. Springer (2013)
14. Shekar, B.H., Bhat, S.S.: Iris recognition using partial sum of second order Taylor series expansion. In: Proceedings of the Tenth Indian Conference on Computer Vision, Graphics and Image Processing, p. 81. ACM (2016)
15. Shekar, B.H., Pilar, B., Sunil, K.: Blockwise binary pattern: a robust and an efficient approach for offline signature verification. Int. Arch. Photogram. Remote Sens. Spatial Inf. Sci. **42**, 227 (2017)
16. Shikha, P., Shailja, S.: Neural network based offline signature recognition and verification system. Res. J. Eng. Sci. **2278**, 9472 (2013)
17. Soleimani, A., Araabi, B.N., Fouladi, K.: Deep multitask metric learning for offline signature verification. Pattern Recognit. Lett. **80**, 84–90 (2016)
18. Yilmaz, M.B., Yanikoglu, B., Tirkaz, C., Kholmatov, A.: Offline signature verification using classifier combination of hog and lbp features. In: 2011 International Joint Conference on Biometrics (IJCB), pp. 1–7. IEEE (2011)

Urban LULC Change Detection and Mapping Spatial Variations of Aurangabad City Using IRS LISS-III Temporal Datasets and Supervised Classification Approach

Ajay D. Nagne, Amol D. Vibhute, Rajesh K. Dhumal, Karbhari V. Kale and S. C. Mehrotra

Abstract An accurate mapping of urban LULC is essential for urban development and planning. Although urban area represents a little portion of Earth surface, which brings an unbalanced impact on its surrounding areas. However urban LULC mapping and change detection is critical issue by traditional methods. Recent advances in Geospatial technology can be used to map built-up areas for detecting the urban growth patterns. In this work IRS LISS-III sensors image data of 2003, 2009, and 2015 of same season were used. The LULC mapping and change detection was carried out by four supervised classifier namely Maximumlikelihood classifier (MLC), Mahalanobis-Distance (MD), Minimum-Distance-to-Means (MDM), and Parallelepiped classifier (PC). Obtained results were examined by considering the efficiency of each classifier to accurately map the identified LULC classes. It is observed that, MLC has given the highest overall accuracy of 73.07, 83.51, and 93.43% with kappa coefficient of 0.64, 0.78, and 0.90 in 2003, 2009, and 2015 respectively, which are superior among others; hence we have used classified layer obtained from MLC for further change detection and analysis from 2003 to 2015.

A. D. Nagne (✉) · A. D. Vibhute · R. K. Dhumal · K. V. Kale · S. C. Mehrotra
Department of CS & IT, Dr. Babasaheb Ambedkar Marathwada University, Aurangabad, Maharashtra, India
e-mail: ajay.nagne@gmail.com

A. D. Vibhute
e-mail: amolvibhute2011@gmail.com

R. K. Dhumal
e-mail: dhumal19@gmail.com

K. V. Kale
e-mail: kvkale91@gmail.com

S. C. Mehrotra
e-mail: mehrotra.suresh15j@gmail.com

P. Nagabhushan et al. (eds.), *Data Analytics and Learning*,
Lecture Notes in Networks and Systems 43,
https://doi.org/10.1007/978-981-13-2514-4_31

Keywords Urban classification · Spatial variation
Maximum likelihood classifier (MLC) · LULC · Change detection

1 Introduction

Land-Use (LU) and Land-Cover (LC) are two separate terminologies, which are often used interchangeably. The LC states the physical characteristics of earth's surface distribution, like rock, soil, water, vegetation, manmade structures and other features of the land. While LU indicates that, the land has been used by humans and their habitat. The LULC pattern is an outcome of natural and socioeconomic factors and their utilization by people [1]. To achieve the increasing demands for basic human needs and welfare; information on LULC and possibilities for their optimal use are essential for the formation and implementation of LU schemes [2].

An accurate temporal LULC change detection of Earth's surface objects are very significant to understand the interactions between humans and natural phenomena, in order to promote better decision-making. New strategies are required for the regular updating of existing databases rather than traditional methods. These are generally in view of field examinations and the visual understanding of Remote Sensing Images [3]. The usual techniques are tedious and costly, and mapping exercises frequently can't stay aware of the pace of urban improvement. Geospatial technology provides an extensive variety of applications which can be used for effective decision-making and future planning [4]. Application of geospatial technology made possible to study the LU pattern and identifies a changes in LC with better accuracy in less time. Under Geospatial approach, we have identified spatial variations of Aurangabad city from 2003 to 2015 using LISS-III data and supervised approach.

This paper is grouped into five sections. The introduction has been presented in Sect. 1. Section 2 illustrates the study area. Section 3 portrays the proposed Methodology for classification of urban LULC. Obtained results and discussions about the observed changes in the classified images are revealed in Sect. 4. At last, conclusion and future works are exhibited in Sect. 5.

2 Study Areas

Aurangabad City (see Fig. 1) is located in the north central part of Maharashtra. This city has many tourist places like Ajanta and Ellora caves, Bibika Maqbara and Deogiri fort, etc. [5].

The atmosphere of the Aurangabad region and its surrounding are identify as a hot summer and dryness atmosphere amid the year except from June to September; whereas post monsoon season start from the month of October to November [6]. The temperatures begin to fall rapidly in November, and December month is considered to be a very coldest and mean min temp is near about 10 °C, and the mean max temp

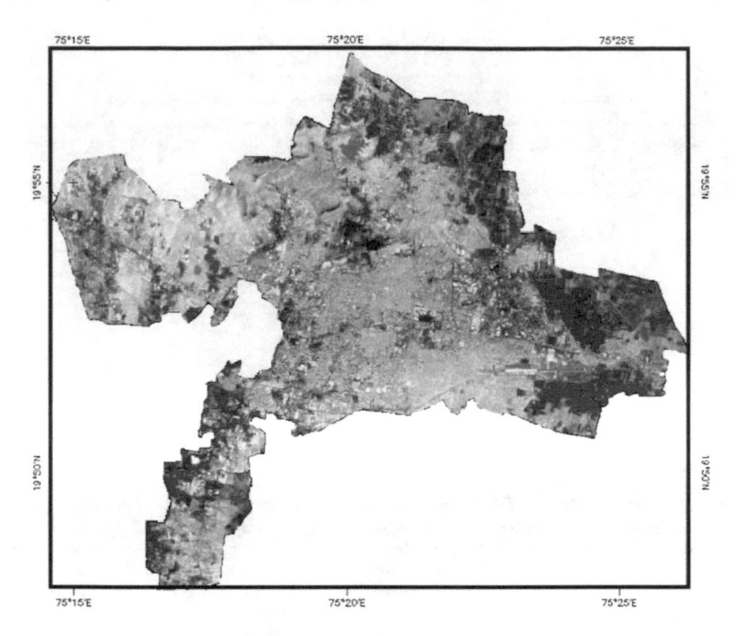

Fig. 1 Area under study

is near about 28 °C. The daily temperature increases from the beginning of March and May and these months are considered to be hottest months with mean minimum temp. of 24 °C and mean max temp. of 40 °C [7]. Therefore December to March is considered to be the best suitable months to identify a LULC of Aurangabad city. According to this, the LISS-III temporal datasets of December-2003, January-2009, and February-2015 have been used for the current study.

3 Proposed Methodology

The literature shows that the multispectral data has been used for the detection of LULC classes. As a result of growing urban improvement and mapping costs city experts keen on viable urban LULC mapping and change detection that can be utilized for improvement in Smart City projects.

The methodology adopted for urban classification of Multispectral images is shown in Fig. 2. As per the study area, the remote sensing LISS-III images were collected from NRSC Hyderabad (India). These multispectral temporal images are captured from same sensor (i.e., LISS-III) with same season and represent same geographical area but it has been captured in different years (i.e., 2003, 2006, and 2015). The preprocessing was performed on this raw LISS-III images and then selection of training data sets, classification and accuracy assessment was done using ENVI tool.

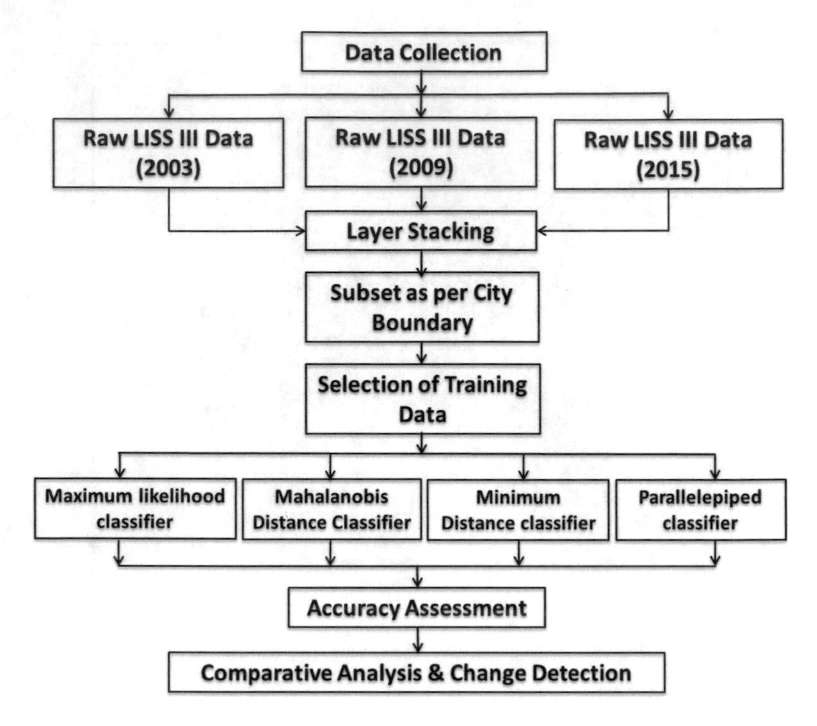

Fig. 2 Proposed methodologies

Table 1 Band specification of LISS-III sensor image

Band	Spectral band (μm)	Spatial resolution (m)
2.	0.52–0.59	23 × 23
3.	0.62–0.68	23 × 23
4.	0.77–0.86	23 × 23
5.	1.55–1.70	23 × 23

3.1 Used Datasets

A Multispectral LISS-3 camera operates in four spectral bands, among these three bands are operates in the Visible-Near-Infrared (VNIR) and 1 operates in the Shortwave-Infrared (SWIR) region [7]. The LISS-III camera provides data with a spatial resolution of 23.5 m. An optical LISS-3 sensor works in four different spectral bands (i.e., Green, Red, NIR, and SWIR). This multispectral sensor covers an area of 141 km wide swath with a moderate spatial resolution of 23 m in all four spectral bands [8]. Table 1 shows a detailed specification of LISS III data [7].

3.2 Preprocessing

Raw LISS III data contains four bands and every band having a separate file. The individual bands for each image are Tagged Image Format File (.TIFF) format. These images are Georeferenced images and having a World Geodetic System (WGS) 1984 Datum (Datum: D_WGS_1984) and projected in UTM Zone 43 N. The raw LISS-III data has four bands which are present in the TIFF format, i.e., Band2 (Green), Band3 (Red), Band4 (NIR), and Band5 (SWIR). The infrared band was utilized in the analysis because its capability detecting vegetation area. Then the Layer Stacking operations were performed and all four bands are comprised in a single image [8].

A single LISS-III image covers near about 24,000 km^2 area of earth surface and Aurangabad Municipal Corporation is our Region of Interest (ROI) and its total area is near about 142 km^2. So, there is need to perform a spatial subset of image and it is achieved by using shape file of Aurangabad corporation area [7]. Spatial Subset (clipping) is a way of reducing the extended area of an image to include just the area that is of interest. In digital image interpretation; most of the times a whole satellite images are not interpreted, only some certain areas of the satellite image were interpreted. In these cases the part we are interested it should be cut (clipped) from the whole image; so it also reduced the size of an image and improve the processing speed and reliability.

3.3 Selection of Training Data

The training pixels have been selected based on their spectral signature for identified LULC classes; with reference to ground truth information which was collected by using GPS-enabled system. The ground truth points were collected during the period of March to April 2015. The collected ground truth points were matched with the 2015 image. The Google earth and visual inspection was used for collection of the old points for 2003 and 2009 images. Moreover, the tested region was well known to us; so the error was less for identifying correct objects and training the data. The six objects were considered for classification namely Residential (Built-up) area, Hill without Vegetation (Rock), Water Body, Vegetation, Fallow Land and Barren Land. The Fig. 3 shows the average spectra of training pixels of selected LULC classes. These selected training pixels were used for classification.

3.4 Classification

A Digital image classification is a mingled process and that requires a different divers factors and it is very challenging task. These factors are selection of RS data, image processing and selection of classification technique. In image classification the prime

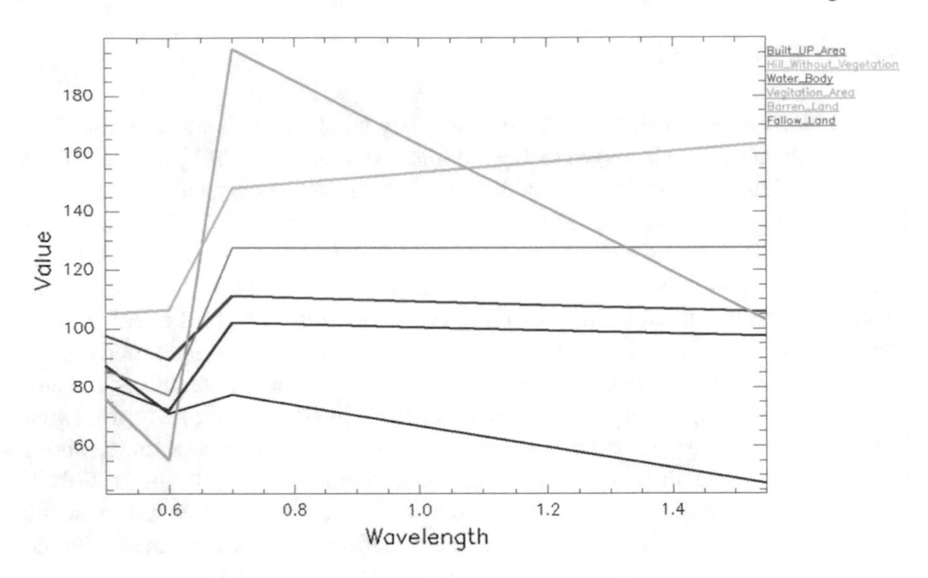

Fig. 3 Average spectra of similar objects used for classification

Table 2 LULC classes along with Anderson LULC code

Sr. No.	Anderson scheme code	Class name	Color
1	1	Residential (built-up) area	Magenta
2	6	Hill without vegetation (Rock)	Cyan
3	5	Water body	Blue
4	2, 4	Vegetation	Green
5	7	Barren land	Yellow
6	3	Fallow land	Red

steps are selection of appropriate classification system, selection of and number of training samples, techniques of image preprocessing, selection of feature extraction technique, post classification processing, and performing accuracy assessment [9]. The motive of classification is to categorize all image pixels into different land-cover classes. This classified data will be used to design thematic maps of the LULC. Table 2 describes the LULC classes for classification of images along with color codes and Anderson LULC code.

3.4.1 Maximum Likelihood Classifier (MLC)

The MLC supervised classification algorithm takes advantages of probability density functions which are used in the classification. It works on overlapping signatures with the help of probability. This classifier is based on Bays theorem, in which an every pixel belongs to the maximum likelihood, are characterized into the related classes [10]. The probability density functions are estimated by two weighting factors through Bayes theorem [11]. Firstly, the user trains the a priori probability or specific signatures of the class in the given image. Second, for each class the cost of misclassification is weighted as per its probability. Both factors outcomes better decreasing the misclassification [12]. This classifier classifies an unknown pixel's spectral response patterns through assessing both the variance and covariance of the class. The user must have the knowledge about the spectral signature or ground truth [13]. This classifier uses a probability density functions to classify undefined pixel by calculating the probability of every pixel value belonging to each category, after this, the pixel would be assigned to the most likely class with highest probability value. If the probability values are all below a threshold then it labeled as "unknown" [14]. The basic discriminant function for pixel X is,

$$X \in C_j \ \text{if} \left(C_j/X\right) \quad \max\left[p(C_1/X), p(C_2/X), \ldots, p(C_m/X)\right], \tag{1}$$

where, the function $\max\left[p(C_1/X), p(C_2/X), \ldots, p(C_m/X)\right]$ returns the highest probability. For identification of pixel X it uses the information class resultant to the probability. The Function $p(Ci/X)$ represents a pixel X's conditional probability as being a member of class Cj. The Bayes's theorem is used to solve this problem [10],

$$P\left(C_j/X\right) = p\left(X/C_j\right) * p\left(C_j\right)/p(X) \tag{2}$$

where, Function $p\left(X/C_j\right)$ is a priori probability or conditional probability, and $p\left(C_j\right)$ is the occurrence probability of class Cj in the input data and $p(X)$ is the probability of pixel X occurring in the input data and written as follows [10]:

$$p(X) = \sum_{j=1}^{m} p\left(X/C_j\right) * p\left(C_j\right) \tag{3}$$

where, $p(X)$ is supposed as normalization constant to confirm $\sum_{j=1}^{m} p\left(C_j/X\right)$ equals to 1 and m is the number of classes [10]. While implementation of this classifier, the user gives the opportunity for stating the probability of each information class, therefore the class of a posterior probability can be written as follows

$$p\left(C_j/X\right) = \frac{p\left(X/C_j\right)p\left(C_j\right)}{\sum_{j=1}^{m} p\left(X/C_j\right)p\left(C_j\right)}, \tag{4}$$

where, $p\left(C_j\right)$ is the prior probability of class C_j and $p\left(X/C_j\right)$ is the conditional probability of observing X from class C_j. Therefore the computation of $p(Cj/X)$ is reduced to determination of $p\left(X/C_j\right)$. The analysis of this function can be expressed as for the remote sensing images [10]:

$$p(X/C_j) = \frac{1}{(2\pi)^{n/2}\left|\sum j\right|^{0.5}} \times \exp\left[-\frac{1}{2}(DN - \mu_j)^T \sum_j^{-1} (DN - \mu_j)\right], \quad (5)$$

where; $DN = (DN_1, DN_2, \ldots, DN_n)^T$ is a vector of pixel with n no. of bands; $\mu_j = (\mu_{j1}, \mu_{j2}, \ldots, \mu_{jn})^T$ is a mean vector of the class C_j and $\sum j$ is the covariance matrix of class C_j which can be written as:

$$\sum j = \begin{bmatrix} \sigma_{11} \; \sigma_{12} \; \ldots \; \sigma_{1n} \\ \sigma_{21} \; \sigma_{22} \; \ldots \; \sigma_{2n} \\ \ldots \; \ldots \; \ldots\ldots \\ \sigma_{n1} \; \sigma_{n2} \; \ldots \; \sigma_{nm} \end{bmatrix} \quad (6)$$

3.4.2 Mahalanobis Distance (MD)

MD classifier is very similar to Minimum Distance-to Means, except that a covariance matrix is used in this Eq. 7 [14]. Mahalanobis classifier is a Manhattan distance classifier which is dissimilar from Euclidian distance due to its simplicity. For classification and analysis of various patterns it's depend on correlations between variables. Also it measures the similar class of an unknown data set using known data. This classifier uses each band to classify the objects which are the statistical summation of the complete dissimilarities between two pixels in the similar band. The differentiations of pixel values are neglected in deriving the difference [10, 15]. The Eq. (7) for the Mahalanobis distance classifier is as follows:

$$D = (X - M_c)^T (Cov_c^{-1})(X - M_c), \quad (7)$$

where

D	Mahalanobis distance
c	a particular class
X	the measurement vector of the candidate pixel
M_c	the mean vector of the signature of class c
Cov_c	the covariance matrix of the pixels in the signature of class c
cov_c^{-1}	inverse of Cov_c
T	transposition function

The pixel is assigned to the class, *c*, for which *D* is the lowest.

3.4.3 Minimum Distance-to Means (MDM)

The MDM is computationally simple which takes the average spectral values within each class signature. It ignores the covariance and standard deviation which are helpful in classification of different unidentified objects of the image [13]. Hence it uses the unknown image data to classes which minimize the distance between the image data and the class in multi feature space for classification with maximum similarity. For all classes the average spectral value in each band is considered. Therefore, the classifier is mathematically simple and computationally able with some drawbacks [10, 14].

The following equation indicates the decision rule behind this classifier:

$$Pixel\,\mathrm{X} \in \mathrm{C_j}\ \mathrm{if}\, \mathrm{d}(\mathrm{C_j}) = \min[\mathrm{d}(\mathrm{C_1}), \mathrm{d}(\mathrm{C_2}), \ldots \mathrm{d}(\mathrm{C_m})] \tag{8}$$

where;

- $\min[\mathrm{d}(\mathrm{C_1}), \mathrm{d}(\mathrm{C_2}), \ldots \mathrm{d}(\mathrm{C_m})]$ is a function for identifying the smallest distance among all those inside the bracket
- $\mathrm{d}(\mathrm{C_j})$ refers to Euclidean distance between pixel X and the center of information class $\mathrm{C_j}$.

It is calculated using the Eq. (9):

$$\mathrm{d(Cj)} = \sqrt{\sum_{i=1}^{n} [\mathrm{DN(i,j)} - \mathrm{Cij}]^2} \tag{9}$$

The above formula is repeated *m* times with all information classes, for each pixel in the input remote sensing data.

3.4.4 Parallelepiped Classifier

The PC is a box classifier whose opposite sides are straight and equivalent. This classifier uses the class limits and stores it in each class to establish whether a specified pixel falls within the class or not. It is a very simple classifier for computation when speed is necessary with less accuracy and most of the pixels overlap in classification [14].

This classifier assigns pixel into one of the predefined information classes in terms of its value in relation to the DN range of each class in the same band.

$$\text{Pixel X} \in C_i \text{ if Min } DN_i \leq DN_x \leq \text{Max } DN_i \tag{10}$$

where:

Ci Information Class
DN_i DN Value of Information Class
X Pixel
DN_x DN Value of pixel X.

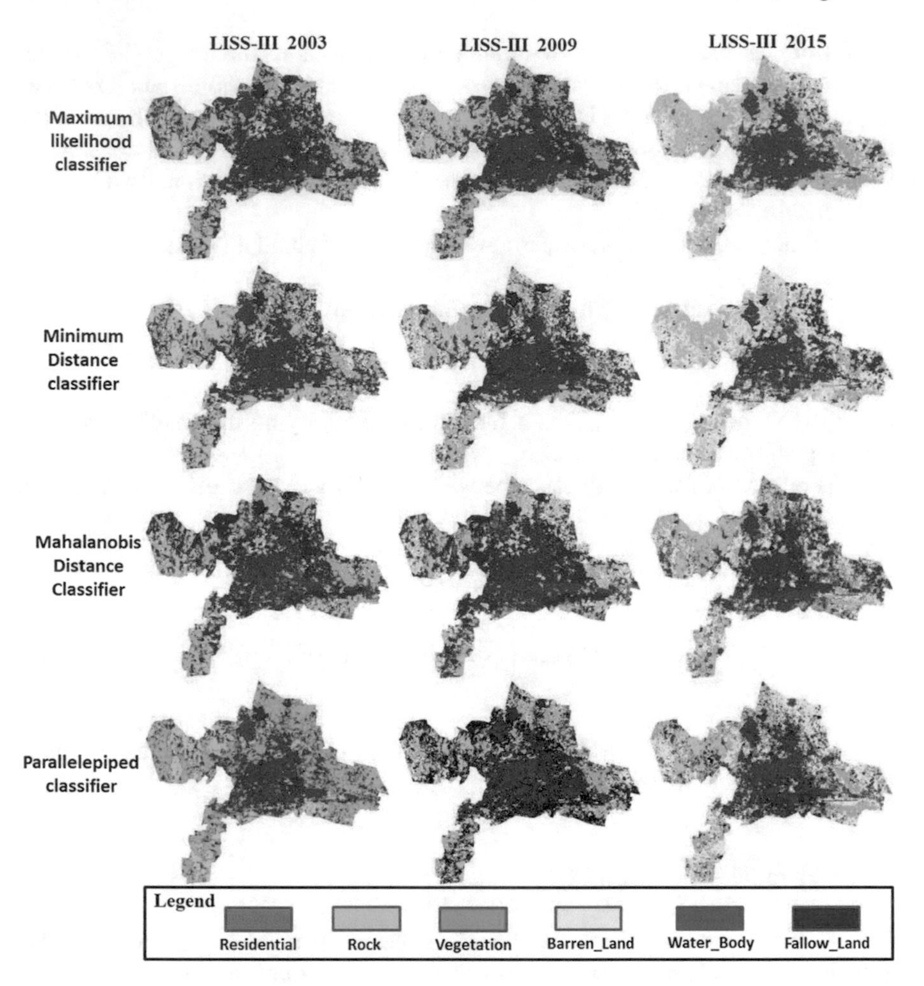

Fig. 4 Classified images of LISS-III multi-temporal data by using four supervised classifiers

The Multitemporal LISS-III datasets of December-2003, January-2009 and February-2015 was classified by using four supervised classifiers. These four classifiers were MLC, MD, MDM and PC and these classifiers are well known and well reported in literature. Figure 4 shows classified images of LISS-III Multitemporal data by using four supervised classifiers.

4 Results and Discussion

The confusion matrix was generated for accuracy assessment of all methods and obtained results are indicated in Tables 3, 4 and 5. The results were evaluated on the basis of overall accuracy, producer's accuracy, user's accuracy and kappa coefficient; computed with reference to confusion matrix. These values are used to compare the results of classification. The said methods have provided the class specific accuracies (Tables 3, 4 and 5). An overall accuracy of the classification is simply defined as the total number of correct classifications divided by the total number of sample points. The producer's accuracy measures how well a certain area has been classified. The producer's accuracy is derived by dividing the number of correct pixels in one class divided by the total number of pixels as derived from reference data. The user's accuracy is therefore a measure of the reliability of the map; it indicates how well the classified map represents what is really on the ground. It is defined as the correct classified pixels in a class are divided by the total number of pixels that were classified in that class [14]. The kappa coefficient measures the agreement between classification and ground truth values. A kappa value of 1 represents perfect agreement; a value of 0 represents no agreement while −1 represents complete disagreement. This is another measure of agreement or accuracy based on the difference between the actual agreement in the error matrix (i.e., the agreement between the remotely sensed classification and the reference data as indicated by the major diagonal) and the chance agreement which is indicated by the row and column total marginal [15].

According to the results of 2015, it is observed that, there were misclassification between barren land and residential area due to its spectral similarity. In fact, all the classes were classified accurately except residential region and barren land using MLC method. However, the MD, MDM, and PC algorithms have given the similar error for residential area, barren land and fallow land due to because of their spectral resemblance. Additionally, the misclassification was found to be normal with vegetation and hill without vegetation using all methods. On the other hand, the results of 2015, 2009, and 2003 were similar for water bodies and vegetation areas. The other classes were also classified well except barren land with 2015, 2009, and 2003 images using all methods. Furthermore, whatever the misclassification were identified, it was due to similarity in reflectance behavior of said LULC classes.

The results clearly indicates that, the MLC has given the highest overall accuracy of 73.07%, 83.51%, and 93.43% with kappa coefficient of 0.64, 0.78, and 0.90 in 2003, 2009 and 2015 respectively. The MLC is a standout among the most mainstream techniques for characterization in remote sensing as compared to other supervised classifiers. The obtained results were compared with others as given in Table 6, and it is found to be better. As per accuracy assessment a MLC was considered for comparing the LULC change detection.

Table 3 Accuracy assessment of LISS-III 2015 classified images

Class name	Maximum likelihood			Mahalanobis distance			Minimum distance			Parallelepiped classifier		
	PA	UA	OA/ƙ	PA	UA	OA/ƙ	PA	UA	OA/ƙ	PA	UA	OA/ƙ
RA	97.44	100	**93.43/0.90**	79.49	68.89	**83.01/0.76**	89.74	100	**88.41/0.84**	79.49	100	**81.46/0.75**
HWV	100	86.72		97.3	87.8		94.59	88.24		98.2	87.2	
WB	100	100		100	100		100	81.82		100	100	
VA	100	100		76.92	100		96.15	100		76.92	100	
BL	5.88	100		23.53	33.33		23.53	28.57		5.88	14.29	
FL	100	100		64.1	78.13		84.62	100		58.97	100	

OA overall accuracy, *PA* producer's accuracy, *UA* user's accuracy, *ƙ* Kappa coefficient, *RA* residential area, *HWV* Hill without vegetation, *WB* water body, *VA* vegetation area, *BL* Barren Land, *FL* Fallow land

Table 4 Accuracy assessment of LISS-III 2009 classified images

Class name	Maximum likelihood			Mahalanobis distance			Minimum distance			Parallelepiped classifier		
	PA	UA	OA/ƙ	PA	UA	OA/ƙ	PA	UA	OA/ƙ	PA	UA	OA/ƙ
RA	94.74	100	**83.51/0.78**	57.89	40.74	**70.87/0.61**	100	61.29	**82.41/0.76**	84.21	100	**79.12/0.73**
HWV	100	44.74		88.24	48.39		100	50.00		94.12	84.00	
WB	100	100		100	100		100	100		97.37	100	
VA	100	100		93.33	100		100	100		93.33	100	
BL	25.00	38.00		12.00	25.00		3.23	100		0.00	0.00	
FL	100	68.57		54.17	39.39		100	96.00		100	88.89	

Table 5 Accuracy assessment of LISS-III 2003 classified images

Class name	Maximum likelihood			Mahalanobis distance			Minimum distance			Parallelepiped classifier		
	PA	UA	OA/ꝁ	PA	UA	OA/ꝁ	PA	UA	OA/ꝁ	PA	UA	OA/ꝁ
RA	89.47	100	**73.07/0.64**	78.95	46.00	**66.85/0.57**	94.74	54.55	**64.84/0.54**	94.74	69.23	**68.13/0.58**
HWV	94.12	59.26		76.47	59.09		100	45.95		94.12	53.33	
WB	94.74	100		92.00	100		84.21	100		94.74	84.71	
VA	86.67	92.86		80.00	100		86.67	100		93.33	63.64	
BL	21.26	30.00		16.45	26.30		5.43	16.79		0.00	0.00	
FL	62.50	35.71		50.00	31.58		25.00	54.55		16.67	22.22	

Table 6 Results comparison with others published work

Author and year	Purpose	Data used	Techniques	Accuracy (%)
Kamrul Islam (2017) [1]	LULC CD	Landsat 5 TM, Landsat 8 OLI/TIRS	MLC	83.96–92.16
Sinha et al. (2015) [16]	Improved LULC classification	Landsat ETM+	MLC	85
Jayanth et al. (2016) [17]	LULC change detection (CD)	LISS-IV, CARTOSAT-1	Support vector machine artificial bee colony (ABC), MLC	62.63–80.4
Butt et al. (2015) [18]	LULC CD	Landsat TM, SPOT-5	MLC, post classification refinement	90
Beuchle et al. (2013) [19]	LULC CD	Landsat TM Landsat ETM	Object based classification	85–90
Our proposed work	LULC CD	LISS-III	MLC, MD, MDM and PC	73.07–93.43

4.1 Change Detection

The purpose of present study is not only to classify the LULC, but also to identify the temporal changes in the region. Effective application of geospatial technology for LULC change discovery to a great extent depends on an appropriate understanding of the study area. Change detection of LISS-III temporal datasets of December-2003, January-2009 and February-2015 shown in Table 7; it shows all results in percentage which was computed by considering the number of classified pixels of each class and total number of classified pixels. As per the results, there is a growth in residential area of 1.24% in 2009 and 11.34% in 2015 as compared to 2003. Since from 2003 the rainfall was remarkably decreased every year in Aurangabad region till 2015, this makes effects on the storage of water bodies in the region and it is clearly observed that, water body area has been decreased by 1.23% in 2009 and 1.9% in 2015 as compared to 2003. This less rainfall make effects on the vegetation area and it has been decreased by 1.96% and 3.69% in 2009 and 2015 respectively. The vegetation's were available with rock surface regions in 2003; however it (vegetation) was reduced in 2009 and 2015. Consequently the rocks were highlighted more in 2009 and 2015 by 1.72% and 3.27% respectively as compared to 2003. Due to fast urban expansion, the fallow Land was decreased by 1.09% and 33.98% in 2009 and 2015 respectively; which is converted into Barren Land, hence barren land is increased by 1.29% and 24.94% in 2009 and 2015 as compared to 2003. Figure 5 shows the Statistical analysis of LISS III Temporal Datasets obtained from MLC.

Table 7 Change detection of LISS-III 2003, 2009 and 2015 by using MLC

Sr. no.	Class name	Color	LISS-III classification			Remark
			2003 (%)	2009 (%)	2015 (%)	
1	Residential	Magenta	17.16	18.40	28.50	Increased
2	Rock	Cyan	18.35	20.07	21.62	Increased
3	Water_Body	Blue	2.20	0.97	0.30	Decreased
4	Vegetation	Green	12.58	10.62	8.89	Decreased
5	Barren_Land	Yellow	10.21	11.50	35.15	Increased
6	Fallow_Land	Red	39.52	38.43	5.54	Decreased

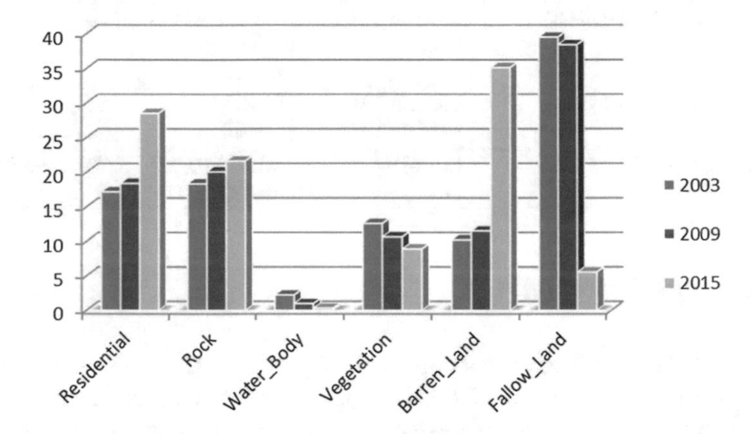

Fig. 5 Statistical analysis of LISS III temporal datasets obtained from MLC

5 Conclusions

LULC is important factor in comprehensible the relations between the human activities and the environment. Satellite Remote sensing systems have an ability to cover a large area. This research reports a Mapping of LULC classification using LISS-III multispectral image datasets of December 2003, January 2009, and February 2015 of the Aurangabad Municipal Corporation (AMC) area. As IIRS LISS-III data sets used in this study having 23.5 m spatial resolution and four spectral bands, this spatial and spectral characteristic gives enough details for LULC mapping and change detection. The LULC classes are derived from a level one classification of the Anderson classification schemes. In the study, MLC has given highest accuracy as compared to other supervised classifiers; hence MLC is the most recommended technique to perform classification and mapping. To achieve the better accuracy with maximum LULC classes (Anderson Level 2), very high spatial resolution Multispectral data can be used.

Acknowledgements Author(s) would like to acknowledge, UGC-BSR Fellowships, DST_FIST and UGC-SAP(II) DRS Phase-I and Phase-II F.No.-3-42/2009 & 4-15/2015/D R S -II for Laboratory Facility to Department of CS & IT, Dr. B.A.M.University, Aurangabad (MS), INDIA.

References

1. Islam, K., Jashimuddin, M., Nath, B., Nath, T.K.: Land use classification and change detection by using multi-temporal remotely sensed imagery: the case of Chunati wildlife sanctuary, Bangladesh. Egypt. J. Remote Sens. Space Sci. (2017)
2. Rawat, J.S., Kumar, M.: Monitoring land use/cover change using remote sensing and GIS techniques: a case study of Hawalbagh block, district Almora, Uttarakhand, India. Egypt. J. Remote Sens. Space Sci. **18**(1), 77–84 (2015)
3. Reis, S.: Analyzing land use/land cover changes using remote sensing and GIS in Rize, North-East Turkey. Sensors **8**(10), 6188–6202 (2008)
4. Lu, D., Mausel, P., Brondízio, E., Moran, E.: Change detection techniques. Int. J. Remote Sens. **25**(12), 2365–2401 (2004). https://doi.org/10.1080/0143116031000139863
5. Nagne, A.D., Dhumal, R.K., Vibhute, A.D., Rajendra, Y.D., Kale, K.V., Mehrotra, S.C.: Suitable sites identification for solid waste dumping using RS and GIS approach: a case study of Aurangabad, MS, India. In: 2014 Annual IEEE India Conference (INDICON), pp. 1–6. IEEE (2014)
6. Balpande, U.S.: Ground_Water Information of Aurangabad District Maharashtra, Ministry of Water Resources Central Ground Water Board, Gov. of India, 1791/DBR/201. http://cgwb.gov.in/District_Profile/Maharashtra/Aurangabad.pdf
7. Nagne, A.D., Dhumal, R.K., Vibhute, A.D., Gaikwad, S., Kale, K., Mehrotra, S.: Land use land cover change detection by different supervised classifiers on LISS-III temporal datasets. In: 2017 1st International Conference on Intelligent Systems and Information Management (ICISIM) (2017)
8. Resourcesat—1 Data User's Handbook, National Remote Sensing Agency, Department Of Space, Govt. Of India NrsaBalanagar, Hyderabad—500037, A.P. India. http://Bhuvan.Nrsc.Gov.In/Bhuvan/Pdf/Resourcesat-1_Handbook.Pdf
9. Hebbara, R., SeshaSaib, M.V.R.: Comparison of LISS-IV MX AND LISS-III+ LISS-IV merged data for classification of crops. ISPRS Ann. Photogramm. Remote Sens. Spatial Inf. Sci. (ISPRS Technical Commission VIII Symposium, 09–12 December 2014, Hyderabad, India) **II**(8) (2014)
10. Vibhute, A.D., Dhumal, R.K., Nagne, A.D., Rajendra, Y.D., Kale, K.V., Mehrotra, S.C.: Analysis, classification, and estimation of pattern for land of Aurangabad region using high-resolution satellite image. In: Proceedings of the Second International Conference on Computer and Communication Technologies, pp. 413–427. Springer, New Delhi (2016)
11. Govender, M., Chetty, K., Bulcock, H.: A review of hyperspectral remote sensing and its application in vegetation and water resource studies. Water SA **33**(2) (2007). ISSN 0378-4738. Water SA (on-line). ISSN 1816-7950. http://www.wrc.org.za
12. Srivastava, P.K., Han, D., Rico-Ramirez, M.A., Bray, M., Islam, T.: Selection of classification techniques for land use/land cover change investigation. Adv. Space Res. **50**(9), 1250–1265 (2012)
13. Murtaza, K.O., Romshoo, S.A.: Determining the suitability and accuracy of various statistical algorithms for satellite data classification. Int. J. Geomat. Geosci. **4**(4), 585 (2014)
14. Gao, J.: Digital Analysis of Remotely Sensed Imagery. McGraw-Hill Professional (2008)
15. Banko, G.: A review of assessing the accuracy of classifications of remotely sensed data and of methods including remote sensing data in forest inventory (1998)
16. Sinha, S., Sharma, L.K., Nathawat, M.S.: Improved land-use/land-cover classification of semi-arid deciduous forest landscape using thermal remote sensing. Egypt. J. Remote Sens. Space Sci. **18**(2), 217–233 (2015)

17. Jayanth, J., Kumar, T.A., Koliwad, S., Krishnashastry, S.: Identification of land cover changes in the coastal area of Dakshina Kannada district, South India during the year 2004–2008. Egypt. J. Remote Sens. Space Sci. **19**(1), 73–93 (2016)
18. Butt, A., Shabbir, R., Ahmad, S.S., Aziz, N.: Land use change mapping and analysis using Remote Sensing and GIS: a case study of Simly watershed, Islamabad, Pakistan. Egypt. J. Remote Sens. Space Sci. **18**(2), 251–259 (2015)
19. Beuchle, R., Grecchi, R.C., Shimabukuro, Y.E., Seliger, R., Eva, H.D., Sano, E., Achard, F.: Land cover changes in the Brazilian Cerrado and Caatinga biomes from 1990 to 2010 based on a systematic remote sensing sampling approach. Appl. Geogr. **58**, 116–127 (2015)

Study of Meta-Data Enrichment Methods to Achieve Near Real Time ETL

N. Mohammed Muddasir and K. Raghuveer

Abstract Data warehouse store historical records on which analysis queries are executed. Data warehouse are populated from the transaction database through a process called as **E**xtract **T**ransform and **L**oad (ETL). Extract is to identify the transaction database records to be moved to the data warehouse. Transform is to make the records normalized to suit the data warehouse environment. Load is to actually store the records in the data warehouse. Today business wants to perform real time analysis of data. Availability of updated records in the data warehouse is necessary for real time analysis. This could be achieved through the process of near real time (ETL), i.e., to improve existing ETL process. Improvement to existing ETL could be achieved in many ways such as to increase the frequency of loads, to identify relevant changes that are required for analysis and move only those changes. Only through increasing the frequency of load would not be useful in many cases. Hence we should identify the changes and then move if they are useful for analysis making it smart ETL. In this paper we study three such techniques. First one is to create a replica of dimension table in data warehouse and move the changes to the replica this reduces the query response time. Second is to load the data in parallel so that loading time could be reduced. Third is to identify changes and trigger the loading process. We have used the first approach to create replica and loaded the data in parallel and observed that by loading in parallel not only does the loading time reduces but also the query response time. Then in the first approach is the query response time reduces to a certain limit methods suggest to move the data from replica dimensions to the original dimensions. Here we bring in right time trigger to move instead of just the query response time. We found that when we have a combined approach to query response time and right time trigger the number of moves are reduced.

N. Mohammed Muddasir (✉)
Department of IS&E, VVCE, Mysuru, India
e-mail: mdmsir@gmail.com

N. Mohammed Muddasir
NIE, Mysuru, India

K. Raghuveer
Department of IS&E, NIE, Mysuru, India
e-mail: raghunie@yahoo.com

P. Nagabhushan et al. (eds.), *Data Analytics and Learning*,
Lecture Notes in Networks and Systems 43,
https://doi.org/10.1007/978-981-13-2514-4_32

Keywords Near real-time ETL · Meta-data · Data warehouse · TPC-DI · Kettle

1 Introduction

Decision support systems require up to date information for making a tactical decision for short-term goals or strategic decision for long-term goals. An illustration of tactical decision in a retail store scenario is to know the demand for a particular category of item in the coming week or a customer base to be served based on events like festivals, holidays, etc. These decisions could be made based on historical information recorded in the transactional databases that could possibly include the purchase patterns of a week before, etc. All these information is stored in a transactional database, these databases are continuously updated (insert, update, delete) and not suitable for query execution. Query execution is required to know the content of the data, if transactional database are used for query execution the response time of the query and the update operations (primary purpose of transactional databases) is affected, they both become slow. Hence we have a process of moving data from transactional database to analysis database. The transactional databases are referred widely as Online Transaction Processing (OLTP) databases and analysis database used for query processing are referred as Online Analytical Processing databases (OLAP). OLTP databases are extensively used for update operations and OLAP databases are used for query processing. Moving data from OLTP to OLAP is done through a process known as Extract Transform and Load (ETL). ETL was previously done in off peak hours like night loads or weekend loads whenever the OLTP is under-utilized. This would be helpful in long term strategic decision making but would not provide updated information and less useful in tactical decision making. Hence today we need to move data from OLTP to OLAP as soon as possible without affecting the performance of OLTP. This is called near real-time ETL. This paper studies some of the methods to achieve near real-time ETL and gives results of comparing the various options available to optimize and achieve near real-time ETL.

We studied the existing work on how to achieve near real time ETL using Meta-data based approach. Out of the various methods available we zeroed down on three things that could possibly optimize ETL or we could call it smart ETL where in the data are moved based on some criteria.

The first such approach was to have a replica of the entire schema in the OLAP database. These replica schemas table were not having any integrity checking constraints like primary key, foreign key, or indexing, etc., that could make loading into the replica schema faster as compared to the original schema. The queries on the original schema would not fetch the records that were recently added simply because the recently updated records were updated on replica schema and not in the original schema. To get the updated results same query has to be run on original schema as well as the replicated schemas and the results needs to be combined using a union operation. This scenario cannot go on for long because if the replica tables grow big they would affect the query response time simply because they were not having any constraints. In case of a high query response time we have to move the data from replicated schema to the original schema to reduce the query response time [1]. If we combine the results from original schema and replica schema using the union operation we call this as union based approach, else if we query only the original table we call this no-union approach.

The second approach is to examine the newly updated data in the OLTP if it is of use in the analysis process. The question of how to know the updated data is useful depends on the individual business case. The idea is not all updates are useful so we wait for certain threshold value to reach before the update could be moved. The method to calculate threshold value could vary from business to business. In [2] they have considered domain specific values such as sales in case of retail store. If the newly added row has sales above a certain value then this row should be moved. Similarly if the no. of rows in the OLTP is above a certain proportionality level then all these rows have to be moved. Proportionality is the ratio of no. of rows in OLTP to the no. of rows in OLAP. Next is the frequency of certain queries, if some OLAP queries are accessed more frequently than others the rows in OLTP for these queries should be moved in priority to the other rows. Also the feedback from the OLAP user if the query were useful is to be considered. They take all the four parameters, i.e., domain specific value, proportionality, frequency and feed back to calculate the threshold. If the threshold has reached a certain level trigger the movement of data from OLTP to OLAP through the ETL processes. This way the number of times ETL is done would be reduced and only required data undergoes ETL.

The third approach is to perform the data flow in parallel. The amount of data moved would be more and the time taken would be less if the flow is partitioned. In [3] they have used quality based metric to partition the flow for getting real-time ETL. They have a router to partition the flow based on quality metric and merger to combine the partition.

To summarize the above discussion we have studied three approaches union/no-union based approach, right time trigger based approach and partition flow based approach to optimize the ETL process. In this paper we have studied the query response time in union/no-union based approach and also query response time in parallel flow versus serial flow of union/no-union based approach. We partition the flow just based on no. of rows and not on any quality metric. We found union based approach with serial as well as parallel flow has a less query response time. Because of this we studied the no. of moves required in case of union based approach for the right time trigger. We devised an algorithm to move the changes after a query has crossed a certain amount of response time and if the threshold value has reached a certain level. We studied various threshold values and found that most of the time there is a reduction in number of moves if we use the union based approach in conjunction with trigger-based approach.

The rest of the paper is organized as related work, then implementation, results, discussion followed by conclusion and future work.

2 Related Work

In [4] they have studied the various optimization options to improve ETL. They consider type of loading technique, query sessions, indexing, referential integrity constraints, refresh of aggregate summaries, and partitions. They used TPC-H bench mark for the evaluation query performance while the loading of data is happening. Also they study the analysis of through put for online versus offline ETLR process. They also report the impact of query session to evaluate the performance of ETLR process. Then they report the impact of data loading strategy, weather bulk, JDBC batch of row by row for online versus offline load. After this they have reported the cost of each activity in ETLR process. They conclude with high loading time and low query response time with the above methods and process a DBMS based partitioning approach for better query response time and index drop while loading and re-created after loading for better loading time.

The best ways to achieve continuous data integration is studied in [5]. They have taken various RDBMS and demonstrate the best methods for continuous data integration pertaining to a specific context. They study three ways of loading viz. bulk load, ODBC load, and internal inserts to study the query performance in each of these cases. They consider loading small tables and loading replica tables for getting good query response time.

Identifying the changes to the OLTP through the log using change data capture feature provided by the database management systems has been studied in [6]. They perform customized movement of data after the changes have been recorder in the log. Also they look into the possible update and transformations that could affect many data warehouse schemas. In their approach Meta-data is generated on the fly there by reducing the cost of maintenance of indices and caches. Movement of data as required by the analysis query is the work done in [7]. They have studied the idea of loading the facts with records when analysis queries are run and later dropping the contents once the queries have finished executing. They have a dice-based approach and normal ETL approach. In the dice-based approach they identify the missing data from the facts and load only those facts so that queries get the updated data and later drop the facts once the analysis is done. In the normal ETL approach its movement of all the changes to the data warehouse. In [8] they identified the frequent patterns in the ETL flow and propose an empirical approach for mining ETL structural pattern. They have used TPC-DI benchmark for the evaluation of their work.

3 Implementation

To study the query response time we have run the data warehouse loading procedure concurrently with the querying of the data ware house. To do this we have used TPC-DI bench mark with postgresql database management system. TPC-DI gives the generated data in the form of flat files under various formats, such as csv, xml, xls, etc.

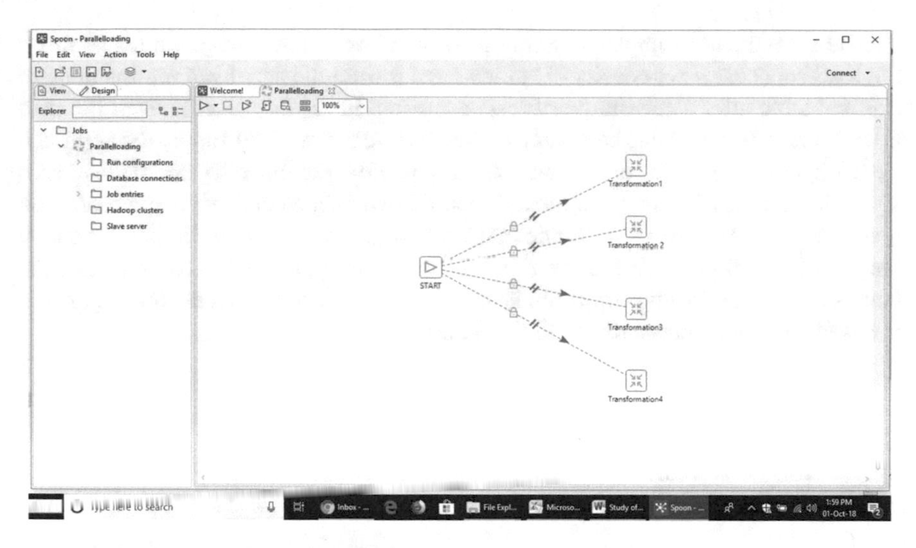

Fig. 1 Parallel loading of transformation

Details of TPC-DI a new bench mark for studying data warehouse and data integration is discussed in [9]. We have implemented the transformation mentioned in TPC-DI benchmark using pentaho kettle data integration tool. We got the implemented transformation from github [10]. We changed the loading of data to run currently with the load parallel feature of pentaho kettle as shown in Fig. 1. We have generated 500 MB of data using the DBGEN utility of TPC-DI. We have used i5 core machine with 4 GB Ram on a windows 10 operating system. The simulation of loading and querying concurrently was done using pentaho repeated load utility (loading every 5 min) and postgresql\watch utility to query every 5 s respectively. We stored the result of query response time along with other parameters into a database table. We have used the loading of dimtrade along with all its transformations. To load dimtrade all other dimensions like dimaccount, dimsecurity, tradetype, etc., were loaded. Dimtrade is both dimension table as well as fact table according to TPC-DI documentation [11]. In postgresql we created replica of dimension table with no indexing or integrity checking constraints. We created two stored procedures, one to query the dimension table dimtrade without union. i.e. this dimension table would contain all the integrity constraints. In the stored procedure we have created variable to record the response time and saved the variable values in database tables. The second stored procedure is to query dimtrade with union where we have a replica table for storing new changes and updated query results are got by performing union of query output from both main dimension table and replica dimension table. This is how we compare the query response time in case of union and no-union based approach. After we assert the union based approach is giving a good query response time we extent that to study the no. of loads, i.e., we use the union based approach to decrease the no. of load.

The second part of implementation is to see if we would reduce the no. of times the load happens. If we consider [1] they have moved the data from replica table to original table after a certain query response time. In [2] they have moved the data after a certain trigger value has reached. We observe that we could apply the approach by [2] in case of [1] we could reduce the no. of time the move happens. Our result show that out of 200 times the moved would have happened if we depend only on query response time, if we combine that with trigger based approach then the move happens only 20 times. We tested this for various threshold value and results were consistent. The following algorithm is used to reduce the no. of loads; this algorithm is based on both response time and threshold.

```
BEGIN
 'query using union operation'
if (tms >80) THEN
RAISE NOTICE 'Query response time crossed set limit val-
ue';
Compute(threshold)
if (threshold >0.25) then
RAISE NOTICE 'MOVING';
ELSE
RAISE NOTICE 'NOT MOVING';
ELSE
RAISE NOTICE 'Query response time within limit';
'query using union operation'
END
```

4 Results

We first show the query response time in union and no-union based approach. The Fig. 2 is the plot of query response time in ms against the no. of row queried. As mentioned in the implementation section we query every 5 s and load every 1 min so there are approx. 12 queries between two loads. Each query would return a different response time depending on the system parameters. We have taken the max query response time and we observe that most of the time union based approach takes less query response time. Further we extent this by loading additional data of 22k, 34k, and 45k in which we have loading 22k in two parallel loads of equal size and 34k in three parallel load and 45k in four parallel loads. The results show that union base approach is better and also we found that query response time reduces in case of parallel load apart from less load time. We have not recorded the load time only the query response time are shown in the Fig. 3, Fig. 4, Fig. 5 respectively for 22k, 34k, 45k, serial and parallel load for union and no-union based approaches.

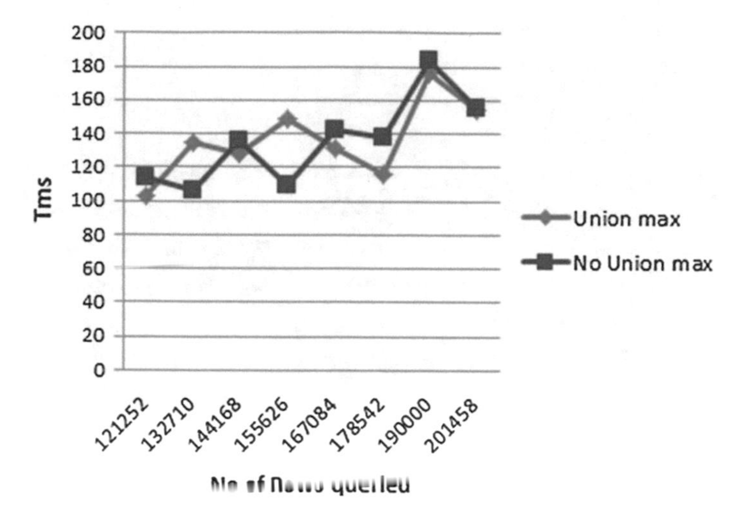

Fig. 2 Query response time union and no-union based approach

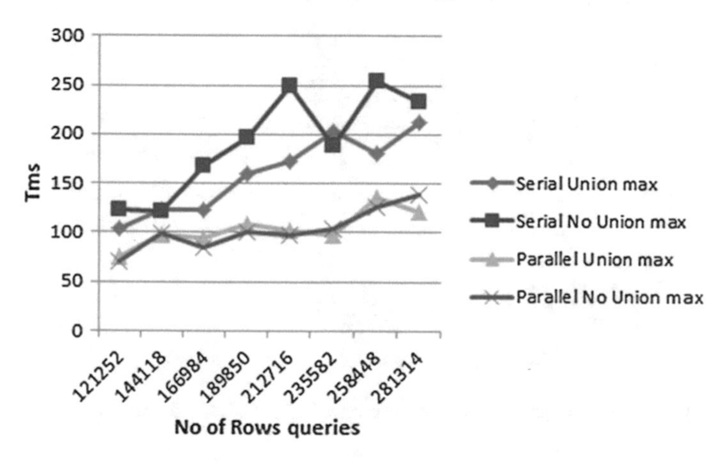

Fig. 3 Query response time union and no-union based approach serial and parallel load 22k rows

Second part of result is to show the reduction in no. of moves after we take both the query response time as mentioned in [1] and right time trigger as mentioned in [2]. The graph shows the query response time plotted against the no. of moves compared with threshold against no. of moves. To calculate the threshold we take ratio of average of quantities in replica table and original table, then the ratio of sales in replica table and original table, then the ratio of no. of rows in replica table and

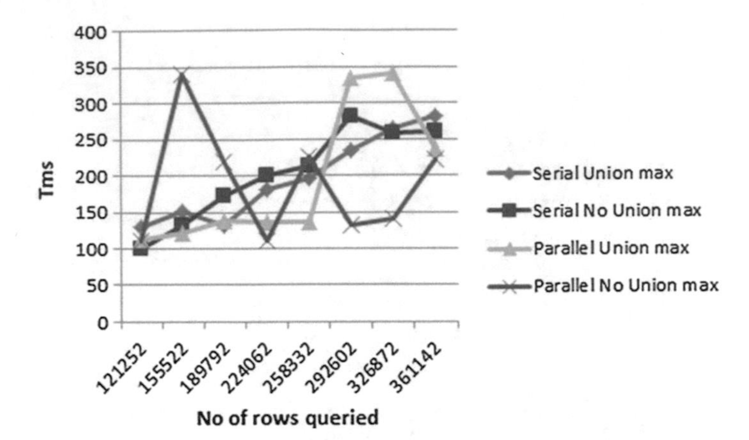

Fig. 4 Query response time union and no-union based approach serial and parallel load 34k rows

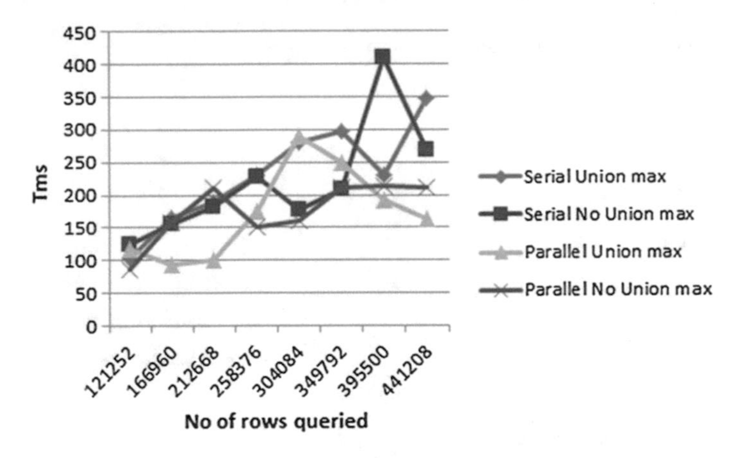

Fig. 5 Query response time union and no-union based approach serial and parallel load 45k rows

original table and lastly the feedback about the query from the OLAP user. Example is shown below

$$\begin{aligned} \text{threshold} = {} & (\text{avgqtytemp}/\text{avgqty}) * 0.30 + (\text{avgsalestemp}/\text{avgsales}) * 0.20 \\ & + (\text{rowcounttemp}/\text{rowcount}) * 0.4 + 0.5 * 0.1 \end{aligned}$$

Here were taking 30%, 20%, 40% and 10% of each of the values to mentioned above to calculate the threshold respectively. This is the first threshold, similarly we have calculate the threshold value varying the percentages for 4 more cases as shown in the below Table 1.

The graphs from Figs. 6, 7, 8, 9 and 10 show the no. of moves for query response time compared with moves based on threshold after the query response time has

Table 1 Threshold values and various percentages

Attempts	avgqtytemp/avgqty (%)	avgsalestemp/avgsales (%)	rowcounttemp/rowcount (%)	Feed back (%)	Threshold value
T1	30	20	40	10	0.55
T2	25	25	25	25	0.62
T3	30	30	20	20	0.69
T4	20	20	30	30	0.55
T5	10	50	10	30	0.74

crossed the set limit value for the threshold from T1 to T5 respectively. If we consider only the query response time then the no. of moves would be about 200 in almost all cases. But if we consider the threshold then the no. of moves would be 20 in almost all cases. In our experiment we have taken query response time to be 80 ms, this could be altered or varied according to the requirement.

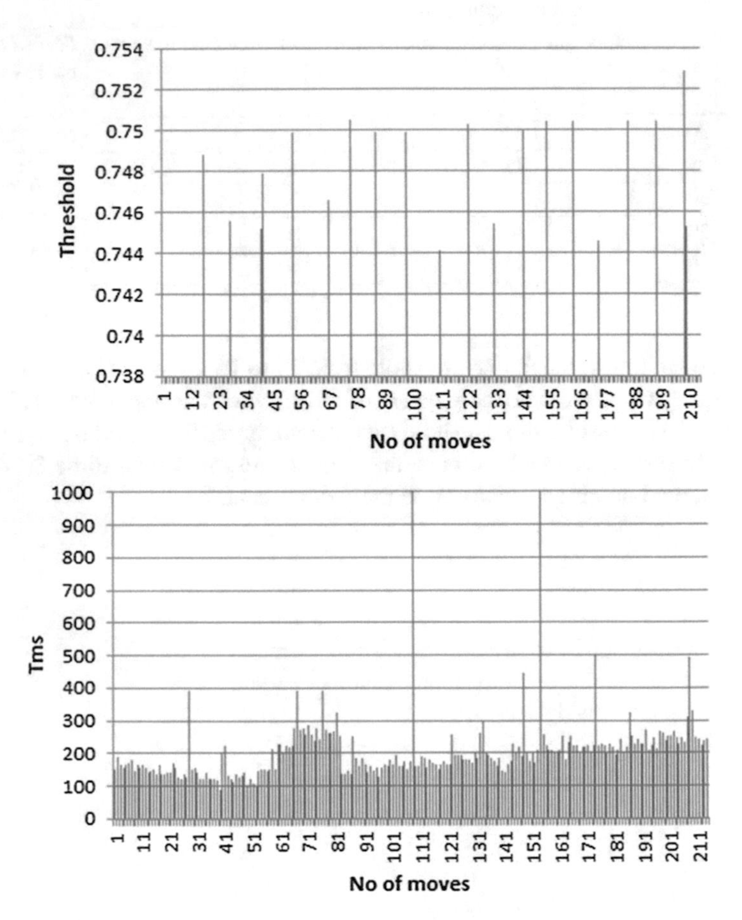

Fig. 6 Threshold T1 data

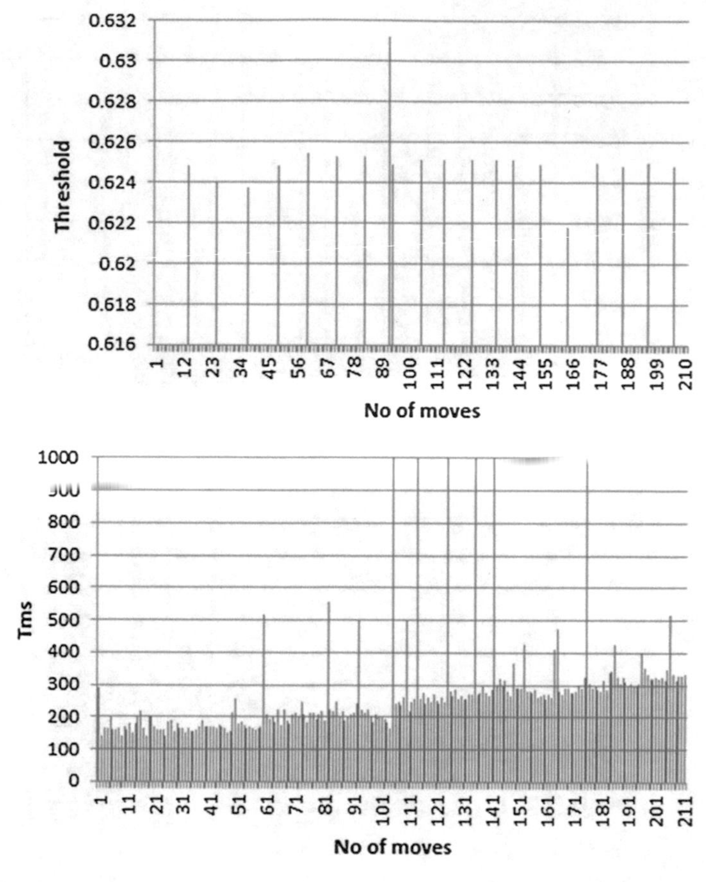

Fig. 7 Threshold T2 data

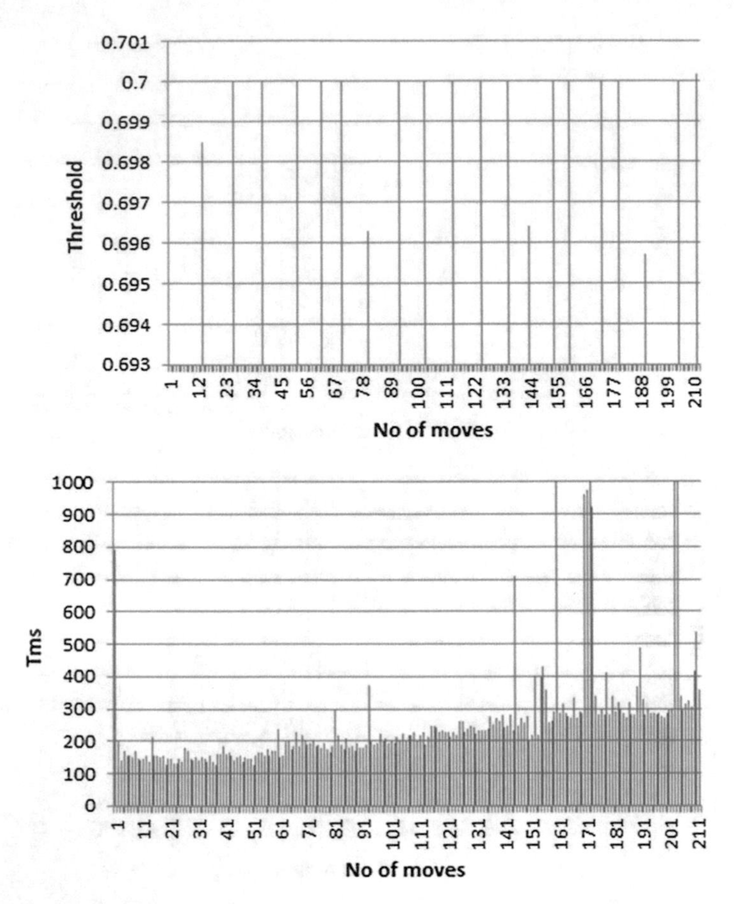

Fig. 8 Threshold T3 data

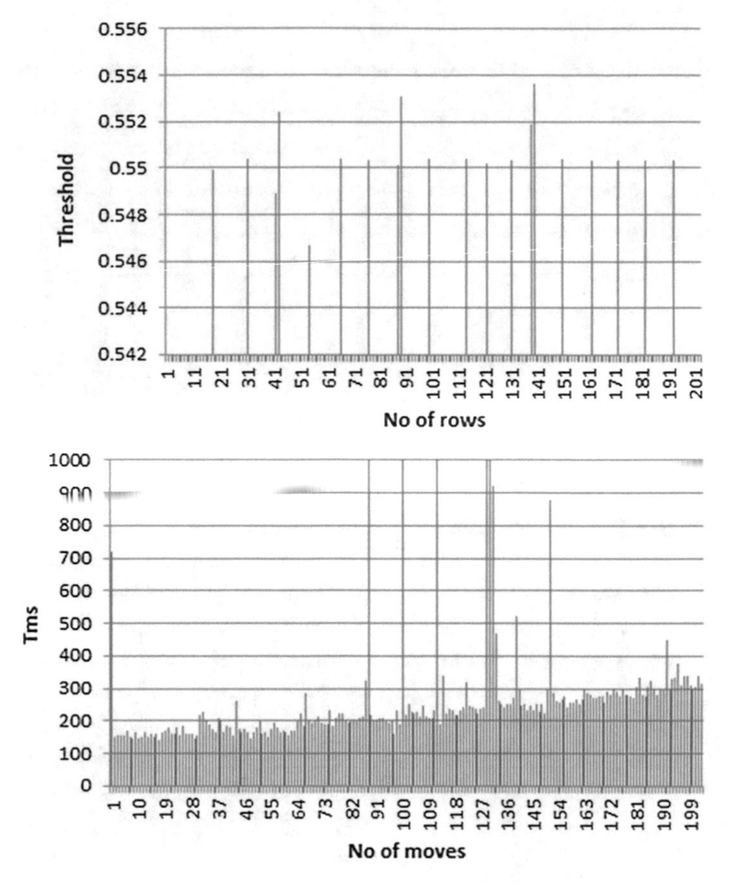

Fig. 9 Threshold T4 data

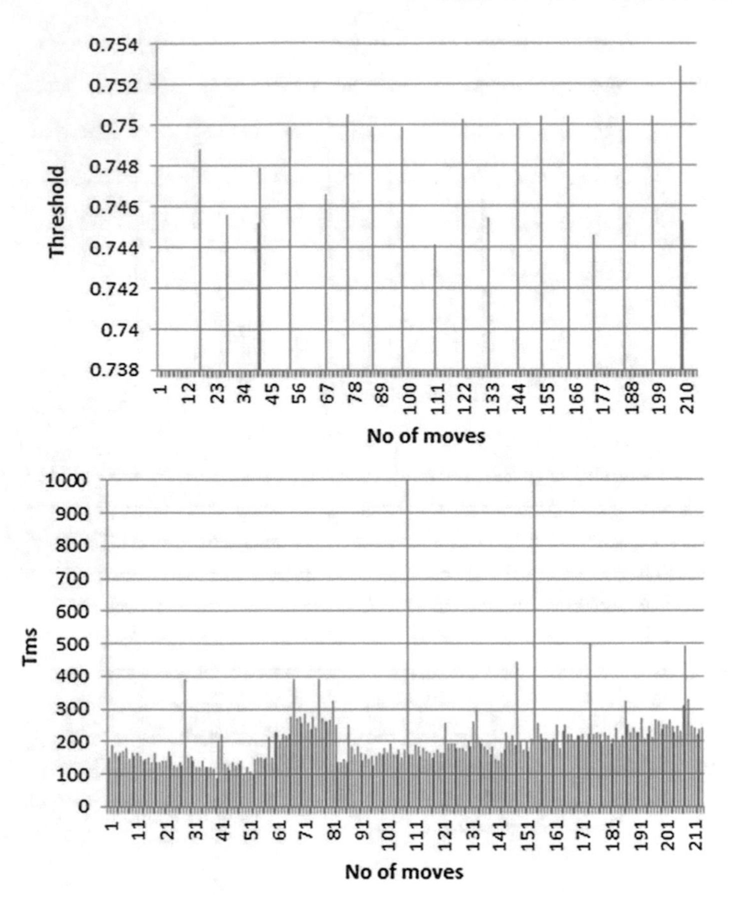

Fig. 10 Threshold T5 data

5 Discussion

Out of the various meta-data based approaches [12–14] that we surveyed for ETL process optimization this study focus more on implementation as a optimization option. Other studies are more on improving the process and planning and optimize the design to improve on ETL and achieve a optimized solution for data warehousing. In our study we have exclusively study the effect of loading data in parallel on query response time and we have found that loading parallel not only reduces the load time but also the query response time. Also we have reduced the no. of moved in a union/no-union based approach based on triggers to move only if the threshold value could reached a certain limit and not only on the query response time.

6 Conclusion and Future Work

In conclusion we have several ways in which ETL could be optimized. In the plethora of optimization possibilities we have studied three of them and found some good results in terms of query response time and no. of times to load the data. This work is part of our ongoing research on study of optimization techniques for near real time ETL. In future we would study some other meta-data based approach to optimize the ETL process.

References

1. Ricardo.: Real-time data warehouse loading methodology. ACM, Coimbra (2008)
2. Rahayu, W., Chen, L.: Towards near real-time data warehousing. IEEE (2010)
3. Simitsis, A., Gupta, C., Wang, S.: Partitioning Real-Time ETL Workflows. IEEE, Long Beach CA (2010)
4. Ferreira, N., Furtado, P.: Near real-time with traditional data warehouse architectures: factors and how to. ACM, New York (2013)
5. Bernardino, J., Santos, R.J.: Optimizing data warehouse loading procedures for enabling useful-time data warehousing. ACM, Cetraro (2009)
6. Donselaar, V.L.: Low latency asynchronous database synchronization and data transformation using the replication log (2015)
7. Rizzi, S., Baldacci, L., Golfarelli, M., Graziani, S., Stefano, R.: QETL: an approach to on-demand ETL from non-owned data sources (2017). Elsvier
8. Lehner, W., Theodoroua, V., Abellóa, A., Thieleb, M.: Frequent patterns in ETL workflows: an empirical approach (2017). Elsvier
9. Poess, M., Tilmann, R., Caufield, B.: TPC-DI: the first industry benchmark for data integration, pp. 1367–1378 (2014)
10. Theodorou, V., Kartashov, A.: Supervisor. https://github.com/AKartashoff/TPCDI-PDI (2016)
11. TPC.: TPC BENCHMARK ™ DI.: TPC (2014)
12. Li, L.: A framework study of ETL processes optimization based on metadata repository. In: 2nd International Conference on Computer Engineering and Technology, pp. 125–129. IEEE, Shanghai (2010)

13. Rahman, N., Marz, J., Akhter, S.: An ETL Metadata Model for Data Warehousing, pp. 1–18 (2012)
14. Lemmik, R.: Model-Driven Development Method of the Virtual Data Warehouse, pp. 197–200. IEEE, Tallinn, Estonia (2010)

BornBaby Model for Software Synthesis

A Program that Can Write Programs

H. L. Gururaj and B. Ramesh

Abstract There have been many improvements on neural networks and neural nets in order to come up with a program that is as intelligent as humans. Recent improvements in natural language processing and machine learning techniques in the field of artificial intelligence has proved that the near future will be very exciting with artificial intelligent robots around us. The BornBaby model proposed is a whole new dimension for Artificial Intelligence and opens up array of possibilities for researchers to come up with natural solutions for problems. Attempts are being made to come up with a software synthesis program, which is able to write programs on its own, like human programmers. BornBaby model is a structured model which defines how a program must learn in order to come up with solutions. Though there are models like Neural networks which emphasizes on the implementation of program to simulate working of human brain, BornBaby model emphasizes on how the program must learn the data based on natural concepts of living beings and the implementation is generic.

Keywords Artificial intelligence · NLP · Bornbaby · Software synthesis

1 Introduction

Artificial Intelligence is the intelligence exhibited by machines. In Computer Science, the field of AI research defines itself as a study of "intelligent agents". Artificial Intelligence includes Machine Learning, Data Mining, and Natural Language Processing.

The sole aim of Artificial Intelligence was to mimic or even outperform human minds. Thus it is important we question the fact whether it has actually been able to

H. L. Gururaj (✉)
Vidyavardhaka College of Engineering, Mysore, India
e-mail: gururaj1711@vvce.ac.in

B. Ramesh
Malnad College of Engineering, Hassan, India

P. Nagabhushan et al. (eds.), *Data Analytics and Learning*,
Lecture Notes in Networks and Systems 43,
https://doi.org/10.1007/978-981-13-2514-4_33

do so. It cannot be ignored that the fact of AI is being implemented especially in the fields of medicine, robotics, law, stock trading, etc. Also some of its applications are homes and big establishments such as military bases and the NASA space station.

BornBaby model uses these techniques and emulates how a brain of baby works. There are 2 kinds in artificial Intelligence: Narrow AI and Broad AI. While BornBaby model is generic, its implementation comes under narrow AI wherein we concentrate on optimizing on one specific task. Besides categorizing as narrow and broad, it is classified into 2 kinds. They are Strong AI and Weak AI [1].

1.1 Strong AI

Strong AI says that machines can equally do tasks like humans. So, in near future, we could expect machines to be working, feel emotions, and understand situations like how humans does and learn from the experienced data. Though it seems impossible, just by studying how a part of the brain functions (Neurons) if we could develop neural networks, what if we get to know the, mystery of brain? Strong AI gets its grip. If this happens, machines would be roaming around us just like humans or be used as personal assistants to do assigned works. This can revolutionize how computing is done.

1.2 Weak AI

The motive behind Weak AI is just the fact that machines can pretend they are intelligent, i.e., they are acting as if they are intelligent. It states that the ability to think can be easily added to computers which can produce better results and tools. For example, when a computer is made to play against humans, it will be taking some moves based on how human is playing. It might behave as if it is intelligent by deciding which move is better. But it is not actually thinking. It is producing the moves based on the inputs or training data already fed into it. It is just trying not to lose the game and eventually wins it by making moves of a professional chess player.

But wait, there is a reason why we are talking about this right now. BornBaby model does not acts as if it is intelligent, but it learns how to be intelligent. Exactly, As the name says, it can generate programs of any kind, of any language just by learning what someone teaches it. If we think deeply on what's intelligence, intelligence is nothing but how one can perform operations to produce results based on what he has learnt.

Well, thanks to the computer. It has got all the memory, processing and devices to take i/o. So, we already have the baby in our hand. What we just have to do, is to train it. Let us see how to train it in the working.

2 Literature Survey

Ali Heydarzadegan, Yaser Nemati, Mohsen Moradi, Islamic Azad University [2]. In this paper evaluation of various machine learning algorithms have been presented. This gives a brief idea about how we structure our data.

Unnati Dhavare, Prof. Umesh Kulkarni, Vidyalankar College Marg [3]. In this paper the applications of natural language processing have been explained briefly as to where the concept can be applied.

Prof. Manish G. Gohil, BCA Mahila College, Gujarat [4]. In this paper it explains about how a computer distinguishes the pauses and punctuations.

Heather Sobeys, Department of Computer Science University of Cape Town [5]. In this paper how current speech recognition is based on hidden Markov model, its advantages and disadvantages.

Hany M. Ammar, Walid Abdelmoez and Mohamed Salah (2012) [6] In their paper authors surveyed on problems associated in the traditional software development such as SBST requires further research and it can be overcome by how the artificial intelligent technique such as KBS, CBR, Fuzzy logic and automated programming tool.

3 Artificial Intelligence and Software Engineering

Software engineering comprises of two kinds of knowledge. It is nothing but the programming knowledge and domain knowledge. Programming knowledge consists of how the language is built, the grammar used in the language, the syntax and semantics of that language and flow control mechanisms built into the language. The domain knowledge is the knowledge gained on the problem statement you are about to solve. If we are going to solve a problem on how to recognize if the color of shirt is blue, we need to know what a shirt is, what a shirt looks like and other properties of shirt in order to recognize it is a shirt. We must also know what the different colors are and how does blue look like.

3.1 Intersections Between AI and SE

Merging these two domains is the main task involved in BornBaby Model for software synthesis. Because, if the AI knows about the domain of what I am asking it to solve, given some rules, syntax and semantics, it can generate a program to solve the problem statement. Traditional software engineering process involves several phases like requirement phase, design phase, development phase and testing phase. Errors can happen in any of these phases. Only thing needed is how you recognize and resolve the error. If these capabilities are built into a machine, it'll be able to write standard code for any problem statement.

The common research area between them is: computational intelligence agents, ambient intelligence [7], and knowledge-based software engineering.

Ambient intelligence help us making an adaptive, sensitive and reactive systems that are informed about the user's emotions, habits, needs, and in order to support them in their work regularly.

4 Sentiment Analysis

Sentiment analysis [8], which is also termed as opinion mining, is a deep concept. Sentiment can be analyzed by the way someone speaks, by the variation in their voice and as such. It is important for us to know learn about opinion mining; because, "How to react to a person who is committing suicide?" is a problem statement based on someone's sentiments and can't be answered based on some given set of questions and answers.

By using various natural language processing techniques, which are advancing very rapidly in today's world is the key for sentiment analysis. If prediction of someone's judgment is a problem, then the solution is to study his attitude, emotional state and present context. Because, we know that one can be convinced. And being convinced is a result of his emotional state, present context, and of course might rely on how he was convinced the previous time.

5 Statistical Language Models

If the baby could learn only from texts, it is impossible for it to understand the sentiment and emotion of those texts. But, learning from texts also plays an important role. If we could train the baby based on the messages and their replies, it can know how humans communicate and how they reply in what state. Emoticons might be digged to know the emotion of that text, roughly.

As it is already said, the BornBaby model is a generic model. But what it exactly means is, it is just a representation of how the machine should work. This representation can be dubbed to generate programs in any language. Also, while we interact with the model, we are going to use our natural languages. Any language in the world can be used to interact with the AI to produce programs. If you guessed it right, the natural language processing techniques of that particular language has to be developed first in order to make the baby learn. After all, everyone who has learnt their mother language, learnt it while they were a baby.

So, now, there would be no need for normal programmers. A person who has better grip on a language and its grammar can interact with this program to generate desired programs. Only thing is, he has to wait until and stuff the baby to learn the necessary things before it is asked to solve the problem. To implement this model, one needs a good insight on how human thinks. For example, when someone asks you to recognize a sign board on the road, you match some rules that sign board must have. Usually, it has some particular colors, shape and size. If that matches, you can say it is a sign board. So, one can now agree to the point that even humans does some matching and mapping process in order to come to conclusions on a problem statement. These matching techniques, for example, regex matching in texts, are also extensive capability to come to solutions for a problem. If the creator knows how his creation must behave, he can take care of making a better creation, which is the simple principle behind the BornBaby Model.

6 The Bornbaby Model

In Bornbaby model there are three phases Hearing Phase, Learning Phase, and Interacting Phase. Hearing phase is input phase, just how a baby starts to hear. Learning phase is the analysing phase, the baby is processing what it has heard about. Interacting phase is how the baby starts to interact with the society, the baby continuously learns as depicted in Fig. 1.

This is a generalized model where in the baby can be made to learn any process in any of the languages. As the language is processed the baby will learn, understand and will possess the ability to solve any desired task. After understanding, the grammar of any desired language must be fed for analytical purpose. This model is a triangular loop of learning, understanding and interacting, because learning is a continuous process. So, to start, the program is setup with a desired language by feeding the dictionary of language and its grammar. Assume the program as a personal assistant to you.

As we talk with the program, it processes the language we talk according to the language we have set it up with. Regex techniques help us to parse the grammar

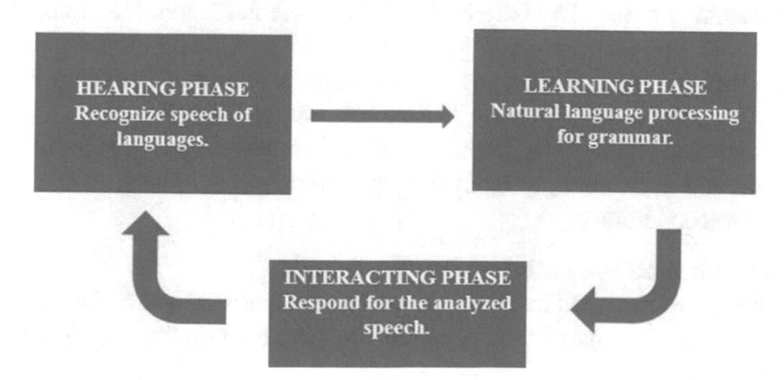

Fig. 1 Phases of Bornbaby model

of the heard sentence. With more advanced natural language processing techniques, more efficient thought process can be setup.

If we ask a baby to find the square of a number, will it answer us? It will, provided that it knows how to multiply a number and has been taught what is square of a number. Fortunately, arithmetic calculations are very easy for a computer. It knows to multiply a number. But we have not taught it to square a number. It definitely cannot find the hypotenuse of triangle if it does not know to square a number. Now, we can agree to the point that things are inter-dependent; only if we know the basics, we can solve complex questions.

In the BornBaby Model, the program synthesizes another program for what it is asked and it stores that program. For example, if we ask "What is the square of 2", the baby does not know yet how to square a number. But, If we teach it saying "Square of a number is number times number", it's now able to square any number and tell us the answer. When we taught it "Square of a number is number times number", it generated a program that takes number as input and returns number*number as result. Now, since hypotenuse needs one to find square of the sides first, the baby can now square the sides using what it has learnt.

7 Result Analysis

To demonstrate the BornBaby model, let us discuss about the implementation we came up with.

The Hearing Phase—The speech data is fed through WebkitSpeech API, which is used for voice recognition based on context. The voice is converted to text and is sent to brain to process it. The baby will be hearing to the world but not listening to it, i.e., the heard data is just in its ear but not processed by its brain. The baby needs a name before we can make it to listen to us. There is a gender detection algorithm which can detect gender of particular name by exploring the web. We can give a name for the baby and then make it to listen to us.

The Learning Phase—The core logic of the implementation, the brain, is implemented in this phase. Regex techniques are used for analysing the input and it is processed through Natural Language Processing techniques. The brain module is implemented using Node.js popular library released by Google Inc.

The Interacting Phase—The interacting phase sends the input from WebkitSpeech to the brain and gets the reply from program, which is converted to speech using responsiveVoice.js library.

1. The program acts as a brain and hence named brain.js.
2. The language processed by the brain are also synthesized as JavaScript modules.
3. All these modules together form the conscious mind with the help of which baby will interact.

As we will see, baby can solve simple arithmetic problems by processing and learning the input text. The JavaScript modules are synthesized by the baby's brain which in turn is also a JavaScript program, justifying the title as "software synthesis". We will follow the classical example of square root and hypotenuse to demonstrate the result.

In Fig. 2, we can see how the program is interacting with a human. When we asked it what the square root is, it replied "I have not learned this yet". Because we've not taught it to square a number. Let's go ahead and teach it how to square a number.

In Fig. 3, It has learnt to square a number as we fed in "Square of number is number multiplied by number".

In Fig. 4, it is producing us the result for the square of 25.

In Fig. 5, we can see thast it is writing the program to square a number. Now that it knows to square a number, using verbal communication, we taught it to find

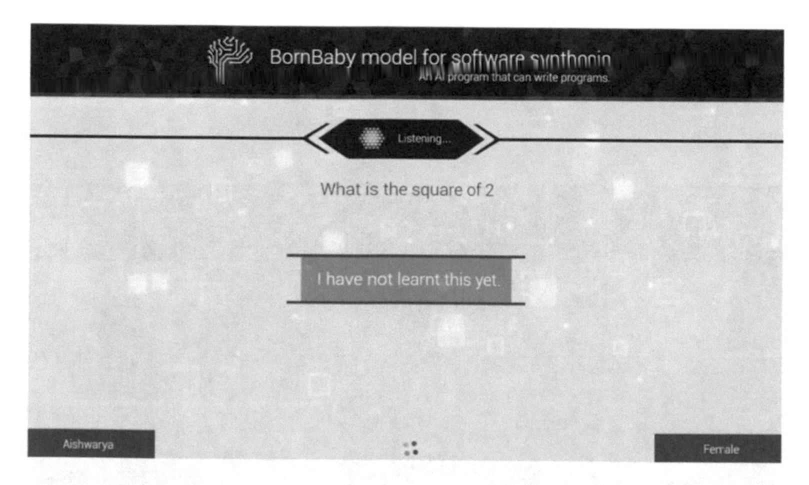

Fig. 2 Initial phase

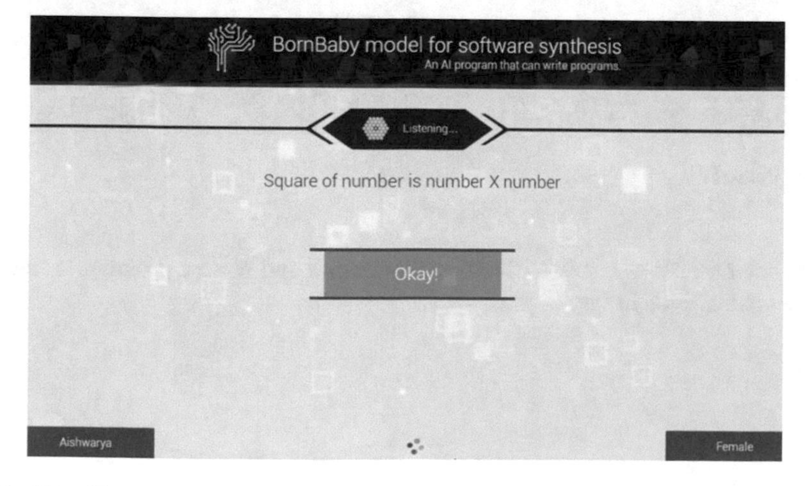

Fig. 3 Phase II

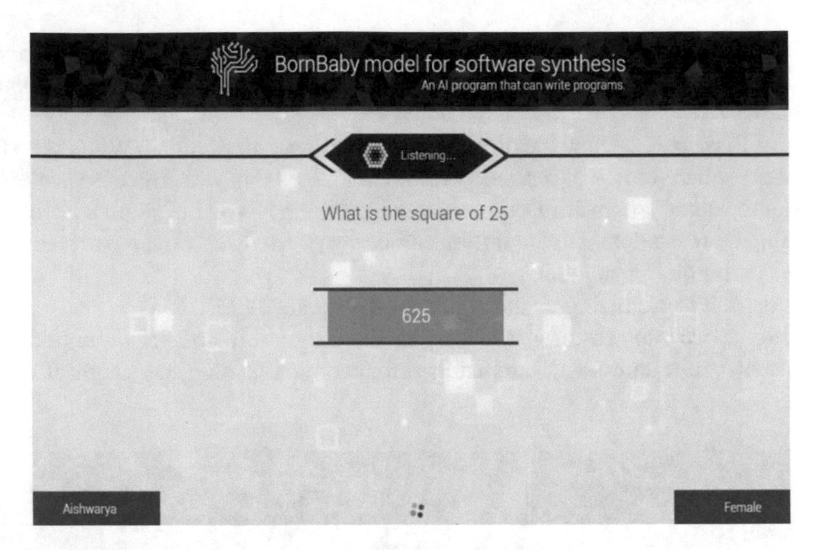

Fig. 4 Phase III

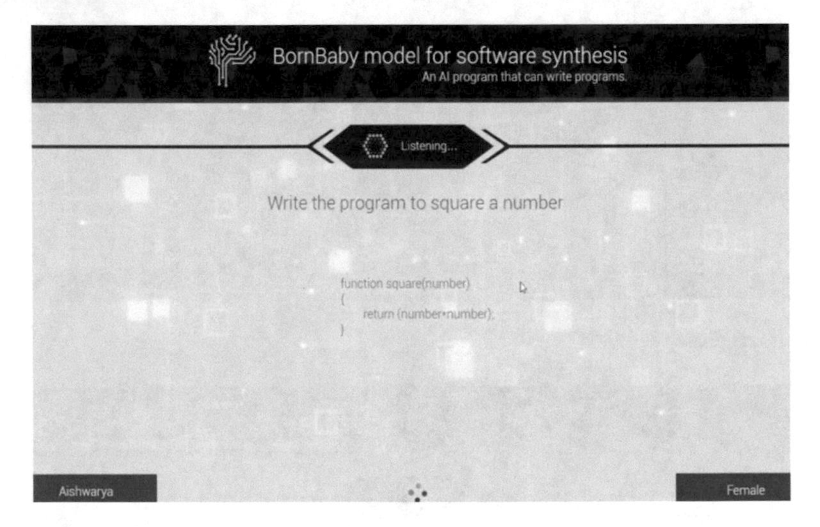

Fig. 5 Phase IV

the hypotenuse. It was able to find the hypotenuse and write a program to find the hypotenuse as seen in Fig. 6.

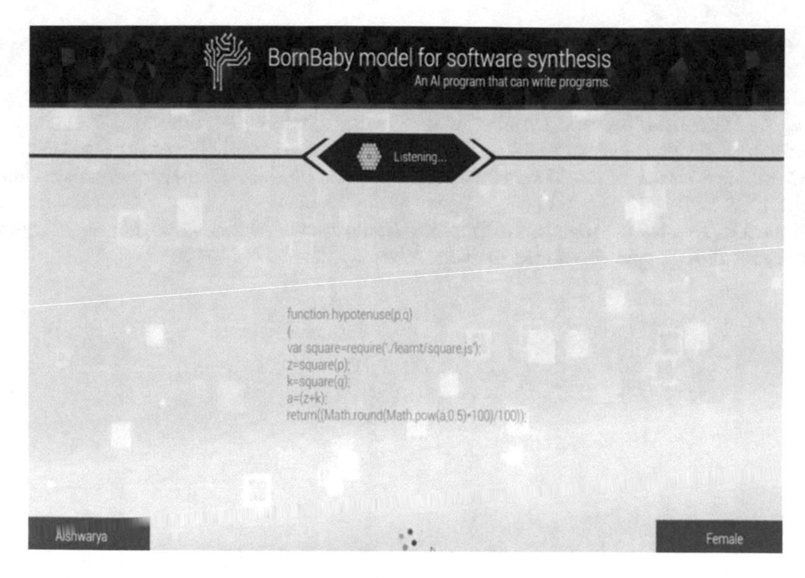

Fig. 6 Phase V

8 Conclusion

We know how to write programs using programming languages, but for the first time this program can write programs using natural communicating languages of daily life. Even a novice can be able to solve complex problems using this model without the knowledge of any programming languages, but Basic English. If children are exposed to this program, they can even get a good grip at grammar and also can know how computers work Different Natural Language Processing techniques and regex simplicity empowered the BornBaby model in the process of software synthesis. BornBaby model will help open a wide area for research interests and introduces a new scope of programming methodology.

References

1. Hu, Y., Chen, J., Rong, Z., Mei, L., Xie, K.: A neural networks approach for software risk analysis. In: Proceedings of the Sixth IEEE International Conference on Data Mining Workshops, pp. 722–725. Hong Kong. Washington DC: IEEE Computer Society (2006)
2. Goldman, S., Zhou, Y.: Enhancing Supervised Learning with Unlabeled Data. Department of Computer Science, Washington University, MO, USA
3. Caruana, R., Niculescu- Mizil, A.: An Empirical Comparison of Supervised Learning Algorithms. Department of Computer Science, Cornell University, Ithaca, NY, USA
4. Lavesson, N.: Evaluation and Analysis of Supervised Learning Algorithms and Classifiers. Blekinge Institute of Technology Licentiate Dissertation Series No 2006:04. ISSN 1650-2140, ISBN 91-7295-083-8

5. Lavesson, N.: Evaluation and Analysis of Supervised Learning Algorithms and Classifiers. Blekinge Institute of Technology Licentiate Dissertation Series No 2006:04. ISSN 1650-2140, ISBN 91-7295-083-8
6. Ng, A.: Deep Learning And Unsupervised, "Genetic Learning Algorithms", "Reinforcement Learning and Control. Department of Computer Science, Stanford University, CA, USA
7. Liu, B.: Supervised Learning. Department of Computer Science, University of Illinois at Chicago (UIC), Chicago
8. Caruana R., Niculescu-Mizil, A.: An Empirical Comparison of Supervised Learning Algorithms. Department of Computer Science, Cornell University, Ithaca, NY, USA

Lung Cancer Detection Using CT Scan Images: A Review on Various Image Processing Techniques

A. P. Ayshath Thabsheera, T. M. Thasleema and R. Rajesh

Abstract These days, image processing techniques are most common in diverse medical applications for the early diagnosis and treatment, predominantly in cancer tumors. Identification of such tumors at the budding stage is a tedious task. Most of the existing methods tests on CT (Computed Tomography) scan images that is having mainly four stages: image enhancement, segmentation, extraction of features, and classification. This paper briefly discusses about different methods already reported in literature for lung cancer detection using CT scan images.

Keywords CT · Image enhancement · Segmentation

1 Introduction

Lung cancer is considered as one of the major cause of cancer death worldwide. However, studies shows that if diagnosed earlier, chance of survival is more. Various imaging modalities such as chest X-ray, positron emission tomography (PET), computed tomography (CT), and magnetic resonance imaging (MRI) are used for the detection of cancer [1]. Since CT image is more sensitive to find the size of the tumor and having less noise content, it is preferred over other imaging modalities [2].

There are many Computer-aided diagnosis (CAD) systems, developed by researchers for the prediction of lung cancer. Such systems help the radiologist to make a final decision by considering the system-generated result along with the clinical diagnosis. Therefore, there are more number of research being carried out in

A. P. Ayshath Thabsheera (✉) · T. M. Thasleema · R. Rajesh
Department of Computer Science, Central University of Kerala, Kasaragod, Kerala, India
e-mail: thabshit4@gmail.com

T. M. Thasleema
e-mail: thasnitm1@hotmail.com

R. Rajesh
e-mail: rajeshr@cukerala.ac.in

P. Nagabhushan et al. (eds.), *Data Analytics and Learning*,
Lecture Notes in Networks and Systems 43,
https://doi.org/10.1007/978-981-13-2514-4_34

this field for lung cancer detection. There are different computational approaches and image processing techniques existing in the literature for the early diagnosis and prediction of lung cancer using CT scan images and this paper gives a short review on such methods used in each stages of the lung cancer detection so that one can have an idea on what methods are being used in this field.

2 Review of Existing Methods

Many researchers have implemented lung cancer detection system on CT images using numerous image processing techniques and approaches. However, all these works involve mainly four steps for the detection of lung cancer: image enhancement, segmentation, extraction of features from segmented images, and finally classifying these into cancerous and non-cancerous categories. Figure 1 shows the basic steps involved in a lung cancer diagnosis.

2.1 Image Enhancement

Main objective of image enhancement is to process an image in order to improve its quality for making it more suitable for a specific application. A large number of image enhancement techniques exists and these techniques are very much problem-oriented. Image enhancement techniques are mainly classified into two: spatial and frequency domain. Many image enhancement approaches proposed in literature such as Gabor Filter [3], Median filter [4], Weiner filter [5], Fast Fourier Transform, Auto

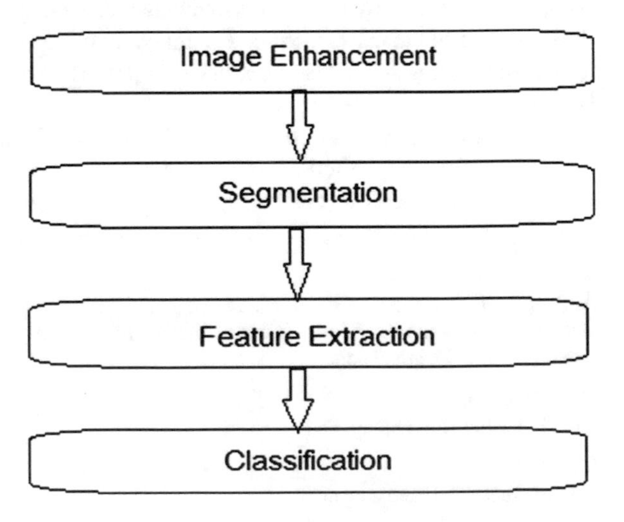

Fig. 1 Procedure for lung cancer detection

enhancement [6], Unsharp filter are used in order to reduce or eliminate noise present in an image and to visually improve the quality of image.

A procedure proposed by Gajdhane and Deshpande to detect the lung cancer in which Gabor filter and watershed segmentation were used in preprocessing stage of CT scan images. Gabor filter is a spatial linear filter depending on the orientation and ridge frequency in an image and used for texture analysis. Use of Gabor filter in enhancement of medical images is very significant. It is mainly used in two-dimensional image like CT scan images [3].

Al-Tarawneh discussed mainly on accuracy and quality of image. In the stage of image enhancement, Gabor filter is used since it is very useful for texture analysis, due to its optimal localization properties in both spatial and frequency domain, it is being mostly used for the enhancement purpose.

Auto enhancement, automatically adjusts and enhances the image brightness, color and contrast to optimum levels which strongly depends on statistical operations such as mean and variance calculation [7]. Another method which is using FFT filter operates on Fourier transform of image [6]. Median filter is often used in image enhancement and it mainly reduces salt and pepper noise [4]. Median filter preserves the edges of image after enhancement. Unsharp filter also enhances edges after enhancement procedure. K-Nearest Neighbors and Weiner filters can also be used for preprocessing. A statistical approach is used by Sharma and Jindal [5] to filter out the noise present in lung CT image using Weiner filter.

Among the above-discussed methods, it is observed that Gabor Filter Enhancement gives comparatively better result to have a good-quality image. Table 1 shows the performance of various image enhancement methods based on Contrast improvement index(CII). It can be represented using the Eq. (1).

$$\text{CII} = \mathbf{I}\text{processed}/\mathbf{I}\text{original} \tag{1}$$

Iprocessed and **I**original indicates contrast value of enhanced and original image respectively. Higher the value of CII better is the performance. CT images are considered and are obtained from NIH/NCI LIDC dataset consist of 244,327 images from 1010 patients and RIDER [8] Lung CT database having 15,419 images from 32 patients are taken from The Cancer Imaging Archive (TCIA) [9].

Table 1 Performance of various image enhancement methods

Dataset	No. of images	Gabor filter	Median filter	Weiner filter	Unsharp filter	FFT
LIDC	244,327	3.19	2.81	2.93	2.64	1.96
RIDER	15,419	2.04	1.96	1.94	1.83	1.34

2.2 Segmentation

In the second stage, the lung region is segmented from surrounding area in the image. Main purpose of segmentation is to decompose the regions in an image for further analysis. There are many types of segmentation approaches such region-based and boundary-based and the segmentation of medical image has many useful applications. Methods such as thresholding [10], marker-controlled watershed segmentation [3, 6], Sobel edge detection [5], Otsu thresholding [4] and morphological operations [7] are used in existing works for the segmentation purpose.

For the segmentation of the image, Marker-controlled watershed method is used by Al-Tarawneh [6]. Main feature of Marker-driven watershed segmentation technique is widely used in medical images that helps identifying touching objects in an image and used to separate such touching objects in an image [10]. Thresholding also can be used for segmentation purpose since the method has many advantages such as less storage space and high processing speed when compared to grey level image. Otsu thresholding [4] and Sobel edge detection method [5] are also well-established approaches used in literature for the segmentation procedure.

Otsu thresholding is a nonlinear clustering-based global thresholding method which converts gray level image into binary image. It is used when an object need to be extracted from its background. Sobel edge detection is useful when we analyze image in structural aspect.

Some morphological operations such as Morphological opening, Edge detection, Morphological closing, and Region Growing Algorithm are also used in order to remove low intensity regions [7]. Figure 2 shows the different steps of image segmentation.

2.3 Feature Extraction

Feature extraction is the next step to be performed, which is essential in order to determine the cancerous nodule [7]. These extracted features are the input for classification

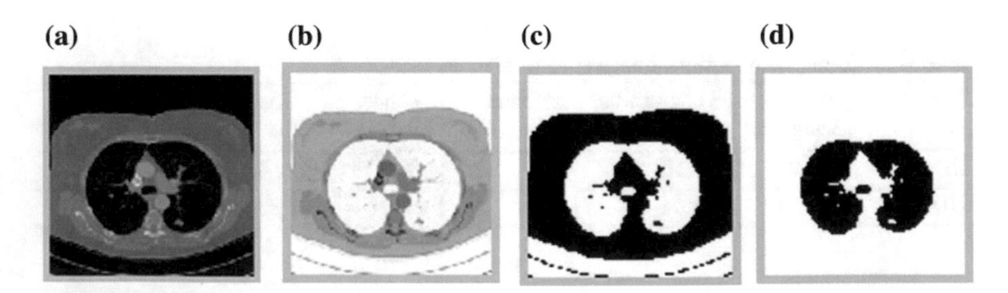

Fig. 2 Steps for segmentation: **a** original image, **b** Gabor enhanced image, **c** thresholded image, **d** segmented image

process. Area, Perimeter, and eccentricity are the features which are considered in the work by Gajdhane and Deshpande [3]. Area gives the actual number of pixels in the region and Perimeter specifies the distance around the boundary of the region by calculating the distance between each adjoining pair of pixels around the border of the region. Eccentricity gives the roundness measure of the region [10].

The method proposed by Tung and Khaing extracts features such as entropy, contrast, correlation, energy, homogeneity, area, perimeter, and eccentricity [4]. These eight features are considered for classification purpose. Tariq et al. considered area, energy, eccentricity, entropy, and mean grey level value of candidate region as the basis for classification [11]. Binarization and masking also used for extracting features [6, 10].

2.4 Classification

Classification is the final step for determining whether a nodule is cancerous or not. Classification using Artificial neural network (ANN) has shown good results (approximately 90%) in classifying lung cancer [4]. Support Vector Machines are also used for classification purpose. An SVM model is a representation of the examples as points in space, mapped so that the examples of the separate categories are divided by a clear gap that is as wide as possible [12]. SVM is preferable when your data has exactly two classes. SVM classifies an input into either class 1—which is normal or class 2—which is cancerous. This classification is performed on the basis of the extracted features.

For the classification, a feed forward back propagation neural network is used by Kaur et al. [13]. Tariq et al. proposed a hybrid classifier based on neural network and fuzzy which uses fuzzy self-organizing network followed by Multilayer Perceptron (MLP) [11].

Miah et al. proposed a multilayer feed forward neural network with supervised learning method for classification of the segmented images [14]. Firmino et al. used a rule-based classifier and Support Vector Machine (SVM) [15]. In [16], Sakthivel et al., proposed an intelligent and dynamic classification method, Intelligent Fuzzy C-means to detect the tumor. To classify the segmented nodule, SVM is used in the work by Makaju et al. [17]. Table 2 shows a brief summary of existing classification methods for lung cancer detection.

From literature review, it is observed that several image processing techniques and computational-based approaches are useful in the prediction of lung cancer. In this paper, we briefly studied on different techniques used for lung cancer detection at its enhancement, segmentation, feature extraction, and classification stages.

Table 2 Summary of existing methods for classification

Author	Classifier	Extracted features	Accuracy
K. M. Tun	Artificial neural network	Area, perimeter, eccentricity, entropy, contrast, correlation, energy, homogeneity	90
Disha Sharma	Diagnostic indicators	Area, shape, size	80
M. S. AL-Taranweh	Masking	Number of black and white pixels	85.7
Tariq	Neuro-fuzzy	Area, energy, eccentricity, entropy, mean and standard deviation	95
Miah	Neural network	Center of image, percentage of black pixels, central moment	96.67
Firmino	Rule-based SVM	Roundness, elongation, energy	94.4
Sakthivel	Intelligent fuzzy C-means	Fused tamura and haralick features	97.6
Makaju	SVM	Area, perimeter, centroid, diameter, eccentricity and mean intensity	92

3 Conclusion

This paper encapsulates a short review on different techniques used for lung cancer diagnosis using CT images. It is observed from the literature that Gabor filter enhancement gives better performance. Features such as area, eccentricity are calculated and given as input to the classifier, which decides whether the lung nodule in the CT image is cancerous or non-cancerous.

As a future work, MRI, X-ray, PET images can be considered for detecting lung cancer. Thus, one can compare and analyze which imaging modality gives better result for the detection of lung cancer. We can also classify the images into several stages of lung cancer using many techniques such as fuzzy or neural network.

References

1. Stark, P.: Use of imaging in the staging of non-small cell lung cancer (2008)
2. Sluimer, C., van Waes, P.F., Viergever, M.A., et al.: Computer-aided diagnosis in high resolution CT of the lungs. Med. Phys. **30**, 3081–3090 (2003)
3. Gajdhane, V.A., Deshpande, L.M.: Detection of lung cancer stages on CT scan images by using various image processing techniques. IOSR J. Comput. Eng. (IOSR-JCE) **16**(5), Ver. III (2014). e-ISSN: 22780661, p-ISSN: 2278-8727
4. Tun, K.M., Khaing, A.S.: Feature extraction and classification of lung cancer nodule using image processing techniques. Int. J. Eng. Res. Technol. (IJERT) **3**(3) (2014). ISSN: 2278-0181
5. Sharma, D., Jindal, G.: Identifying lung cancer using image processing techniques. In: International Conference on Computational Techniques and Artificial Intelligence ICCTAI (2011)

6. Al-Tarawneh, M.S.: Lung cancer detection using image processing techniques. Leonardo Electr. J. Pract. Technol. **20** (2012). ISSN 1583-1078
7. Chaudhary, A., Singh, S.S.: Lung cancer detection on CT images using image processing. Int. Trans. Comput. Sci. **4** (2012)
8. Armato III, S.G., McLennan, G., Bidaut, L., McNitt-Gray, M.F., Meyer, C.R., Reeves, A.P., Clarke, L.P.: Data from LIDC-IDRI. The Cancer Imaging Archive (2015)
9. Zhao, B., Schwartz, L.H., Kris, M.G.: Data from RIDER_Lung CT. The Cancer Imaging Archive (2015)
10. Patil, B.G., Jain, S.N.: Cancer cells detection using digital image processing methods. Int. J. Latest Trends Eng. Technol. (IJLTET) **3**(4) (2014)
11. Tariq, A., Akram, M.U., Younus Javed, M.: Lung nodule detection in CT images using neuro fuzzy classifier, ©IEEE (2013)
12. Cortes, C., Vapnik, V.: Support-vector networks. Mach. Learn. **20**(3), 273–297 (1995). https://doi.org/10.1007/BF00994018
13. Kaur, J., Garg, N., et al.: Segmentation and feature extraction of lung region for the early detection of lung tumor. Int. J. Sci. Res. (IJSR) ISSN (Online), **3**(6), 2319–7064 (2014)
14. Miah, M.B.A., Yousuf, M.A.: Detection of lung cancer from CT image using image processing and neural network. Electr. Eng. Inf. Commun. Technol. (ICEEICT) (2015)
15. Firmino, M., Angelo, G., Morais, H., Dantas, M.R., Valentim, R.: Computer-aided detection (CADe) and diagnosis (CADx) system for lung cancer with likelihood of malignancy. BioMed. Eng. OnLine (2016)
16. Sakthivel, K., Jayanthiladevi, A., Kavitha, C.: Automatic detection of lung cancer nodules by employing intelligent fuzzy c means and support vector machine. Biomed. Res. (2016)
17. Makajua, S., Prasad, P.W.C., Alsadoona, A., Singh, A.K., Elchouemic, A.: Lung cancer detection using CT scan images. In: 6th International Conference on Smart Computing and Communications, ICSCC (2017)

Deep Learning Approach for Classification of Animal Videos

N. Manohar, Y. H. Sharath Kumar, G. Hemantha Kumar and Radhika Rani

Abstract Advances in GPU, parallel computing and deep neural network made rapid growth in the field of machine learning and computer vision. In this paper, we try to explore convolution neural network to classify animals in animal videos. Convolution neural network is a powerful machine learning tool which is trained using large collection of diverse images. In this paper, we combine convolutional neural network and SVM for classification of animals. In the first stage, frames are extracted from the animal videos. The extracted animal frames are trained using Alex Net pre-trained convolution neural network. Further, the extracted features are fed into multi-class SVM classifier for the purpose of classification. To evaluate the performance of our system we have conducted extensive experimentation on our own dataset of 200 videos with 20 classes, each class containing 10 videos. From the results we can easily observed that the proposed method has achieved good classification rate compared to the works in the literature.

Keywords Convolutional neural network (ConvNet)
Support vector machine (SVM) · AlexNET · GPU · Animal classification

---

N. Manohar (✉) · G. H. Kumar
Department of Computer Science, University of Mysore, Mysuru 570006, India
e-mail: manohar.mallik@gmail.com

G. H. Kumar
e-mail: ghk.2007@yahoo.com

Y. H. Sharath Kumar
Department of Computer Science and Engineering, Maharaja Institute of Technology, Belawadi, Srirangapatna Tq, Mandya 571438, Karnataka, India
e-mail: sharathyhk@gmail.com

R. Rani
SBRR Mahajana First Grade College, Mysuru, India
e-mail: radhika.rani435@gmail.com

P. Nagabhushan et al. (eds.), *Data Analytics and Learning*,
Lecture Notes in Networks and Systems 43,
https://doi.org/10.1007/978-981-13-2514-4_35

1 Introduction

In recent days convolution neural network has got a great success in object recognition. Implementing of convolution neural network has becoming more and more in the computer vision field to make an attempt of improving the original architecture of the system to achieve better accuracy. In this paper, we tried to classify animals using convolution network. Identification and classification of animals is a very challenging task. In early days the biologist used to manually classify the animals to study their behavior, which is a tedious and time-consuming task. Also animal classification has got various applications like, avoiding animal-vehicle collisions, anti-theft system for animals in zoo, restricting animal intrusion in residential areas, etc. Designing an automated system for animal classification is an effortful job since the animals posses large intra and inter class variations (shown in Fig. 1). Also the videos considered are in real time with complex background, different illumination, different postures, occlusion, and different view (shown in Fig. 2) makes the problem of animal classification a complex job. Further the usage of high-resolution cameras increase the size of the dataset which also increase the complexity in processing frames. Traditional methods involve some of the major steps in any object recognition problem: preprocessing steps like enhancement, segmentation, feature extraction, and classification. Since the animals often covered with more greenery, complex background which makes the task of fully automatic segmentation a difficult one. Without performing of segmentation, extracting features and classification leads to inefficiency. All these challenges made us to implement convolution neural network to classify animals.

Here we present some of the recent works carried towards animal classification and also presenting work carried in convolution neural network.

Fig. 1 Sample images with inter and intra-class variations

Fig. 2 Sample images with different pose, illumination, view point and occlusion

Like all typical traditional computer vision algorithm, even animal classification also got three stages: preprocessing, feature extraction, and classification. Lahiri et al. [1] introduced a method for identification of individual zebra in the wildlife footage. In this work, images are segmented manually, extracted color features and mapped to the dataset for classification. In [2], we find a method for identifying elephants. Here, segmentation is done using semi-automatic algorithm, shape features are extracted for classification. Raman et al. [3] proposed a method to track and classify animals in videos. Segmentation is done through semi-automatic algorithm using texture features. For classification of animals texture features are extracted and SVM and KNN classifiers are used. Zeppelzauer [4] proposed a method to detect elephants in wildlife video. Images are preprocessed to segment the elephant using color information. For the detection of elephant they have used shape and color features. Kumar et al. [5] deployed a method to identify and classify animals. For segmenting animals semi-automatic graph cut method is introduced, extracted color, and texture features to classify animals using KNN and PNN classifiers. Since animal possess high intra- and inter-class variations Manohar et al. [6] introduced both supervised and unsupervised method for classifying animals. In this work, semi-automatic region merging based segmentation algorithm is used to extract animals from the images, later gabor features are extracted. For classification symbolic classifier is used in supervised approach and k-means clustering is used in unsupervised approach.

Simonyan and Zisserman [7] used convolution neural network to increase the accuracy in large-scale image recognition. They used a very small (3×3) convolution filters to classify the images of ILSVRC-2012 dataset which consists of 1.3 million images with 1000 classes. In [8] proposed a method to classify high-resolution images in the ImageNet LSVRC-2010 dataset with 1.2 million images, 1000 classes using deep convolution neural networks. Burney and Syed [9] implemented the concepts of deep learning for crowd video classification. Zhang et al. [10] proposed a model to process real-time video basically for object detection using deep learning. Deep learning has proved very efficient in classifying images by modeling data using multiple processing layers. Training images using deep learning requires very large datasets and which is a time consuming task. So it is recommended to use pre-trained neural network, which will be used to extract relevant features [11, 12].

From the literature it is understood, there are plenty of works carried towards identification of animals and classification of animals, but all of them have used shallow features. Also most of the works are carried for a very small dataset. In addition, all these approach needs preprocessing stage like enhancement and segmentation. Since animals images are with very complex background it is not possible to come up with a fully automatic segmentation algorithm. With this, in some of the methods they segment the animals manually and in rest of the work they make use of semi-automatic segmentation algorithm. As the size of the dataset increases both manual segmentation and semi-automatic segmentation algorithm consumes more time and a tedious task. Further, for the purpose of classification a simple classifier such as nearest neighbor, support vector machine have been used. In this work, we propose a system to classify animals using convolution neural network. Frames are trained

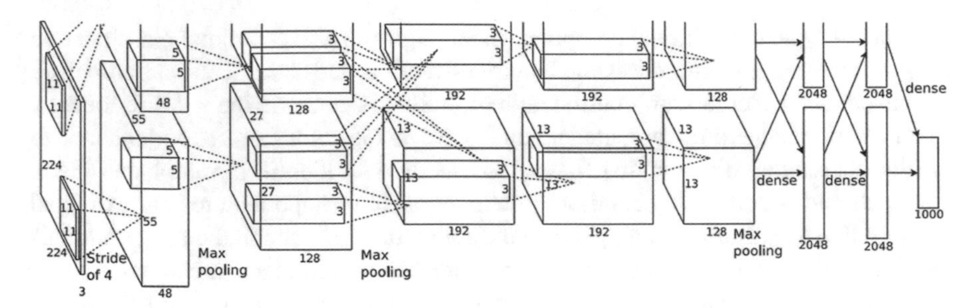

Fig. 3 General architecture of ConvNet [8]

using pre-trained neural network which will be used to extract relevant features and a multi-class SVM which is trained on these features is used to classify the animals.

Section 2 presents the proposed methodology, Sect. 3 briefs about the experimental setup and experimental results and the paper is concluded in Sect. 4.

2 Convolution Neural Network Configuration

In this section we describe about the general layout of ConvNet using general principles described by [7, 8]. The architecture of the ConvNet is summarized in Fig. 3. It is composed of eight layers, first five are convolutional layers and other three are fully connected layers. The first convolution layer filters the input image of size 224 $\times$ 224 $\times$ 3 with 96 kernels which is of size 11 $\times$ 11 $\times$ 3 with a stride of 4 pixels. The second convolution layer takes input from first layer and filters with 256 kernel size of size 5 $\times$ 5 $\times$ 48. Similarly, third convolution layer has 384 kernels with size 3 $\times$ 3 $\times$ 256 connected to the output of second convolution layer. The fourth layer has 384 kernels of size 3 $\times$ 3 $\times$ 192, the fifth layer has 256 kernels with size 3 $\times$ 3 $\times$ 192. The output of every convolution layer and fully connected layer are passed through ReLU nonlinearity. Finally the fully connected layers have 4096 neurons. The output of the last fully connected layer is sent to 1000-way soft-max layer.

3 Proposed Methodology

Our video dataset clearly suggests that, the presence of the animal is there thought out the video so, we randomly extracted 100 frames from each video. The proposed method has two stages namely training and testing. The system is trained using convolution neural network and classification is done using multi-class SVM. The architecture of the proposed system is shown in Fig. 4.

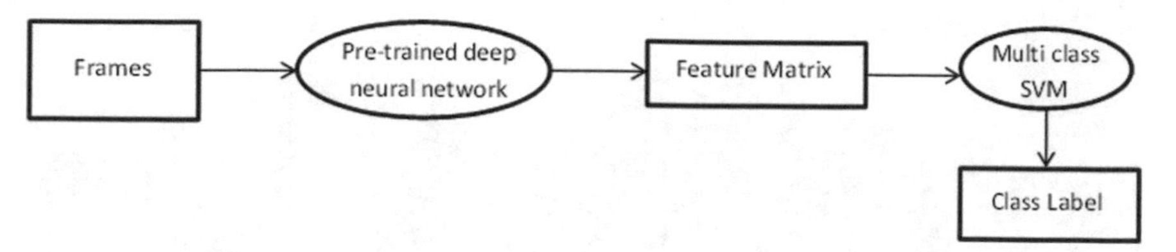

Fig. 4 Architecture of the proposed model

3.1 *CNN with SVM-Based Classification*

Previously machine learning approach uses traditional features for images like color histograms, SIFT, SURF, local binary patterns. But from the literature we can observe ConvNet outperform these features. Convolutional neural network learns these features automatically from images in a hierarchical manner. Lower layer learns features like edges, corners wherein middle layer learns features like color, shape, etc., and higher layer layers represents the object in the image. Thus convolutional neural network suits well for image classification problems. In convolutional network only few will train the entire system from the scratch due to the lack of sufficient dataset. So it is common to use a pre-train convolutional neural network instead of training the system from the scratch. The pre-trained networks have learned rich features representations for a wide range of natural images. We can apply these features to classification problems using transfer learning and feature extraction. The pre-trained convolutional networks are trained on more than a million images and classified the images into 1000 object categories. Some of the popular pre-trained networks are AlexNet, VGG-16 and VGG-19, GoogLENET, Inspection-v3, etc. In this work we train the images using AlexNet pre-trained convolutional neural network [8]. As like general ConvNet, AlexNet also has eight layers: Five layers are convolution layers and other three layers are fully connected layers. For feature extraction we consider only the first seven layers, i.e., five convolution layers and two full connected layers, the last fully connected layer is removed. For our experimentation the images are resized to $227 \times 227 \times 3$. AlexNet will compute 4096 dimension vector for every image which are used as features to classify the images using some linear classifier as showed in Fig. 5. In our work we have used multi-class SVM for classification.

3.2 *Multi-Class SVM Classifier*

Basically SVM is a discriminant-based classifier which is considered as the most powerful classifier amongst the classifiers exists in the literature [13]. Hence, in this work, we recommend to use SVM as a classifier for labeling an unknown animal image. But, the conventional SVM classifier is suited only for binary classification [14]. Here, we are handling with multiple species of animal images. Hence, we

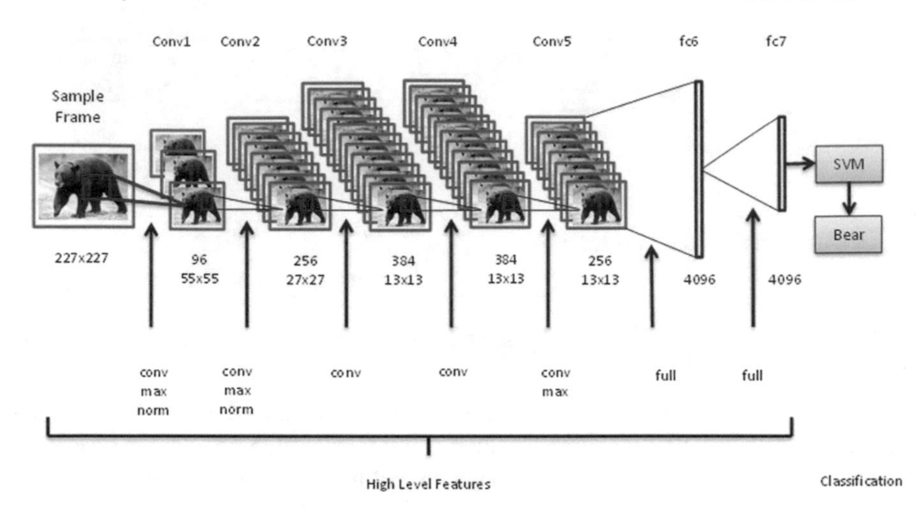

Fig. 5 Proposed AlexNetConvNet architecture with SVM classification

recommend the multi-SVM classifier for classification purpose. The multi-SVM classifier is designed based on two standard approach viz., one-versus-all (OVA) and one-versus-one (OVO) [14]. In this work, we have adopted the former approach for multi-SVM classification. For further details on multi-SVM classifier can be found in [14].

During training, the multi-SVM classifier is trained with 'N' samples with 'k' number of features and the support vectors associated with each class are preserved in the knowledgebase. During testing, a query sample with the same 'k' features is projected onto to the hyper-plane to classify the query sample as a member belongs to any one of the available classes.

4 Experimentation

In this section, we briefly explain about the datasets created for the purpose of the experimentation to test and evaluate our proposed methodology, experimental setup, and also we explain the detailed results obtained from the extensive experimentation.

4.1 Dataset

As convolution neural network works effectively only if the dataset is large and also there is no standard dataset of animals are available, we have created our own dataset for the experimentation. Our dataset is composed of 200 videos with 20 different class, 10 videos in each class. In our dataset, we can see that animal is present

Fig 6 Sample frame of animals from each class

throughout the video so from each video 100 frames are extracted randomly. So in total 20,000 frames are considered for experimentation. To analyze the effectiveness of the proposed methodology we have considered the video with large intra and inter class variations and also some of the challenges like videos with different illumination, occlusion, and animals captured in different pose and different view. Figure 6 shows the sample video frames randomly selected from each class.

4.2 Experimental Setup

Our implementation is done on the above dataset with training and testing phases. For training and testing, the extracted frames are resized into 227 × 227 and implemented on 1 GB Radeon HD 6470 M GPU processor. Alex Net Pre-trained weights are used to train the frames during training phase and extracted 4096 features. During testing phase, an unlabeled animal frame is classified by extracting the same set of features using multi-class SVM classifier. The linear kernel is used for SVM training.

4.3 Results

The proposed model has been evaluated using standard validity measures such as precision, recall and F-measure. These measures are computed from a confusion matrix obtained during classification of animal images. The definitions of these measures are given below.

Table 1 Classification results in terms of average precision, recall, and F-measure under varied train-test percentage of samples

Train–test (%)	Accuracy	Precision	Recall	F-measure
30–70	72.14	74.73	74.74	74.73
50–50	81	83.17	83.17	83.17
70–30	83.33	85.83	85.83	85.83

Precision is the fraction of retrieved documents that are relevant to the search

$$Precision = \frac{|\text{correctly classified frames}|}{|\text{classified frames}|}$$

Recall is the fraction of the documents that are relevant to the query that are successfully retrieved.

$$Recall = \frac{|\text{correctly classified frames}|}{|\text{expected frames to be classified}|}$$

F-measure is a measure that combines precision and recall is the harmonic mean of precision and recall.

$$F\text{-}measure \frac{2 * \text{precision} * \text{recall}}{\text{precision} + \text{recall}}$$

To evaluate the efficacy of the proposed system we have conducted experiments on our own dataset and reported the result for different training and testing percentage of samples.

Figure 7 shows the confusion matrices obtained for varied train-test percentage of samples and Fig. 8 shows the class wise performance analysis in terms of precision, recall, and F-measure for different training and testing percentage of samples. Table 1 shows the classification results in terms of average Precision, Recall, and F-Measure under varied train-test percentage of samples.

5 Conclusion

In this work a system to classify animals using convolution neural network is designed. The animals are fed into pre-trained system, i.e., AlexNet to extract the features. Multi-class SVM classifier is used for classification purpose. For experimentation we have created our own dataset of 20,000 frames from 200 videos with

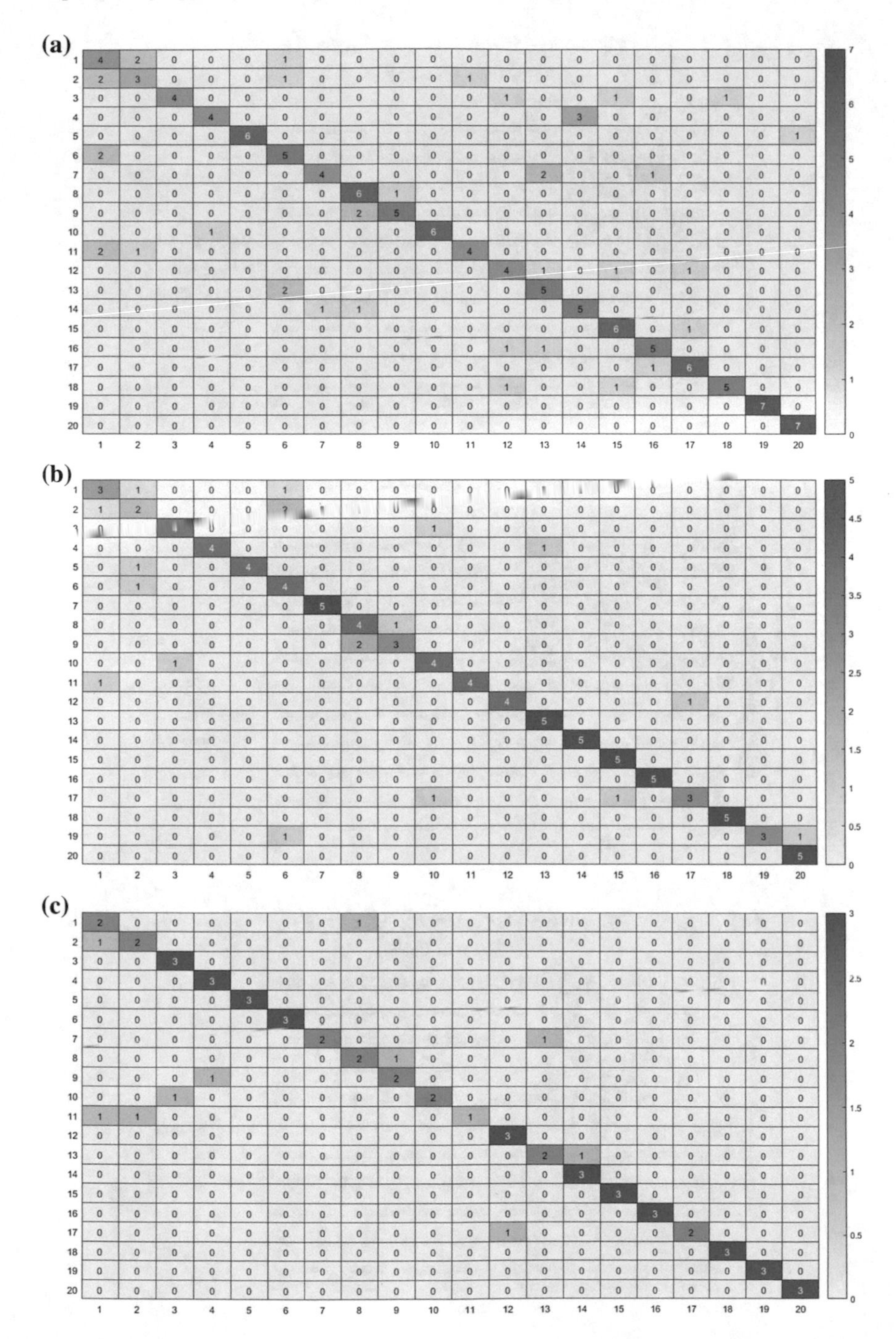

Fig. 7 Confusion matrices obtained for varied train-test percentage of samples, **a** 30–70%, **b** 50–50%, **c** 70–30%

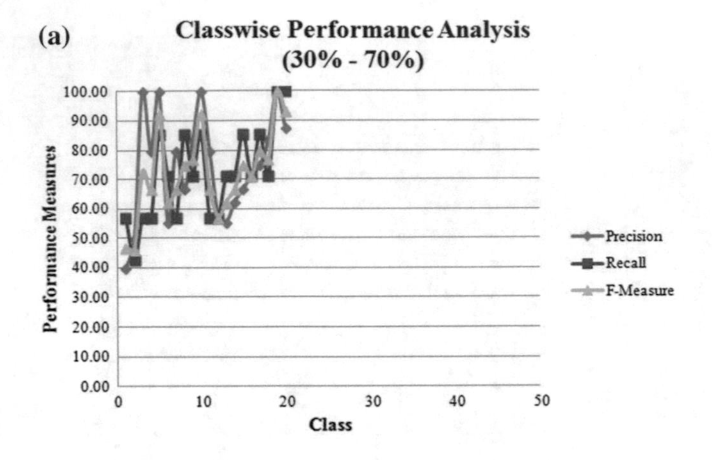

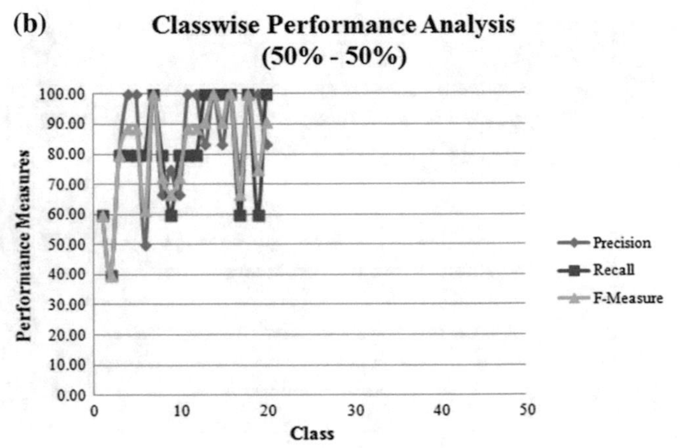

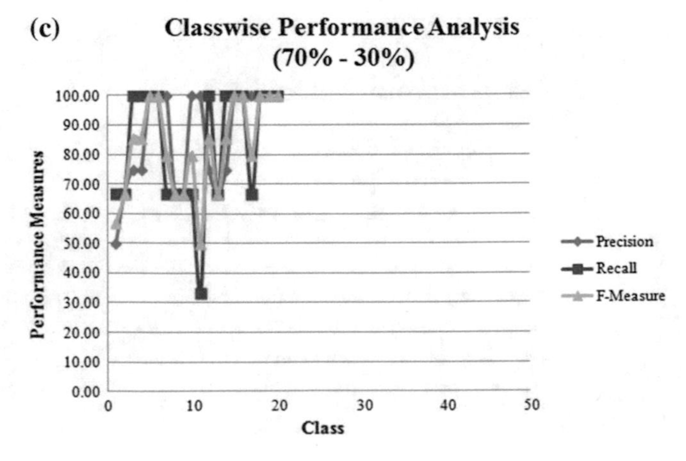

Fig. 8 Class-wise performance analysis in terms of precision, recall, and F-measure for **a** 30–70%, **b** 50–50%, and **c** 70–30% train-test samples

20 different classes. Experimental results show the effectiveness of the proposed methodology, which has high accuracy, precision values. The proposed system can automatically classify animals in videos with considering many challenges in the dataset.

References

1. Lahiri, M., Tantipathananandh, C., Warungu, R.: Biometric animal databases from field photographs: identification of individual zebra in the wild. In: Proceedings of the ACM International Conference on Multimedia Retrieval (ICMR) (2011)
2. Ardovini, A., Cinque, L., Sangineto, E.: Identifying elephant photos by multi–curve matching. Pattern Recogn. 1867–1877 (2007)
3. Ramanan, D., Forsyth, D.A., Barnard, K.: Building models of animals from videos. IEEE Trans. Pattern Anal. Mach. Intell. **28**(8), 1319–1334 (2006)
4. Zeppelzauer, M.: Automated detection of elephants in wildlife video. J. Image Video Process. (2010)
5. Kumar Y.H., Manohar, N., Chethan H.K., Kumar, H.G.: Animal classification system: a block based approach. In: International Conference on Information and Communication Technologies (ICICT) (2014)
6. Manohar, N., Kumar, Y.H.S., Kumar, G.H.: Supervised and unsupervised learning in animal classification. In: International Conference on Advances in Computing, Communications and Informatics (ICACCI), pp. 156–161, Jaipur (2016)
7. Simonyan, K., Zisserman, A.: Very deep convolutional networks for large-scale image recognition. In: ICLR (2015)
8. Krizhevsky, A., Sutskever, I., Hinton, G.E.: ImageNet classification with deep convolutional neural networks. In: Proceedings of the 25th International Conference on Neural Information Processing Systems—Volume 1, pp. 1097–1105 (2012)
9. Burney, A., Syed, T.Q.: Crowd video classification using convolutional neural networks. In: 2016 International Conference on Frontiers of Information Technology (FIT), pp. 247–2515, Islamabad (2016)
10. Zhang, W., Zhao, D., Xu, L., Li, Z., Gong, W., Zhou, J.: Distributed embedded deep learning based real-time video processing. In: 2016 IEEE International Conference on Systems, Man, and Cybernetics (SMC), Budapest, pp. 001945–001950 (2016)
11. Elleuch, M., Maalej, R., Kherallah, M.: A new design based-SVM of the CNN Classifier architecture with dropout for offline Arabic handwritten recognition. ELSEVIER Proc. Comput. Sci. **80**, 1712–1723 (2016)
12. Nagi, J., Di Caro, G.A., Giusti, A., Nagi, F., Gambardella, L.M.: Convolutional neural support vector machines: hybrid visual pattern classifiers for multi-robot systems. In: 11th International Conference on Machine Learning and Applications, pp. 27–32 (2012)
13. Duda, R.O., Hart, P.E., Stork, D.G.: Pattern Classification, 2nd edn. Wiley-Interscience (2000)
14. Hsu, C.-W., Lin, C.-J.: A comparison of methods for multiclass support vector machines. IEEE Trans. Neural Netw. **13**(2), 415–425 (2002)

Building Knowledge Graph Based on User Tweets

Seyedmahmoud Talebi, K. Manoj and G. Hemantha Kumar

Abstract For any social network, due to its vast amount of content for each user, is the best source for personalization and recommendation systems In this paper, we work on building users' knowledge graph, which uses Wikipedia as its knowledge base and users' tweets as input data. The knowledge graph contains the relevant articles matched accordingly with the keywords extracted from users tweets. This structure is helpful for a recommendation in E-commerce and social network analysis.

Keywords Social network · Big data · Knowledge base · User knowledge graph
Text similarity · Distance matching

1 Introduction

User profiling has been focused to generate and extract useful information from users for recommendation and e-commerce. Therefore, a social network is the best source to access this information for each user.

Social networks play an important role nowadays in our life. Among various social networks, Twitter is considered one of the best, with almost 330 million active users as of 2017. Twitter is a microblogging service that allows a user to post messages (each of maximum 280 characters) in order to follow friends and to communicate with them.

In this work, knowledge graph is generated from users' tweets. To achieve the model, we need to use a knowledge base to generate proper output therefore among

S. Talebi (✉) · G. Hemantha Kumar
Department of Computer Science, University of Mysore, Mysuru 560007, India
e-mail: sm.talebi1@gmail.com

G. Hemantha Kumar
e-mail: ghk.2007@yahoo.com

K. Manoj
Bengaluru, India
e-mail: manoj.krishnaswamy@gmail.com

P. Nagabhushan et al. (eds.), *Data Analytics and Learning*,
Lecture Notes in Networks and Systems 43,
https://doi.org/10.1007/978-981-13-2514-4_36

all encyclopedias, Wikipedia is helpful because it is open source and covers most of the recent topics.

In Wikipedia, three different types of pages are identified as

- *Disambiguation pages*—The title of articles which is similar to other pages but with a different meaning.
- *Redirect pages*—The content of these pages link to another article which has an almost same title and to avoid repetition of contents.
- *Articles*—The main pages of Wikipedia that explain all the details regarding the title.

After finding the proper pages of Wikipedia relates to user tweets then we extract the links connected to each page of Wikipedia. To find the proper links between users' tweets and Wikipedia pages we use natural language processing techniques to remove noise, extract keywords and find similarity between texts. Therefore, with the above steps, we generate users' knowledge graph.

2 Literature Survey

Personalization is started by analyzing web documents that users visit in order to generate user's interests [1, 2]. Recently, the increasing amount of data which are available at social networks such as Twitter has made the personalization systems to analyze user activities by different methods such as Bag of Words [3], Topic Models [4, 5] or Bag of Concepts [6–8]. Wikipedia has been used as the base for generating concept graph in our proposed approach. Other approaches have utilized it for tasks such as ontology alignment [9], and clustering [10], classification of tweets [11].

3 Algorithm Design

Knowledge source is the entire Wikipedia itself. The goal is to find the topics related to users' interests and based on that to generate knowledge graph. Therefore, the entire Wikipedia dump (English version) is used. Wikipedia has all sorts of information in which many parts can be wholly ignored. To have a better and suitable structure to our work, we convert it to a different structure, which covers only the title, id, and links connected to each article. The links are other related pages mentioned in the content of any articles. This new format of Wikipedia is labeled as "wiki_converted" in this work.

In the next step, users' tweets are processed from a CSV file format as shown in Fig. 2. First, repeated tweets are first removed and then noise such as HTML links, emoji's, etc., are also eliminated. In the next step, keywords based on Rapid Automatic Keyword Extraction (RAKE) method are extracted. In the last step, the

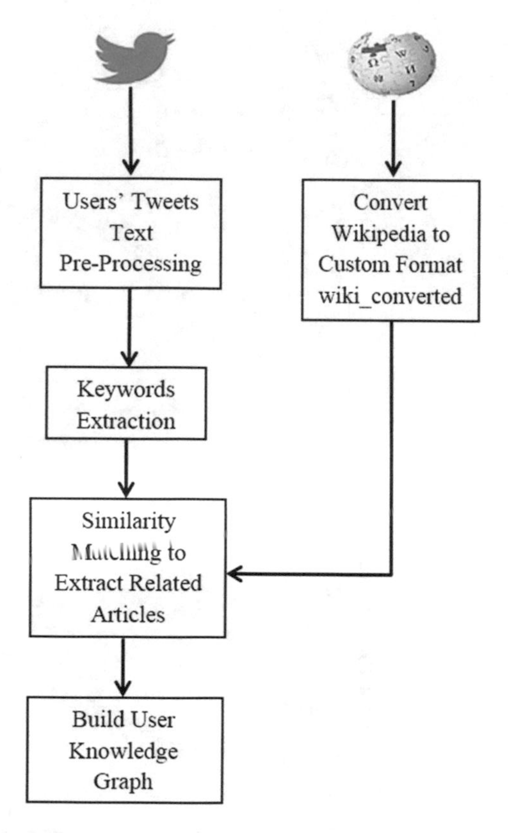

Fig. 1 Proposed algorithm

id	created_at	text
8.71889E+17	06-06-17 0:39	Mimas is a speck in the upper right as our @CassiniSaturn craft gazes from above the planet's northern hemisphere:â€¦ https://t.co/AqcTG90X4D
8.71888E+17	06-06-17 0:34	Apply to join experts as we preview this summerâ€™s total solar eclipse at a #NASASocial at the @Newseum on June 21:â€¦ https://t.co/LGJdrGeYI8
8.71874E+17	05-06-17 23:37	Two planetsâ€¦same size & temp and orbit around nearly identical starsâ€¦then why are their atmospheres so different?â€¦ https://t.co/PzHtvkcCyd
8.71857E+17	05-06-17 22:31	This newly discovered Jupiter-like world is so incredibly hot that itâ€™s being vaporized by its own star! Details:â€¦ https://t.co/7PbF9ahhA4
8.71854E+17	05-06-17 22:19	RT @Astro2fish: We finally caught a #Dragon by the tail and weâ€™re not letting go! It has a ton of great @ISS_Research aboard (& maybe a litâ€¦
8.71854E+17	05-06-17 22:19	RT @Astro2fish: It is pretty when a #Dragon spreads its wings! 1st return visitor to @Space_Station since Atlantis in 2011. Will reuse rockâ€¦
8.71842E+17	05-06-17 21:29	A newly arrived experiment on @Space_Station looks to better understand pesticide poisoning and improve antidotes:â€¦ https://t.co/XYmiV0LuTQ
8.71825E+17	05-06-17 20:24	New research on @Space_Station will test the bone-forming molecule NELL-1 to see if it can prevent bone loss. Watchâ€¦ https://t.co/DxIto9P6BO
8.71794E+17	05-06-17 18:19	Among the research that arrived on @Space_Station today is a study on bone density & muscle wasting in microgravityâ€¦ https://t.co/B12bxEFsXa
8.71793E+17	05-06-17 18:15	RT @Space_Station: #Dragon bolted to station at 12:07pm ET. Crew will unload new @ISS_Research, spacewalk gear and crew supplies. https://tâ€¦

Fig. 2 Tweets of User C

whole wiki_converted is parsed to find the appropriate matches with the users' keywords. Pages of Wikipedia that are related to the keywords have a connection to other pages, which helps to expand the concepts of users' interests. Finally, User Knowledge Graph or UKG is generated by comparing with Levenshtein distance and Cosine similarity matching methods. Figure 1 shows the general review of these steps. Below are steps of the proposed algorithm

Step 1: Convert the entire Wikipedia dump to custom format by removing redundant information and labelled as wiki_converted. This is done only once.

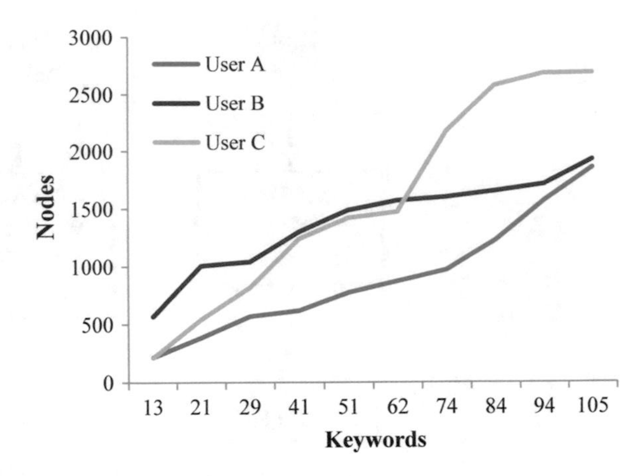

Fig. 3 UKG nodes comparison of Users A, B, C based on Cosine similarity matching

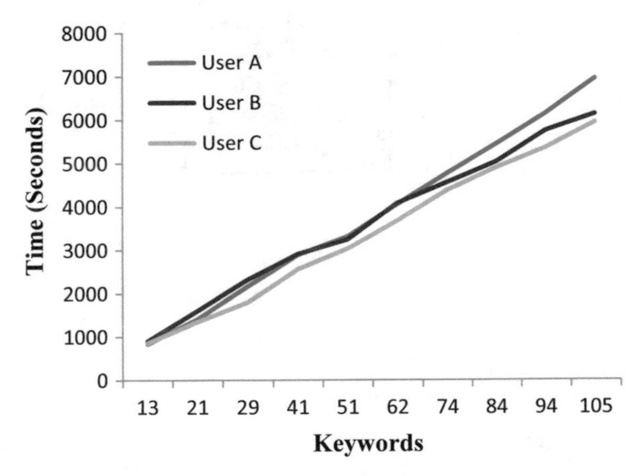

Fig. 4 Time taken for Users A, B, C based on Cosine similarity matching

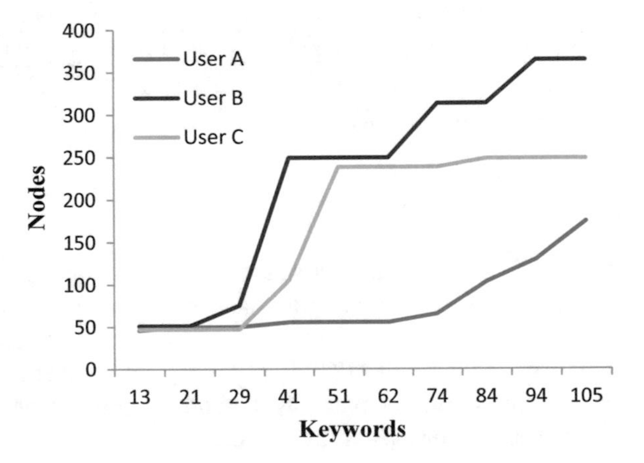

Fig. 5 UKG nodes comparison of Users A, B, C based on Levenshtein distance matching

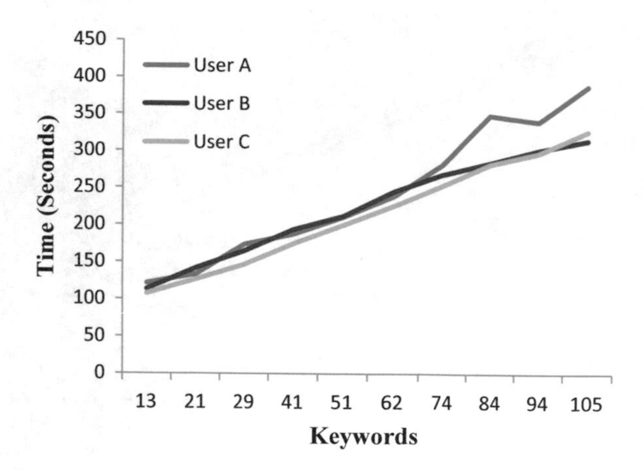

Fig. 6 Time taken for users A, B, C based on Levenshtein distance matching

Step 2: Extract users' tweets and remove noise.
Step 3: Extract Keywords from users' tweets.
Step 4: Perform similarity matching with Levenshtein distance and Cosine similarity methods with wiki_converted in order to find proper articles that match with the extracted keywords.
Step 5: Store the matched articles and extract the links by regular expression into a graph structure.

4 Results and Analysis

Wikipedia 2017 dump version has been used which is a 59.1 GB file in XML format. It has 5 million+ pages and each page have information such as the author, date of creation, title and the text. Out of 5 million+ pages, some of them redirects to other pages, which is removed due to its irrelevance in knowledge graph building and also, in order to reduce unnecessary parsing time. The reduced wiki_converted is set to usable 3 million+ pages. In this work, only the links directly connected to other pages have been used (single level depth). Therefore, the wiki_converted contains pages, which have title, id, links tags and size is reduced to 5.57 GB XML file.

The entire process of this work is on python 3 under Ubuntu 16.04 OS with Intel Xeon 3.10 GHz and 8 GB RAM hardware configuration. Twitter's Tweepy API is used in order to access users' tweets. The keyword extraction performed by using RAKE (Rapid Automatic Keyword Extraction) algorithm has good and fast performance.

For matching between each keyword and Wikipedia pages, Levenshtein distance and Cosine similarity matching methods have been used. After finding the appropriate pages, the links connected to each page are extracted and output in a graph structure.

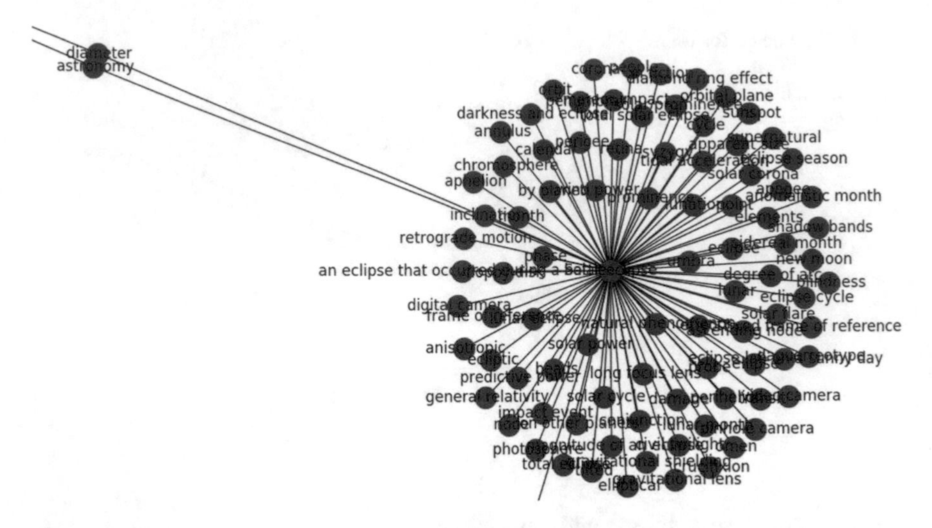

Fig. 7 Part of user C's UKG based on Cosine similarity matching

Table 1 Experimentation results of User A by Cosine similarity matching

Twitter posts	Extracted keywords	Linked articles	Nodes in graph	Time taken (s)
10	13	12	217	832
20	23	26	385	1410
30	36	81	569	2160
40	48	86	617	2890
50	56	90	774	3310
60	69	96	872	4046
70	80	100	971	4763
80	93	103	1228	5439
90	105	110	1575	6132
100	117	123	1861	6934

A different number of tweets for users' are extracted to track statistics such as number of keywords and total time for the entire process.

For a specific Twitter poster NASA, which is denoted as User C in this work, initially ten number of posts, is used to generate the knowledge graph. Figure 7 is the knowledge graph representation of 10 posts, 13 Wikipedia articles matches with the extracted keywords. Due to the limitation of representation, only the partial graph is shown here.

In this work, the generated knowledge graph is generic and covers all the categories that exist in Wikipedia and it is not based on any specific topic or keywords. To have a better analysis of the performance of these methods and the sizes of the graph are represented in Table 1 through 6.

Table 2 Experimentation results of User B by Cosine similarity matching

Twitter posts	Extracted keywords	Linked articles	Nodes in graph	Time taken (s)
10	14	13	569	899
20	26	24	1005	1599
30	37	28	1039	2314
40	49	47	1297	2908
50	56	51	1486	3244
60	67	55	1570	4073
70	76	57	1601	4553
80	82	73	1656	5030
90	92	77	1717	5742
100	102	130	1932	6121

Table 3 Experimentation results of User C by Cosine similarity matching

Twitter posts	Extracted keywords	Linked articles	Nodes in graph	Time taken (s)
10	13	9	214	854
20	21	14	542	1346
30	29	18	817	1795
40	41	34	1236	2560
50	51	40	1420	3031
60	62	42	1472	3667
70	74	53	2172	4374
80	84	81	2578	4895
90	94	94	2685	5347
100	105	97	2692	5932

To demonstrate our output and performance, three users are selected from varied categories like technology, politics, and science. For each user, we processed 10, 20, 30, …, 100 number of posts based on Cosine Similarity and Levenshtein distance matching methods.

Tables 1, 2 and 3 corresponds to three different Twitter users experimentation results based on Cosine similarity matching. Tables 4, 5 and 6 corresponds to same users experimentation results based on Levenshtein distance matching method. The number of keywords is the size of keywords extracted from the corresponding twitter posts. The number of articles is the number of Wikipedia pages that matched with the extracted list of keywords. Each Wikipedia page has connections to other pages, therefore, the number of nodes represent the size of the user knowledge graph and time taken indicates the total process including loading CSV file of tweets until generating the knowledge graph.

Table 4 Experimentation results of User A by Levenshtein distance matching

Twitter posts	Extracted keywords	Linked articles	Nodes in graph	Time taken (s)
10	13	1	46	121
20	23	2	50	133
30	36	2	50	173
40	48	4	55	187
50	56	4	55	210
60	69	4	55	238
70	80	5	65	280
80	93	6	103	347
90	105	7	129	338
100	117	9	174	386

Table 5 Experimentation results of User B by Levenshtein distance matching

Twitter posts	Extracted keywords	Linked articles	Nodes in graph	Time taken (s)
10	14	1	51	113
20	26	1	51	141
30	37	2	75	164
40	49	4	249	193
50	56	4	249	211
60	67	4	249	244
70	76	6	313	267
80	82	6	313	283
90	90	7	364	300
100	103	7	364	313

Table 6 Experimentation results of User C by Levenshtein distance matching

Twitter posts	Extracted keywords	Linked articles	Nodes in graph	Time taken (s)
10	13	2	47	107
20	21	2	47	126
30	29	2	47	146
40	41	3	105	174
50	51	5	238	199
60	62	5	238	225
70	74	5	238	252
80	84	7	248	281
90	94	8	248	296
100	105	8	248	325

As an example, for User C below is the extracted keywords of ten posts, as shown in Fig. 2:

['prevent bone', 'incredibly hot', 'better understand pesticide poisoning', 'total solar eclipse', 'return visitor', 'great aboard maybe', 'bone density muscle wasting', 'finally caught', 'orbit around nearly identical', 'newly discovered world', 'newly experiment', 'upper right', 'size temp']

Below is the portion of user knowledge graph of User C based on Cosine similarity method from the above-extracted keywords:

<UKG> … <2017 Total Solar Eclipse stamp>*adhesive; astrophysicist; astrophysics; ceremony; cut; day of issue; full moon; ink; silver dollar; solar eclipse; stamp; summer solstice; the total solar eclipse on the stamp; ultraviolet* </2017 Total Solar Eclipse stamp>

<Solar eclipse>*an eclipse that occurred during a battle; anisotropic; annulus; anomalistic month; aphelion; apogee; …; total solar eclipse; transit; umbra; video camera; wind power* </Solar eclipse> … </UKG>

Below is the portion of user knowledge graph of User C based on Levenshtein distance method with same keywords extracted from the text

<UKG>…<Field experiment>*anthropology; data; discrimination; education; experiment; external validity; health care; inference; laboratory; negative income tax; precision and accuracy; prototype; psychology; randomization; randomize; research; sample size; selection bias; situ; standard deviation* </Field experiment>

<Belle experiment>*calorimeter; calorimetry; collider; detector; effect; electron; experiment; fb; hermetic; integrated luminosity; iodide; kaon; luminosity; matrix; meson; oscillation; particle detector; particle physics; physics; pion; plot; positron; resonance; solid angle; tau particle; violation* </Belle experiment>… </UKG>

From Tables 1, 2 and 3 as compared to Tables 4, 5 and 6, shows that Cosine similarity matching takes almost three times longer as compared to Levenshtein distance matching but could find better matches related to keywords and also matches many more number of articles from Wikipedia, which are related to users' tweets. The more it matches the articles the more it covers data and knowledge from users; therefore, it could be more reliable and relevant for many practical purposes such as targeted advertisements.

Figures 4 and 6 shows that both methods are consistently increasing w.r.t. time taken but Cosine similarity method as compared to Levenshtein distance method takes more time. Figures 3 and 5 shows that the number of nodes matched in Cosine similarity is almost ten times more than Levenshtein distance method. In Fig. 8 as compared to Fig. 7, the size of UKG is much smaller which shows that Levenshtein distance method matches less number of articles as compared to Cosine similarity method.

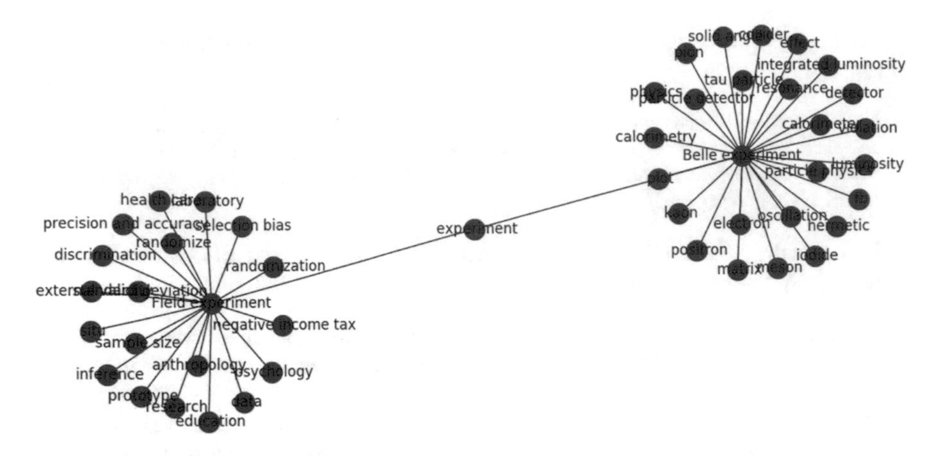

Fig. 8 Part of User C's UKG based on Levenshtein distance matching

5 Conclusion

The whole process of generating knowledge graph is standalone, since the entire Wikipedia dump is processed offline. Wikipedia access to its API is very limited if performed directly online, so the proposed method bypasses such restrictions. In addition, multiple users can be analyzed simultaneously with good performance as shown, since all the resources are available on the local machine itself.

The proposed methods are generic and can be used for large-scale extraction and analysis of user-based tweets. In our experimentation, Cosine similarity matching yields better results than Levenshtein distance matching for generating the user knowledge graph, albeit the time taken is more.

The generated knowledge graph has various practical applications in real life such as in suggesting friends, events, shopping, etc. It also gives a good indicator of a particular user's interests and it covers multiple aspects and domains. The user knowledge graph is also useful in artificial intelligence systems that require such automated processing of human data.

References

1. Godoy, D., Amandi, A.: Modeling user interests by conceptual clustering. Inf. Syst. **31**(4), 247–265 (2006)
2. Ramanathan, K., Kapoor, K.: Creating user profiles using wikipedia. Concept. Model. ER **2009**, 415–427 (2009)
3. Mislove, A., Viswanath, B., Gummadi, K.P., Druschel, P.: You are who you know: inferring user profiles in online social networks. In: Proceedings of the Third ACM International Conference on Web Search and Data Mining pp. 251–260 (2010)

4. Harvey, M., Crestani, F., Carman, M.J.: Building user profiles from topic models for personalised search. In: Proceedings of the 22nd ACM International Conference on Conference on Information & Knowledge Management, pp. 2309–2314 (2013)
5. Ramage, D., Dumais, S.T., Liebling, D.J.: Characterizing microblogs with topic models. In: ICWSM, vol. 10, p. 1 (2010)
6. Abel, F., Gao, Q., Houben, G.J., Tao, K.: Analyzing user modeling on twitter for personalized news recommendations. In: User Modeling, Adaption and Personalization, pp. 1–12 (2011)
7. Abel, F., Gao, Q., Houben, G.J., Tao, k.: Semantic enrichment of twitter posts for user profile construction on the social web. In: Extended Semantic Web Conference, pp. 375–389 (2011)
8. Orlandi, F., Breslin, J., Passant, A.: Aggregated, interoperable and multi-domain user profiles for the social web. In: Proceedings of the 8th International Conference on Semantic Systems, pp. 41–48 (2012)
9. Jain, P., Hitzler, P., Sheth, A.P., Verma, K., Yeh, P.Z.: Ontology alignment for linked open data. In: International Semantic Web Conference, pp. 402–417 (2010)
10. Xu, T., Oard, D.W.: Wikipedia-based topic clustering for microblogs. Proc. Assoc. Inf. Sci. Technol. **48**(1), 1–10 (2011)
11. Genc, Y., Sakamoto, Y., Nickerson, J.: Discovering context: classifying tweets through a semantic transform based on wikipedia. In: Foundations of Augmented Cognition. Directing the Future of Adaptive Systems, pp. 484–492 (2011)

Interoperability and Security Issues of IoT in Healthcare

M. Shankar Lingam and A. M. Sudhakara

Abstract Internet of Things (IoT) gadgets being utilized now can help to overcome certain limitations that inhibit their use in medical and healthcare systems. Security and Interoperability are particularly affected by these constraints. In this paper, the current issues faced are discussed, which incorporate advantages and challenges, and additionally methodologies to evade the issues of utilizing and coordinating IoT devices in medical and healthcare systems. Also, with regards to the REMOA project, which focuses on a solution for tele-monitoring of patients who suffer from chronic ailments.

Keywords Internet of things (IoT) · Healthcare · Security

1 Introduction

Internet of things (IoT) constitutes a group of technologies that allow an extensive variety of devices, appliances, and objects (or basically "things") to associate and convey among each other utilizing networking advances. Individuals constitute to the major part of the substance and information found on web up until this point, but considering the case of IoT, the components that actively provide data are small devices. IoT has numerous applications in healthcare systems, and this is the primary consideration of this paper. Healthcare systems utilize an arrangement of devices that are interconnected to make an IoT network committed to healthcare evaluation, for example, tracking patients and consequently identifying circumstances wherein there is a requirement for medical intervention. For patients who are chronically sick, example with hypertension, breathing infections or diabetes there is a requirement of clinical, health center, and emergency services swiftly and more frequently than

M. Shankar Lingam (✉) · A. M. Sudhakara
University of Mysore, Manasagangotri, Mysore 570006, India
e-mail: shankumacharla@gmail.com

A. M. Sudhakara
e-mail: sudhakara_am@yahoo.com

P. Nagabhushan et al. (eds.), *Data Analytics and Learning*,
Lecture Notes in Networks and Systems 43,
https://doi.org/10.1007/978-981-13-2514-4_37

normal patients [1]. Information and communication technologies are among the instruments that can be used to ease a part of the issue related with growing older populations, accelerated rates of perpetual sicknesses, and deficiency of healthcare experts, and, in the meantime, encourage service restructuring. The latest measuring gadgets, which incorporate, weight, blood pressure, and movement sensors, include correspondence capacities. These can make IoT networks applied for home account observations. These indistinguishable gadgets are those normally utilized by medical specialists to check the general state of chronically ill patients.

2 REMOA Project

REMOA is a venture which aims at providing home solutions for care or for observing of chronically ill patients. Distinctive systems and protocols for exchange of information among monitoring gadgets like movement sensors and blood pressure screens have been considered. The project likewise incorporates the design and implementation of devices that are dedicated to health care which function as dedicated middleware. The general framework gathers data from various detecting gadgets through a middleware that gives interoperability and security required with regards to IoT for social insurance. Wireless network systems link monitoring devices. The observation application front-end, which is similar in function to a network supervisor, is used to for store, aggregate, consolidate, and analyze at the collected data against previously arranged data and when the limit is reached, a flag is raised and a particular procedure is executed. At the point when breaking points are crossed, contingent upon limit controls, alerts are brought about to allow medicinal specialists to specifically respond to well-being-related occasions. The topology of the standard home story checking social insurance organize incorporates a middle person handling intermediary that advances detecting information to a remote server which is responsible for information examination, union, and basic occasions location. In this way, encompassed by more intelligent condition that utilizes correspondence and data advancements, natives (particularly older folks) and medicinal groups discover better conditions to continue with legitimate locally situated medications.

3 Devices to Be Connected

Considering necessities of home story monitoring system for chronically ill patients with ailments such as blood pressure, etc., a few gadgets have been chosen (wireless 1) to be utilized as a part of the REMOA project, so as to screen a few parts of patient's health remotely. These devices have interface with Wi-Fi and elements that allow interoperability and information transmission. Gadgets chosen to be utilized as a part of the venture are—A Panasonic BL-C230A Wi-Fi IP digital camera that is used to detect movements of patients under observation—the chosen gadget can identify motion in view of three special resources of information: (1) image; (2) body heat; and (3) sound. When a movement is detected, it sends forward a ready message that incorporates the images which are transferred to a FTP (wi wireless I Wi-Fi wireless transport Protocol) server;—A remote body scale equipped with a Wi-Fi interface that is manufactured by Witlings, operational with a software program capable of asserting the patient's percentage of fat, bulk, and body mass index A Witlings blood pressure gadget with this feature that its operation relies on the association with an Apple device like iPod, iPhone or iPad contact). The operation is like the scale, mostly from a similar manufacturer, with the distinction that this gadget is worked by means of an application program introduced in an Apple gadget. The application program is additionally in charge of sending information of the measured quantities to the remote server and reporting to the patient about any progress, cautioning of conceivable errors in managing the gadget, for example, negative stance or sudden movements that influence precise estimation of the blood pressure of the patient.

Threading information sent from observing gadgets and patient's data will be gotten to using a web browser on a normal PC and in Apps on smart cellphones or tablets. Medical staff can make use of these gadgets to monitor patient's well-being and by patients who can get to data about their own treatment and may have connection with medical staff and doctors and different patients using social networking and different exercises standard of the setting of patient 2.0 where patients participate effectively of their own treatment [2]. The gadget in charge of bringing together the information transmission of all chosen gadgets is the wireless access point (WAP) which supports Open WRT or DD-WRT. The use of the above technique empowers the change and sending of additional software, and supports prominent protocols such as IPv4, NAT, IPv6, SNMP proxy and also serves as a gateway for other monitoring devices.

4 REMOA—Interoperability Issues

The Project was intended to deliver a monitoring domain in light of Wi-Fi (802.11). To begin with, it can be understood that different monitoring gadgets would have the capacity to interact with remote servers essentially through Wi-Fi access points that are capable of acting as a gateway between monitored environment and Internet. This approach, even though relevant, has confined the assortment of medical gadgets accessible, especially as a result of the inclination to utilize Bluetooth innovation in human health gadgets. This pattern is because of the minimal effort and low power utilization of Bluetooth-empowered gadgets. Regardless of its crucial accessibility, a couple of solutions have been found which empower the interaction between internet and Bluetooth (802.14)/ZigBee (802.15.4) systems. This factor was testing the Wi-Fi and the challenges in transmitting information gathered by healthcare gadgets to a remote server over the group.

Considering the case of an environment which is completely Wi-Fi-based, the interoperability issue could be understood through a Wi-Fi/Bluetooth gateway, giving all administration abilities required, by either using Simple Network Management

Protocol (SNMP) or by using web administrations. Be that as it may, even healthcare devices which has built in Wi-Fi interfaces are not completely suitable for use in an adaptable observing environment. It has been found that the vast majority of such gadgets perform through communicating with servers and proprietary systems, just like the instance of the Witlings body scale. This issue required a stay away from answer for be executed in the correspondence framework. This workaround is trotted in a access point (AP) running a Linux (Open WRT/DD-WRT), and accordingly permitting the expansion of the product to the AP. In spite of the fact that the transmitted information arrangement is proprietary, by and large, similar to the Witlings body scale, and on analyzing the traffic it was found that the information is transmitted in unquestionable content and the use of the JSON design, which represents weak point as it grants 'eavesdropping' during the connection.

5 REMOA—Security Issues

The way that individual private information can be obtained through tele-monitoring infers the requirement for techniques and components to guarantee sufficient security and protection. As described in [3], "having each aspect" associated, latest security and protection issues emerge, authenticity, confidentiality, and integrity of the sensed data and exchanged by 'things'. According to the author, standard security necessities are: Resilience to attacks—The framework needs to evade single points of failure and ought to regulate itself to avoid node disasters; Data authentication—As a rule, recovered addresses and object data must be validated; Access control—data organizations must have the capacity to execute access control on the information provided; Customer privacy—Measures should be taken that only the device supplying information can reason from observing utilizing the lookup system.

However, in perspective of the portability setting where health specialists will be able to access patient's information while going to their homes, it suggests the requirement for "security enhancing technologies," as said in [4], to assure that the personal data of the patient is ensured against non-approved access. The solution to interoperability and administration of home care network has been anticipated to utilize intensified network administration ideas in order to comprise of web of things administration disengaging the capacities and components of a question from the forcing innovation, for example, WLAN or RFID as presented in [5]. Thus, the administration and security models should be arranged to monitoring events to guarantee the security of clinical gadgets and cell phones and further utilize a 'bound together authentication' design for the entire arrangement of networks, i.e., medical devices and infrastructures. The deliberate confirmation administrations utilize Shibboleth, a middleware layer for validation and access control evolved as an Internet2 project [6]. By utilizing Shibboleth it is plausible to exploit a Federated Authentication benefit effectively embedded by method for RNP (Rede National de Encino e Pesquisa—the Brazilian research and training system).

CAFe ("Comunidade Acadêmica Federada"), an alliance of Brazilian research and training establishments acting as recognizable proof and service provider. CAFe permits each client to maintain just one client account in his/her beginning organization this is legitimate for all administrations provided by the federation, accordingly discarding the need of different passwords and numerous registration forms (single sign-On SSO). In any case, in the IoT setting other security issues ascend, information integrity and client privacy since cell phones are incapable to handle malignant attacks, which encourage enhancements as far as research and security countermeasures. In account of social insurance situations, keeping away from fruitful strikes to the framework or possibly alleviating them are essential objectives since disappointments or information holes can cause harm to the patients. As opposed to different areas that can take in a few costs of system abuse, medicinal services frameworks cannot. When delicate information about a patient's medical issues is exposed and social harm is done, it is difficult to invalidate the information [7]. Aggressors have well Wi-Fi desires when representing considerable authority in cell phones. Ordinarily, assaults reason at taking client's information, assaulting gadgets assets, or even closing down a few applications [8]. There are numerous dangers encompassing cell phones; the greater part of the general population of these dangers are acquired from conventional computing systems and adjusted for cell phones. Be that as it may, a few dangers get more consideration on account of the potential issues they can bring about to the frameworks. Cases of these strings are: (i) man-in-the-center; (ii) Routing redirection assault; (iii) Denial of administration expecting to bring about a Battery Exhaustion in the device. These strikes, when connected solely on cell phones, speed up given variables, for example, (a) characteristic utilization of broadcast for correspondence; (b) absence of certification sources; (c) utilization of batteries as power supply; and (d) portability. Subsequently, to oversee such a colossal assortment of cell phones mapping and alleviating the most significant threads in a given setting is a fundamental test. In this specific circumstance, devices that perform in a wake-up and input information approach, as the Witlings body scale utilized as a part of this venture, are less defenseless against battery discharge strikes in light of the fact that the introduction time comes down a few moments in which the gadget is associated with Wi-Fi interfaces and service is not required. Also, interception and association inconveniences hold on and the absence of an infrastructure that deal with server/service authentication to the destination to which information will be dispatched and the shortage of clarity of substance are security crevices that needed a circumvent arrangement in REMOA project.

6 Conclusion

The examination expressed on utilizing off-the-rack IoT gadgets for an application of health wireless tele observing at home demonstrated that albeit conceivable, the rising business sector still does not give adaptable items that might be effectively adjusted for use in settings other than the offered by the producer and that grants just access to pre-configured servers now and again. These focuses on the way that IoT interoperability inconveniences are still nascent paying little mind to not being viewed as an issue to build up an information exchange gadget associating social insurance suppliers with patients [9] and the utilization of such arrangements may turn into a pickle with regards to incorporating IoT gadgets in a more extensive setting. A few middleware proposition utilize Service-Orientated Architectures (SOA) instruments as reason for a middleware design in embedded networks [10] however regardless there is a requirement for models to improve interoperability of gadgets particularly on account of healthcare devices.

Required gauges ought to incorporate open APIs, desire of interfaces for interconnection, and various configuration options of the modes of operation of the monitoring/control device, including additional security mechanisms. The embedded middleware that has been proposed in this paper offers a solution to interoperability improvement and security administration in a setting utilizing a special kind of internet of things gadgets for wireless health monitoring using Wi-Fi interfaces. Such gadgets cannot be associated directly to the internet for security and interoperability reasons. With a middleware it can be possible to give an improved AAA (Authentication, Authorization and Accounting) supplier that it is especially essential in this setting as maintained in [4]. For this situation, it was viewed as simpler to make the middleware to manage and moderate the correspondence between the "things" than to search for an answer that may regulate the "things" because of the accessibility of the access point helping the middleware programming. Since the product utilized is free, this solution can be discharged as an item at no more cost.

References

1. Paré, G., Moqadem, K., Pineau, G., St-Hilaire. C.: Clinical effects of home telemonitoring in the context of diabetes, asthma, heart failure and hypertension: a systematic review. J. Med. Internet Res. http://www.jmir.org/2010/2/e21/. https://doi.org/10.2196/jmir.1357 (2010)
2. Bos, L., Marsh, A., Carroll, D., Gupta, S., Rees, M.: Patient 2.0 empowerment. In: International Conference on Semantic Internet and Internet Services, pp. 164–167 (2008)
3. Babar, S., Mahalle, P., Stango, A., Prasad, N., Prasad, R.: Proposed security model and threat taxonomy for the internet of things (IoT). In: Communications in Computer and Information Science, 89 CCIS, pp. 420–429 (2010)
4. Interneter, R.H.: Internet of things—new security and privacy challenges. Comput. Law Secur. Rep. **26**(1), 23–30 (2010)
5. Sedlmayr, M., Prokosch, H., Münch, U.: Towards smart environments using smart objects. In: Paper presented at. Studies in Health Technology and Informatics, vol. 169, pp. 315–319 (2011)
6. Internet2, Shibboleth. http://shibboleth.internet2.edu/
7. Katzenbeisse, S., Petkovi, M.: Privacy-preserving recommendation systems for consumer healthcare services. In: Paper presented at the ARES 08—3rd International Conference on Availability, Security and Reliability, Proceedings, pp. 889–895 (2008)

8. Delac, G., Silic, M., Krolo, J.: Emerging security threats for mobile platforms. In: MIPRO, 2011 Proceedings of the 34th International Convention, pp. 1468–1473, May 2011
9. Ladyzynski, P., Wojcicki, J.M., Foltynski, P.: Effectiveness of the telecare systems. In: IFMBE Proceedings, vol. 37, pp. 937–940 (2011)
10. JesúsSalceda, I.D., Touriño, J., Doallo, R.: A middleware architecture for distributed systems management. J. Parallel Distrib. Comput. **64**(6), 759–766 (2004)

Author Index

P. Nagabhushan et al. (eds.), *Data Analytics and Learning*,
Lecture Notes in Networks and Systems 43,
https://doi.org/10.1007/978-981-13-2514-4

Printed in the United States
By Bookmasters